Redefining the Social Policy Discourse

Emmanuel Ndhlovu • Clement Chipenda
Editors

Redefining the Social Policy Discourse

African Perspectives for a Sustainable Continental Future

Springer

Editors
Emmanuel Ndhlovu
School of Public Management, Governance and Public Policy, College of Business and Economics
University of Johannesburg
Auckland Park, South Africa

Clement Chipenda
South African Research Chair in Social Policy, College of Graduate Studies
University of South Africa
Pretoria, South Africa

ISBN 978-3-032-24066-8 ISBN 978-3-032-24067-5 (eBook)
https://doi.org/10.1007/978-3-032-24067-5

This Springer imprint is published by the registered company Springer Nature Switzerland AG
The registered company address is: Gewerbestrasse 11, 6330 Cham, Switzerland

Preface

The African continent faces unprecedented development challenges that demand urgent attention and innovative responses. Despite its abundant natural resources, Africa contends with deepening poverty, widening inequality, violent conflicts, environmental degradation, political polarisation, and the lingering socioeconomic impacts of the COVID-19 pandemic. These crises are further exacerbated by unjust global trade structures, conditional aid regimes, and debt arrangements that severely constrain African countries' policy autonomy. The resulting dynamic has produced what Thandika Mkandawire aptly described as 'incapacitated states' that demonstrate greater accountability to external forces than to their own citizens.

This volume emerges from the recognition that conventional social policy approaches often transplanted from the Global North with little adaptation to African realities have proven inadequate in addressing these challenges. The disconnection between economic and social policies, characterised by an aversion to socialised provisioning and universal approaches, has resulted in largely residual interventions that fail to address structural inequalities or promote transformative change. There is, therefore, an urgent imperative to redefine the social policy discourse in contemporary Africa.

Contributions in this volume bring fresh methodological perspectives that break from traditional approaches. Rather than presenting an exhaustive methodological overview, the analyses in this volume benefit from interdisciplinary insights that bridge academic disciplines and practice. The research presented draws from both empirical evidence and conceptual innovations, with scholars employing tools appropriate to their specific inquiries while maintaining a commitment to both methodological rigour and contextual relevance.

A distinguishing feature of this collection is its commitment to centring African agency and knowledge systems while maintaining critical engagement with global debates. Rather than merely critiquing dominant paradigms, the contributors offer constructive alternatives rooted in African realities. Across six thematic areas which include theoretical foundations, contemporary challenges, stakeholder dynamics, sectoral innovations, cross-cutting themes, and regional perspectives, they highlight both the persistent impact of colonial legacies and innovative pathways being forged by communities, institutions, and governments.

From Tanzania's COVID-19 response to Zimbabwe's land reform, from healthcare reforms in Kenya to food security interventions in Senegal, the chapters reveal how locally grounded policies can deliver meaningful outcomes despite structural constraints. The analyses of informal economies, community banking initiatives, and resource governance models demonstrate both institutional barriers and transformative possibilities.

As the editors, we view this volume not as a definitive statement but as an urgent call for collaborative action towards creating more effective, equitable, and contextually appropriate social policy solutions. We hope it stimulates further research, policy dialogue, and practical interventions that honour Africa's rich diversity while addressing shared aspirations for inclusive and sustainable development. Most importantly, we hope it contributes to ongoing efforts to decolonise social policy thinking and practice in Africa at this critical juncture in the continent's developmental journey.

Auckland Park, South Africa Emmanuel Ndhlovu
Pretoria, South Africa Clement Chipenda

About the Book

This volume, *Redefining the Social Policy Discourse: African Perspectives for a Sustainable Continental Future*, provides a comprehensive examination of social policy approaches across the African continent, critically engaging with both theoretical foundations and practical implementations. It brings together diverse perspectives that challenge conventional social policy paradigms while proposing contextually appropriate alternatives rooted in African realities and epistemologies. This book interrogates social policy from multiple dimensions, from theoretical debates and historical trajectories to sectoral innovations and emerging challenges. The intention is to showcase transformative pathways for inclusive development through effective policies and practical implementation. This book outlines key aspects and issues which must be understood and addressed to develop effective and equitable social policies capable of responding to Africa's unique developmental challenges. Attention is given to both persistent structural constraints and innovative opportunities for transformation, resilience, and sustainability in social provisioning. Through case studies spanning healthcare, education, food security, resource governance, land reform, informal economies, and youth development, it highlights how locally grounded approaches can deliver meaningful outcomes despite significant constraints. The volume advances the Transformative Social Policy (TSP) framework as a paradigmatic shift that conceptualises social policy as a multidimensional instrument for societal transformation through production enhancement, social protection, redistribution, reproduction, and social cohesion functions. It provides foundational perspectives into the underpinnings of Africa's social policy landscape, the lingering impact of colonial legacies, and opportunities for transformative change that centres African agency and knowledge systems. This book is of interest to researchers, social policy scholars, development practitioners, civil society organisations, and policymakers committed to reimagining and implementing more effective, equitable, and contextually appropriate social policy solutions across the African continent.

Contents

1 **Redefining the Social Policy Discourse in Contemporary Africa** 1
Emmanuel Ndhlovu and Clement Chipenda

Part I Theoretical Foundations

2 **Beyond External Paradigms: Theorising Transformative Social Policy Through Thandika Mkandawire and Jimi Adesina's Vision for Postcrisis African Development** 27
Clement Chipenda and Tom Tom

3 **Rethinking Gender, Social Policy and Development with Thandika: A Transformative Social Policy Perspective** 53
Newman Tekwa

4 **Rethinking and Unthinking Social Policy in Africa: Towards Decolonisation and Prospects for Inclusive Development** 67
Emmanuel Ndhlovu and Neo Mokone

Part II Sectoral Innovations

5 **From Denial to Engagement: Tanzania's Pandemic Response and Lessons for Inclusive Health Policy Development** 89
Ambrose T. Kessy

6 **Political Economy of COVID-19 Social Protection in Southern Africa** 115
Emmanuel Ndhlovu and Daniel F. Meyer

7 **In Pursuit of a Future Healthy Nation: Post-COVID-19 Kenyan Healthcare Social Policy Review and Re-alignment** 139
Jeremiah M. Muneeni

8 **Ethical and Technological Innovations for Social Protection in Africa** 153
Theshaya Naidoo

9 **Exploring the Interplay Among Food and Nutrition Social Policy Actors in Senegal from 1918 to 2021** 171
Alex Nadège Ouedraogo

10 **Agricultural Market Participation and Welfare in Rural Zimbabwe: Land Reform as a Social Policy Instrument** 193
Clement Chipenda and Jimi Adesina

Part III Cross-Cutting Themes (Civil Society, Informal Economies and Gender)

11 **Civil Society as a Catalyst for Inclusive Governance: Examining the Role of Non-state Actors in Shaping Social Policies in South Africa** 211
Otshepeng Mazibuko

12 **The Unseen Architects: Civil Society's Role in Social Policy Innovation Across Africa** 233
Takudzwa Musekiwa

13 **Informal Economies and Transformative Social Policy in Sub-Saharan Africa: Bridging Critical Gaps** 249
Neo Mokone and Emmanuel Ndhlovu

14 **Loan Acquisitions and Repayments Among Rural Women Participating in Informal Economics of Village Community Banks (VICOBAs) in Busega District, Tanzania** 283
Deborah Andrew Ngusa

15 **Transforming Masculinity and Mental Health in Rural Southern Africa: A Comparative Analysis of Gender Transformation and Social Policy in Zimbabwe and South Africa** 305
Winnet Hokoza and Clement Chipenda

16 **From Policy to Practice: Interventions for Learners with Disabilities in South Africa** 335
Raymond Chirowamhangu

Part IV Vulnerable Populations and Social Justice

17 **The Interface of Social Development and Drug Abuse Among Youths in Southern Africa: A Rehabilitative Approach** 357
Edward Muzondo, Wilson Zvomuya, Ntokozo Nyathi, and Tariro Katsomba

18 Transformative Pathways to Children's Rights: A Social Policy Analysis in Rural Zimbabwe 377
Primrose Hove and Clement Chipenda

Part V Synthesis and Future Directions

19 Conclusion: Towards Reframing Social Policy for Inclusive Development 409
Emmanuel Ndhlovu and Clement Chipenda

Editors and Contributors

About the Editors

Emmanuel Ndhlovu is a Research Fellow at the University of Johannesburg, South Africa. He holds a Ph.D. (2021), MA (2017) (Cum Laude), and BA Honours (2013) degrees in Development Studies from the University of South Africa. He also holds a BA Honours degree in English studies (2008) from the University of Zimbabwe. His research interests cover a wide range of topics, including agrarian political economy, food systems, food security, land reform, agriculture and agribusiness, and social policy. Emmanuel is the author of the book *Agriculture, Autonomous Development and Prospects for Industrialisation in Africa*. He has also published over 40 journal articles and 50 book chapters and has co-edited eight books.

Clement Chipenda, PhD, is a Postdoctoral Research Fellow at the University of South Africa's College of Graduate Studies, where he contributes his expertise to the SARChI Chair in Social Policy. He is a co-investigator on the research projects 'The Social Policy Dimensions of Land and Agrarian Reform in International Perspective' and 'Social Policy in the Post Covid-19 Pandemic Era: Investigating a Transformative and Gender-Sensitive Framework in South Africa'. His research interests span a diverse yet interconnected range of topics, including agrarian political economy, agro-food systems, transformative social policy, children's rights and child protection, social security, urban and rural sociology, as well as youth and gender studies.

About the Contributors

Jimi Adesina, PhD, is a Professor and the DSI/NRF SARChI Chair in Social Policy at the College of Graduate Studies, University of South Africa (Unisa) in South Africa. A past President of the South African Sociological Association

(2004–2006), Professor Adesina was elected to the Academy of Science of South Africa (ASSAf) in 2005. He served on the board of the UN Research Institute for Social Development, Geneva (2013–2019), and on the board of RC19 of the International Sociological Association (2014–2018). His research interests include sociology, social policy, and the political economy of Africa's development. He has published widely in these areas.

Raymond Chirowamhangu, PhD, is a postdoctoral fellow at University of Pretoria, Department of Sociology. His current research project explores child marriages and harmful cultural practices in Eastern and Southern Africa. He holds a doctorate in politics from the University of Pretoria and a Master's in Development Studies from the University of Fort Hare. He is a RISC-RISE consortium, University of Johannesburg, and University of Luxembourg scholar on human rights and sustainability. His research interests include human rights, governance, gender studies, and international relations. He has several publications on diverse scholarly discourses.

Winnet Hokoza is a PhD candidate and a Lecturer at the University of Zimbabwe in the Department of Community and Social Development, with an unwavering passion for community empowerment, gender issues, and social justice. Winnet's research and academic pursuits are driven by her deep commitment to addressing pressing social challenges and promoting transformative change. Her research focuses on the complex intersections of gender, social justice, and community development, with a particular emphasis on marginalised populations. Winnet's work explores innovative approaches to fostering gender transformation, challenging deeply entrenched societal norms, and empowering individuals and communities to become agents of change.

Primrose Hove is currently a PhD student at the University of Pretoria in the Department of Sociology. She is also a Lecturer at the University of Zimbabwe (UZ) in the Department of Community and Social Development. She holds BSc and MSc degrees in Sociology and Social Anthropology, M&E, and a Postgraduate Diploma in Social Work. Her research revolves around migration, gender, livelihoods, climate change, and social policy.

Tariro Katsomba is a Junior Lecturer in the Department of Social Work at the University of Zimbabwe. She holds a Bachelor of Social Work Honours Degree and a Master of Science Degree in Social Work and Social Policy from the University of Zimbabwe. Her research interests are social protection, child protection, and disability issues, with a commitment to promoting inclusive and equitable social policies. As an academic and researcher, she is passionate about contributing to the development of knowledge and practice of social work, particularly in the context of Southern Africa.

Ambrose T. Kessy, PhD, is a Professor at the Law School of Tanzania. He possesses a Bachelor's degree in Political Science and Public Administration and a Master's in Public Administration from the University of Dar es Salaam, culminating in a PhD in Development Policy and Management from the University of Manchester in the UK. He served in the Department of Political Science and Public Administration at the University of Dar es Salaam from 2000 to September 2018, then joined the University of Dodoma, where he was appointed Director of Research, Publications, and Consultancy in 2019. Currently, he holds the position of Deputy Principal for Planning, Finance, and Administration at the Law School of Tanzania, a role he has occupied since 2023. His research interests include local governance, human resources management, strategic planning, climate change, political economy, and education management.

Otshepeng Mazibuko is an emerging researcher pursuing a PhD in Development Studies at the University of Pretoria, building on a foundation that includes a Master of Social Sciences and a Bachelor of Social Sciences with honours. Her academic journey reflects a commitment to excellence and a passion for development issues. With presentations at the HSRC Emerging Researcher Conference, the CAPSI Wits PhD Writing and Methodology Workshop, and the CODESRIA College of Mentors, she engages in the discourse on research knowledge, policy decision-making, and governance. She has published a manuscript in an international journal (Routledge) and co-authored several academic articles, demonstrating her dedication to contributing to the academic community.

Daniel F. Meyer is a Professor in the College of Business and Economics (CBE) at the University of Johannesburg (UJ), South Africa. He is a National Research Foundation (NRF)-rated researcher. Daniel is a development economist and a specialist in regional and local economic development analysis and policy development. He has developed various innovative measurement tools, indices, and scales to analyse regional economies. He also has a research focus on macroeconomics and the linkages with good governance. He has authored more than 150 international peer-reviewed research papers since 2015 and has also presented more than 60 international conference papers, including a number of keynote addresses.

Neo Mokone, PhD, is a Postdoctoral Fellow in the Faculty of Economics, Development, and Business Sciences at the University of Mpumalanga. He holds a PhD in Development Studies and an MSc in Agricultural Economics from the University of South Africa (Unisa). He is currently pursuing a Master of Management in African Philanthropy at Wits University. With over 10 years of experience in local government, his research focuses on economic and social policy, African philanthropy, governance, sustainability, and policy analysis, with a particular interest in development issues and socioeconomic transformation. Dr. Mokone has published in peer-reviewed journals such as the *African Journal of Development Studies* and the *Journal of Local Government Research and Innovation*, contributing to interdisciplinary research on development and sustainability challenges in Africa. He is actively engaged in policy research, sustainability, and governance, bridging research and practice to drive sustainable development.

Jeremiah M. Muneeni, PhD, holds a Bachelor of Education in English, linguistics, and literature (Egerton), a Master of Arts (literature), and a Doctor of Philosophy (literature), both from Kenyatta University. He teaches literary studies in the Department of Literature at Kenyatta University. His research interests include gender studies, specifically the novelistic representation of women by writers from both genders and how the same can be tapped to develop policies that promote equality and equity, a shift from Eurocentrism to Afrocentrism using Afrocentric language(s), the use of imagery in poetry to communicate ideologies, and social policies, specifically health care policies during the pre-, during, and post-COVID-19 periods.

Takudzwa Musekiwa, PhD, is a Postdoctoral Research Fellow at the University of Pretoria and a researcher specialising in governance, development, and public policy in Africa. Holding a PhD in Development Studies from the University of Johannesburg, he has researched climate resilience, digital inclusion, and service delivery. Dr. Musekiwa bridges academic research and policy, with a strong focus on state-civil society relations, co-production in governance, and inclusive economic development. Dr. Musekiwa's interdisciplinary approach and commitment to bridging theory and practice aim to make a significant impact on contemporary African governance frameworks.

Edward Muzondo is a licensed social worker, published researcher, and educator. He holds a Bachelors' Degree Honours in Social Work and a Master of Social Work from the University of Zimbabwe. Currently, he is serving as a social work lecturer at Eswatini Medical Christian University in Eswatini. His research interests lie in the areas of social work decolonisation, cultural competence, disability right advocacy, mental health, SRHR, and drug and substance use. Edward advocates for the adoption of African indigenous knowledge systems for social development. He is the founder of Pandima Wellness Center, a mental health organisation seeking to advance the mental wellness of people of all ages.

Theshaya Naidoo is a PhD (Law) candidate at the University of KwaZulu-Natal, researching the implications of neurotechnology on the South African judicial process. She holds an LLM in Medical Law, an LLB (cum laude), and a Bachelor of Social Science (cum laude) majoring in law, criminology, and forensic sciences. Beyond academics, she serves as the Central President of UKZN Golden Key (2025), advocating for academic excellence and leadership. With expertise in health law, human rights, and emerging technologies—AI, neurotech, and cryptocurrency—her research seeks to address the ethical and legal challenges these innovations present. Looking forward, Theshaya Naidoo is focused on developing AI and technological solutions to address social justice challenges in South Africa, particularly gender-based violence (GBV). She is currently building legal-tech and EdTech startups aimed at improving equality and democratising South Africans' access to legal services and education.

Deborah Andrew Ngusa, PhD, is a Senior Lecturer at the University of Dodoma, where she has built a distinguished career focusing on community development. With a deep commitment to advancing knowledge and practice in this field, her professional expertise spans rural development, gender and development, and broader development studies. Dr. Ngusa conducts research and teaching that make significant contributions to understanding and addressing the challenges and opportunities in community development. Her rigorous analytical approach and innovative methodologies have established her as an expert in the field, making her an invaluable asset to both her institution and the wider academic community.

Ntokozo Nyathi is a registered social worker and a mental health practitioner. He holds a Bachelors' Degree Honours in Social Work from the University of Zimbabwe. His research interests lie in the areas of mental health, maternal healthcare, and Sexual and Reproductive Health Rights. He is the founder of the Youth Support Network, a youth mental health organisation.

Alex Nadège Ouedraogo, PhD, is a scholar-practitioner with a doctorate in Economic and Social Sciences from the University of Bremen and an MA in Political Science from the Université Catholique de Louvain. She has extensive field experience in West Africa, focusing on migration, human rights, gender, and social policy. Her doctoral research explored food security and social protection in Senegal, examining both colonial and postcolonial contexts. Alex Nadège has worked with LARTES-IFAN in Dakar, UNHCR in Burkina Faso, the Ethnographic Museum at the University of Zurich, and now GIZ in Cameroon. She supports displaced populations and marginalised groups through participatory diagnostics, civil servant training, and gender-inclusive programmes.

Newman Tekwa, PhD, is a postdoctoral research fellow at the South African Research Chair in Social Policy, University of South Africa. His research interests centre on issues of gender, transformative social policy, inequalities, and land and water sector reforms, areas in which he is well published. Currently, Newman convenes a GETSPA Southern Africa Reading Group covering 15 SADC countries within the continent-wide GETSPA project. The GETSPA research project which covers 31 countries across the continent seeks to transform the cultures of social policymaking in Africa in addition to building a cohort of African social policy scholars.

Tom Tom, PhD, is a Postdoctoral Research Fellow at the DSI/NRF SARChI Chair in Social Policy, College of Graduate Studies, University of South Africa and a Senior Lecturer in the Department of Development Studies, Zimbabwe Open University. He focuses on inclusive and transformative social policy and development.

Wilson Zvomuya is a Social Worker with over a decade of frontline child protection experience across Zimbabwe, Eswatini, and the United Kingdom, where he practices as an Advanced Practitioner. He holds a Master of Social Work from the University of Zimbabwe. Wilson has authored and co-authored numerous articles and book chapters on ubuthu Social Work, the elderly, environmental and climate-focused social work, child protection, and social development. His creative footprint also extends to contributions in local peotry.

Chapter 1
Redefining the Social Policy Discourse in Contemporary Africa

Emmanuel Ndhlovu and Clement Chipenda

1.1 Introduction: Africa's Contemporary Context and Development Challenges

The mid-2020s present a critical juncture for African social policy discourse, as the continent confronts an unprecedented convergence of global disruptions and local imperatives that demand transformative responses. With the 2030 Agenda for Sustainable Development approaching its terminal phase, Africa faces the dual challenge of accelerating progress towards the Sustainable Development Goals (SDGs) while simultaneously preparing for post-2030 development frameworks. This convergence occurs within a twenty-first-century global landscape characterised by interconnected crises, including climate change, rising inequality, and democratic instability. These have exposed the limitations of traditional social policy approaches and necessitate more integrated and adaptive frameworks (UNRISD, 2023).

Social policy emerges as a crucial instrument for achieving multiple SDGs simultaneously, especially SDG 1 (No Poverty), SDG 2 (Zero Hunger), SDG 3 (Good Health and Well-being), SDG 4 (Quality Education), SDG 5 (Gender Equality), SDG 8 (Decent Work and Economic Growth), and SDG 10 (Reduced Inequalities). In addition, it contributes to broader objectives of sustainable development and peaceful societies. However, Africa's present positioning differs markedly from previous decades, presenting both significant opportunities and persistent

E. Ndhlovu
School of Public Management, Governance and Public Policy, College of Business and Economics, University of Johannesburg, Auckland Park, South Africa

C. Chipenda (✉)
South African Research Chair in Social Policy, College of Graduate Studies, University of South Africa, Pretoria, South Africa
e-mail: chipec@unisa.ac.za

E. Ndhlovu, C. Chipenda (eds.), *Redefining the Social Policy Discourse*,
https://doi.org/10.1007/978-3-032-24067-5_1

challenges. The continent's SDG implementation presents a mixed picture, with significant progress in some areas offset by persistent challenges in others. While improvements in child mortality (SDG 3.2) and primary education enrolment (SDG 4.1) demonstrate the potential for rapid advancement, persistent extreme poverty (SDG 1.1) and widening inequality (SDG 10.1) highlight the need for more comprehensive approaches (UN, 2024a). These challenges are acute given that Africa confronts poverty, inequality, violent conflict, environmental degradation, political polarisation, and the lingering effects of the COVID-19 pandemic simultaneously.

The continent's performance on SDG 16 (Peace, Justice and Strong Institutions) has been particularly concerning, with democratic backsliding and conflict affecting multiple countries. Similarly, SDG 13 (Climate Action) and SDG 15 (Life on Land) face significant challenges as Africa, despite contributing minimally to global emissions, bears disproportionate costs of climate change impacts (UN, 2024a, 2024b). Moreover, the continent's social policy landscape is characterised by generally low coverage of health insurance systems, persistent food security challenges, and the dynamics of land distribution amid accelerating urbanisation. These features create obstacles to achieving universal health coverage (SDG 3.8), ending hunger (SDG 2.1), and ensuring sustainable urbanisation (SDG 11.1) while simultaneously presenting opportunities for innovative policy approaches that can address multiple SDGs through integrated social policy interventions.

Despite these challenges, the continent faces new opportunities that could significantly reshape its development trajectory. The African Continental Free Trade Area (AfCFTA) exemplifies this potential, with projections suggesting it could lift 30 million people out of extreme poverty and raise incomes for 68 million others living on less than $5.50 per day (Yaya et al., 2025; World Bank, 2020). However, realising this potential requires social policy frameworks that can simultaneously accelerate SDG achievement while laying foundations for post-2030 development priorities that emphasise resilience, sustainability, and inclusive transformation. Recent economic indicators suggest a more dynamic picture, with the African Development Bank's 2025 African Economic Outlook projecting economic growth improving from 3.4% in 2024 to 3.7% in 2025, reflecting recoveries in key economies and ongoing regional integration efforts (AfDB, 2025). Nevertheless, this growth trajectory occurs against a backdrop of significant aid cuts by major development partners, creating funding pressures for low-income African countries heavily dependent on international development assistance (AfDB, 2025).

The COVID-19 pandemic exposed the limitations of narrowly conceived social policies and highlighted the dangers of divorcing economic policy from social considerations. This disconnection, characterised by an aversion to socialised provisioning and universal approaches, has resulted in largely residual social policies that inadequately address systemic challenges (Adesina, 2021; Tom, 2024). The pandemic's impact on SDG progress has been severe in Africa, affecting poverty reduction (SDG 1), education (SDG 4), and health outcomes (SDG 3). However, the pandemic's aftermath has also catalysed innovative policy responses, with several African countries experimenting with digital social protection systems and emergency cash transfer programmes that demonstrate both the continent's capacity for adaptive governance and the potential for social policy to accelerate SDG recovery.

These emerging policy innovations must be understood within the broader context of structural constraints that have historically limited African states' policy autonomy. This situation has been further complicated by persistent neoliberalism and neocolonialism that continue to shape both social policy discourse and implementation across the continent (Chipenda & Tom, this volume; Tekwa, this volume; Adesina, 2025; Tom, 2024; Chipenda, 2024). The post-independence trajectory of African governments reveals a dynamic pattern of social policy development heavily influenced by donor agendas and external policy recommendations, often resulting in fragmented approaches that prioritise donor preferences over locally defined priorities. This has created what Mkandawire (2001) identifies as states more accountable to external forces than to their citizens. Thus, while the pandemic has demonstrated African governments' capacity for innovative social policy responses, the sustainability and transformative potential of these initiatives remain constrained by the same structural dependencies that have historically undermined comprehensive social policy development. This is occurring in a striking paradox that characterises Africa's present situation. Despite possessing abundant natural resources including minerals, forests, arable land, and human capital, the continent continues to witness increasing numbers of people falling into deep poverty. This contradiction has intensified socio-economic and political tensions across the continent (Gumede & Ndhlovu, 2023). The situation is further exacerbated by unjust trade engagement rules, conditional aid, and debt structures promoted by powerful countries in the global North, which continue to regulate how African countries envision their development agenda (Gumede & Ndhlovu, 2023).

The continent's stability is further compromised by a complex web of political conflicts, insecurity, and economic challenges that significantly impact social policy implementation. For instance, ongoing civil conflict in Sudan has created a humanitarian crisis that strains existing social protection systems. The Sahel region presents a concerning picture, marked by successive military coups in Mali (2020/2021), Burkina Faso (2022), and Niger (2023), while persistent jihadist insurgencies have severely undermined state capacity to deliver social services (Danboyi, 2025). These security challenges are further complicated by xenophobic tendencies, especially in South Africa, and the risks associated with migration patterns across the continent.

Against this backdrop of multifaceted challenges and emerging opportunities, it becomes essential to examine the theoretical foundations that can inform more effective social policy approaches while drawing on comparative experiences from other world regions to identify pathways for transformative change in Africa's development trajectory.

1.2 Theoretical Foundations and Comparative Perspectives

The complexity of Africa's development challenges necessitates a deeper understanding of their historical and structural roots while positioning African experiences within the broader global social policy discourse. While foundational analyses by scholars like Mkandawire (2004, 2010) and Adesina (2011) remain relevant, contemporary circumstances demand engagement with theoretical frameworks that can address emerging realities while drawing insights from comparative experiences across world regions.

1.2.1 Lessons from Latin American Social Policy Development

International comparative experiences provide critical perspectives for analysing social policy development trajectories, especially regarding the intersection of political dynamics and institutional formation. Latin America's welfare state evolution offers essential frameworks for understanding how institutional development pathways influence welfare outcomes and create enduring patterns of inclusion and exclusion. Barrientos' (2024) framework shows that understanding social protection institutions requires examining their embedded stratification mechanisms, as these institutions both constrain and enable actors' behaviours within specific social contexts.

The evolution of social policy frameworks in Latin American nations has been shaped by diverse economic, political, and social transformations across recent decades. This transformation represents a dynamic nexus of historical inheritances, neoliberal restructuring, and contemporary movements towards more inclusive policy arrangements. In contrast to the comprehensive welfare systems developed by early industrialising nations, Latin American countries constructed social protection mechanisms that operated largely separate from public service provision. This created important and distinctive institutional patterns wherein social protection maintained selectivity despite the universalisation of educational provision throughout the region (Barrientos, 2024). The developmental trajectory unfolded across multiple phases, beginning with the primary export era when Latin American states assumed limited responsibilities, concentrating on elementary education and urban sanitation infrastructure (Filgueira et al., 2023). The subsequent import substitution phase established contributory welfare systems that predominantly served formal sector workers while systematically excluding informal sectors, resulting in highly stratified social security arrangements (Filgueira et al., 2023).

Mesa-Lago's (1978) seminal scholarship demonstrated how diverse pressure groups with varying levels of influence became the primary drivers in the historical establishment of stratified social security systems in Latin America, creating significant inequalities in service distribution. This stratification dynamic, wherein social protection institutions perpetuate rather than address existing inequalities, offers crucial cautionary lessons for African policy development. These institutional configurations established what Barrientos (2024) theorises as a spectrum of social protection institutions, extending from occupational insurance funded through payroll contributions and limited to contributing members to social assistance financed through budget allocations with entitlements determined by need and residence. Mesa-Lago (2004) distinguished three primary models of social security development: first, where pressure groups secure state recognition of occupational funds; second, where states co-opt pressure groups through social security provision; and, third, involving collaborative arrangements between pressure groups and political parties influencing state provision.

The interconnection between economic development patterns and welfare state formation provides crucial understanding of institutional trajectories. Research indicates that welfare institutions evolved along two clusters during 1930–1970: one characterised by robust economic development, limited openness, and democratic or left-wing party governance (including Southern Cone countries and Costa Rica) and another displaying weak economic development, high openness, and restricted democratisation (including Andean and Central American countries) (Segura-Ubiergo, 2007). This pattern challenges theories emphasising responses to trade openness uncertainties while supporting industrialisation and power resource explanations.

Huber and Stephens' (2012) extension of power resources theory demonstrates how Latin America's more diversified class structure compared to early industrialisers accounts for limited redistribution and creates what they characterise as a sui generis development pathway for welfare institutions. Their analysis reveals that Latin America's class structure is less conducive to class organisation and political mobilisation compared to Western Europe, with significantly more constrained opportunities for union organisation. These structural characteristics have similarly hindered democratic development, weakened civil society density, and thus reduced the potential for autonomous political mobilisation of lower-class groups (Huber & Stephens, 2012:25–26). Haggard and Kaufman's (2008) concept of critical realignments provides valuable analytical instruments for understanding institutional development. These disruptions in political domination patterns and the formation of new ruling coalitions help explain how worker organisation incorporation led to social insurance coverage expansion for organised urban working classes but failed to create new political opportunities for rural poor populations (Haggard & Kaufman, 2008:346). This process generated the urban bias that characterised Latin America's stratified welfare systems, reflecting the urban-centred priorities of political coalitions that challenged oligarchic governance. These biases were subsequently reinforced through the shift towards import substitution (Haggard & Kaufman, 2008:79).

The 1980s and 1990s represented a significant transition towards market-oriented reforms, characterised by state withdrawal and increased dependence on market mechanisms in social policy, which intensified social stratification (Cortés, 2008; Cruz-Martínez et al., 2021). Countries pursued divergent approaches: Cuba implemented fully socialised economies, while Chile adopted radical neoliberal reforms, establishing contrasting models for social policy frameworks (Mesa-Lago, 2004). By 1980, only a few countries maintained comprehensive social safety nets, primarily benefiting formal sector workers (Huber & Ponce de León, 2019). The post-Washington Consensus era witnessed renewed emphasis on inclusive policies, driven by economic growth and democratic consolidation, seeking to expand coverage and benefits (Filgueira et al., 2023). The 2000s commodity boom enabled increased public expenditure, facilitating expanded coverage for informal workers and enhanced cash assistance programmes (Huber & Ponce de León, 2019). Nevertheless, substantial inequalities continue in service quality and access, especially between contributory and non-contributory programmes (Huber & Ponce de León, 2019).

Contemporary Latin American social policy debates remain confronted by tensions between universalism and targeting, with significant implications for social inclusion and equity. The continuation of segmented labour markets has produced dual and stratified welfare systems, with substantial populations remaining excluded from comprehensive social protection (Cruz-Martínez et al., 2021). This situation exemplifies what Barrientos (2024) identifies as the predominance of three institutional forms in the region: occupational insurance, individual retirement savings, and social assistance, each incorporating different forms of social stratification. The erosion of social protection policies in countries like Brazil demonstrates the vulnerabilities of regional welfare states, influenced by both global trends and local institutional weaknesses. The COVID-19 pandemic necessitated urgent social assistance measures throughout the region, emphasising the requirement for targeted support for vulnerable populations (Diyakova, 2022). Simultaneously, it demonstrated the resilience and adaptability of existing institutional frameworks.

These experiences provide valuable lessons for African social policy development. Barrientos' (2024) comprehensive analysis of social protection institutions demonstrates that understanding these regional patterns contributes to advancing theoretical frameworks for social protection among late industrialisers, with relevance for countries still developing their welfare architectures.

1.2.2 Asian Welfare Development: Diversity and Innovation

The Asian experience presents a multifaceted developmental trajectory that offers important perspectives for African social policy development. It highlights how diverse historical legacies, cultural foundations, and political economies have shaped different welfare architectures across the region. The evolution of Asian social policy reveals a dynamic tension between traditional welfare systems rooted in family structures and modern state-driven approaches, creating hybrid models that challenge conventional Western welfare state typologies.

The historical foundations of Asian welfare systems diverged markedly from both Latin American and European models, with East Asian countries initially relying on informal networks and what has been termed the developmental model, which emphasised self-reliance and community support rather than state provision (Lin & Chan, 2015). This approach was deeply influenced by Confucian values that prioritised family responsibility and social harmony, creating welfare architectures where familial support systems formed the primary safety net rather than formal state institutions (Rakšnys et al., 2022). The enduring influence of these cultural foundations has continued to shape contemporary social policy development, distinguishing Asian approaches from the individualistic assumptions underlying Western welfare models.

The post-war period witnessed significant diversification in Asian welfare development paths. Countries like Japan, Taiwan, and South Korea developed comprehensive welfare states alongside their industrial and democratic transformations through the institutionalisation of social insurance models influenced by democratisation and economic growth (Wong, 2025). However, this development occurred within a broader framework that de Haan (2011) characterises as exhibiting two approaches: China's productivist model influenced by economic reforms since 1978 and India's welfarist model with decentralised public management. These divergent paths reflect how different political ideologies and economic strategies have shaped welfare development across the region, even as economic globalisation has prompted some convergence towards Western models.

The contemporary Asian welfare landscape demonstrates what Tiwari (2013) identifies as a three-stage developmental progression comprising early-stage systems in underdeveloped countries, take-off stage systems in developing countries, and drive to maturity stage systems in developed countries. This typology reveals varying degrees of state commitment across the region, with some nations successfully implementing universal education and health policies while others continue to struggle with implementation and resource constraints. These implementation challenges have led many countries to rely heavily on non-contributory welfare systems and social safety nets for vulnerable populations, creating a patchwork of welfare provision that varies significantly in coverage and quality (Tiwari, 2013).

The characteristics of Asian welfare states thus reveal significant differences from Western models, especially in their emphasis on productivity and familial support systems (Rakšnys et al., 2022). This productivist orientation, which Lin and Chan (2015) identify as prioritising economic growth and viewing welfare as linked to employment, reflects the developmental state logic that has dominated East Asian policymaking. However, contemporary trends show the emergence of hybrid approaches that blend productivist elements with redistributive policies, responding to growing economic uncertainty and social inequality across the region.

Contemporary social policy frameworks in Asian countries are characterised by a dynamic nexus of economic growth, rising inequality, and crisis management imperatives. The key features of these frameworks encompass a strategic mix of social assistance, social insurance, and labour programmes aimed at targeting disadvantaged groups (Saunders & He, 2017). This multi-pronged approach includes extending social security systems, addressing inequality through policy innovations, and integrating human capital development with social protection measures (Croissant, 2004; Mok et al., 2017). However, these efforts encounter significant challenges, including budget constraints, institutional weaknesses, and the imperative for more inclusive policies.

Several features characterise contemporary Asian social policy frameworks. Some Asian countries have implemented ambitious programmes to extend social security systems, aiming to mitigate the social impact of economic crises (Croissant, 2004). There is a growing trend of implementing social rights-based public policy innovations, such as working family support mechanisms, to address rising inequalities and evolving welfare expectations (Mok et al., 2017). In addition, policies in South and Southeast Asia often integrate human capital formation with social protection measures to address chronic poverty and demographic challenges. This reflects a holistic approach to welfare development that links immediate protection with long-term capacity building (Mok et al., 2017).

Despite these innovations, Asian countries face significant challenges in addressing inequality through social policy. These challenges arise from chronic poverty, worsening social inequality, and inadequate budget support and accountability mechanisms (Saunders & He, 2017). Technological change, globalisation, and market-oriented reforms have led to significant distributional consequences, favouring capital owners and skilled workers, thus exacerbating inequality (Zhuang et al., 2014). Despite sustained economic growth, disparities in income and social opportunities persist and disproportionately affect vulnerable populations such as the poor, youth, women, and migrants (Saunders & He, 2017). Unequal access to opportunities due to institutional weaknesses and social exclusion compounds the impact of economic forces, leading to increased spatial disparities (Zhuang et al., 2014).

Crisis management has further complicated the social policy landscape, especially in the wake of the COVID-19 pandemic. The pandemic has exacerbated inequality, with social policy frameworks characterised by limited equitable public services, inadequate progressive taxation, and insufficient labour rights protections, especially for women (Martin, 2022). Many governments remain constrained by growing debt burdens and post-COVID austerity measures, limiting their ability to combat rising inequality effectively (Martin, 2022). While some governments have enhanced public services and labour rights, especially for women, most have not taken significant steps to address inequality during the pandemic period. This has revealed the varied and often inadequate policy responses across the region (Martin, 2022).

The Asian experience thus demonstrates that while countries are making considerable strides in developing social policy frameworks, the challenges of rising inequality and crisis management remain formidable. South Korea's experience exemplifies many of these challenges, with rapid population aging, increasing income inequality, and high relative poverty necessitating improved social insurance systems like the National Pension Scheme and National Health Insurance (Jones, 2012). In addition, the South Korean case highlights the need for targeted social programs, reduced labour market dualism, and addressing low social spending compared to other OECD countries. It also emphasises the reliance on hagwon for education and the necessity to raise tax revenues without hindering economic growth (Jones, 2012).

The diversity in policy responses and the varying degrees of success across the region highlight the need for tailored approaches that consider the unique socio-economic contexts of each country. Like the Latin American experiences, the Asian experiences offer valuable lessons for African social policy development in navigating the complexities of welfare state construction in developing economies.

1.3 Transformative Social Policy: A New Paradigm for Africa

Central to this chapter's analytical framework is the concept of transformative social policy (TSP). TSP offers a paradigmatic shift from conventional welfare approaches by conceptualising social policy as a multidimensional instrument for societal transformation rather than merely a mechanism for addressing immediate vulnerability (Mkandawire, 2007; Adesina, 2011). At its core, TSP understands social policy as '…collective interventions that directly affect social welfare, social institutions, and social relations' (Mkandawire, 2010). Unlike residual approaches that position social policy as a safety net deployed for market failures, TSP views social policy as integral to development processes through several interconnected functions. These include production enhancement, social protection, redistribution, reproduction, and social cohesion or nation-building (Adesina, 2011). This multifunctional understanding recognises that effective social policy simultaneously addresses immediate needs while transforming structural constraints that perpetuate inequality and supports community wellbeing across diverse contexts (Adesina, 2011).

A distinguishing feature of TSP is its normative foundation in principles of equality and solidarity, coupled with the rejection of artificial divisions between social and economic policy domains. Mkandawire (2004) noted that social policy should work in tandem with economic policy to achieve socio-economic development. This integration creates mutually reinforcing relationships where economic development generates resources for social policy expansion, while social investments enhance human capabilities that drive economic productivity and strengthen community resilience. When economic and social policies share normative commitments to equality and solidarity, they can address the interconnected challenges of poverty, inequality, and environmental sustainability while promoting collective welfare within communities (Adesina, 2025).

The TSP framework provides valuable perspectives for addressing what Mkandawire (2009) terms the 'national question' and the 'social question', which present interconnected challenges that are relevant in post-colonial and post-apartheid contexts (Mkandawire, 2009). The social question addresses '…problems engendered by social differentiation along class, ethnicity, gender and other social cleavages that arise or are unresolved within a nation', while the national question concerns challenges of nation-building and social cohesion in diverse societies with historical divisions (Mkandawire, 2009:1–2). Social policy thus plays a vital role in what Mkandawire describes as the 'nation-building project', which creates shared citizenship rights and responsibilities that transcend historical divisions while fostering community wellbeing (Mkandawire, 2009). By centring African intellectual traditions and focusing on context-specific policy solutions, TSP also provides a theoretical foundation for decolonising social policy knowledge and practice. This approach recognises that effective social protection systems must emerge from local political, economic, and social realities rather than externally imposed templates (Adesina, 2011; Gumede & Ndhlovu, 2023; Chipenda, 2023).

Unlike Latin America's institutional path, which created stratified systems that later required reform towards greater inclusion, or Asia's productivist approach that integrated social policy with economic development strategies, Africa has the opportunity to build universalistic social protection systems from the outset. Drawing on the TSP framework, Africa can potentially avoid the stratification trap that characterised Latin American development by designing inclusive systems from the beginning rather than trying to reform exclusionary institutions later. The TSP approach provides the normative foundation for such inclusive design, emphasising equality and solidarity as core principles rather than market-based stratification.

1.4 Features of African Social Policy

Africa's social policy development occurs within a context that presents both unique challenges and unprecedented opportunities. Several features differentiate African social policy contexts from other world regions and create specific imperatives for policy innovation.

1.4.1 Food Security as a Transformative Challenge

Unlike other world regions where food security has been largely resolved through agricultural modernisation and social protection systems, Africa continues to face acute food insecurity challenges. The continent's performance on SDG 2 (Zero Hunger) has been concerning, with the number of undernourished people increasing rather than decreasing in many countries (Atukunda et al., 2021). The COVID-19 pandemic and recent conflicts have further exacerbated these challenges, threatening progress on multiple nutrition-related targets.

The continent's agricultural sector, which employs approximately 60% of the population, remains vulnerable to climate shocks, limited technological adoption, and inadequate infrastructure (Le Roux & Cilliers, 2025). As argued by Chipenda and Adesina (in this volume), a TSP perspective is critical in addressing food security challenges that require integrated approaches to simultaneously enhance agricultural productivity, strengthen social protection for rural populations, and build resilient food systems that can withstand environmental shocks. This multi-functional approach recognises food security not merely as a welfare issue but as a foundation for economic development and social cohesion, with direct implications for achieving SDG 2 while contributing to poverty reduction (SDG 1), health outcomes (SDG 3), and sustainable agriculture (SDG 15). As Africa prepares for post-2030 development frameworks, food system transformation will likely become even more critical, requiring social policy innovations that can address climate adaptation, rural development, and nutrition security simultaneously.

1.4.2 Health Insurance Coverage and Universal Access

Africa's health insurance coverage remains among the lowest globally, with most countries achieving less than 20% population coverage through formal health insurance schemes (Odipo et al., 2024). This represents a significant barrier to achieving SDG 3.8 (universal health coverage), which requires that all people have access to needed health services without financial hardship. The continent's slow progress on this target contrasts sharply with achievements in other regions and highlights the urgent need for innovative financing mechanisms and service delivery models.

This situation contrasts sharply with both Latin American countries, where health insurance expansion has been central to social policy development since the 1990s, and Asian countries that have increasingly integrated health protection within their developmental strategies. The predominance of out-of-pocket health expenditure in Africa creates vulnerabilities that social policy frameworks must address through innovative financing mechanisms and service delivery models. The TSP framework suggests that health policy must be understood not as a separate sector but as integral to broader development objectives, requiring coordination between health investments and economic policy to create mutually reinforcing outcomes. This integration becomes important as post-2030 development frameworks may emphasise health system resilience and preparedness, requiring social policy innovations that can address health security as a foundation for sustainable development.

1.4.3 Land (Re)distribution and Urbanisation Dynamics

The dynamic relationship between land tenure systems, rural-urban migration, and urbanisation patterns creates complex social policy challenges. Africa's rapid urbanisation, proceeding at rates that exceed most other regions, creates both opportunities and challenges for SDG 11 (Sustainable Cities and Communities). While urbanisation can drive economic growth and improve access to services, unplanned and rapid urban growth often results in increased inequality, environmental degradation, and social tensions. Unlike other world regions where urbanisation preceded or accompanied industrial development, Africa's rapid urbanisation is occurring without corresponding industrialisation, creating large informal urban populations with limited access to formal social protection.

The dynamics of land governance in Africa are pronounced in post-settler contexts, where colonial legacies continue to shape contemporary land distribution patterns and social policy imperatives. In countries such as Zimbabwe, South Africa, and Namibia, the historical concentration of productive agricultural land in the hands of white minority populations has created enduring structural inequalities that demand transformative social policy interventions (Chipenda & Adesina, 2025). These contexts present unique challenges where land redistribution efforts must navigate between addressing historical injustices, ensuring agricultural productivity, and maintaining social cohesion. Across these post-settler contexts, elite capture of land redistribution processes emerges as a significant challenge undermining social policy objectives. Political elites and their networks often access redistributed land through informal channels, displacing intended beneficiaries and perpetuating inequality. This elite capture phenomenon is evident in Zimbabwe, where political connections frequently determine access to productive agricultural land (see Mkodzongi, 2022; Chipenda, 2022), and in South Africa, where there have been reports of elite capture of the land redistribution process (see Mtero et al., 2023). The result is the reproduction of elite-controlled agricultural systems that fail to address rural poverty or food security challenges effectively.

Large-scale land acquisitions by foreign investors, commonly termed 'land grabs', represent another dimension of contemporary land governance challenges. These acquisitions, often justified through food security or biofuel production narratives, frequently displace smallholder farmers and undermine local food systems. Countries such as Ethiopia, Ghana, and Mozambique have experienced significant foreign land acquisitions that have generated limited benefits for local populations while constraining agricultural development opportunities (see also Kebede et al., 2021). The social policy implications of these land grabs extend beyond immediate displacement to include long-term impacts on rural livelihoods, food security, and social cohesion.

The intersection of customary land tenure systems with formal property rights creates additional complexities for social policy design in post-settler contexts. In South Africa and Namibia, communal land areas governed by traditional authorities operate alongside freehold agricultural land, creating fragmented land governance systems that complicate policy implementation. Women's land rights remain constrained within customary tenure systems, limiting their participation in agricultural development and perpetuating gender inequalities (Commission for Gender Equality, 2024).

The challenge for TSP lies in developing integrated approaches that can navigate these dual tenure systems while promoting inclusive agricultural development. The persistence of customary land tenure systems alongside formal property rights creates additional dynamics for social policy design that differ from both the Latin

American latifundia systems and Asian land reform experiences. These dynamics are particularly complex in post-settler contexts where colonial land policies created enduring structural inequalities that continue to shape contemporary social policy challenges. Unlike Latin American systems or Asian land reform experiences, African post-settler contexts must navigate the intersection of racial inequalities, customary tenure systems, and formal property rights within single national frameworks.

The TSP framework emphasises the need for integrated approaches that can address rural-urban linkages while ensuring that urbanisation contributes to inclusive development and SDG achievement. In post-settler contexts, this integration becomes important as land redistribution efforts must be coordinated with urban development policies to prevent the reproduction of spatial inequalities and ensure that rural land reform contributes to broader social transformation. The failure to achieve meaningful land redistribution in countries like South Africa has contributed to urban migration pressures and informal settlement growth, highlighting the interconnected nature of land and urban policies. As post-2030 frameworks may place greater emphasis on spatial planning and territorial development, social policy innovations that can manage urbanisation as a transformative force will become increasingly important. This will require attention to post-settler contexts where historical land inequalities continue to shape contemporary development patterns and social policy imperatives.

1.4.4 Continental Integration as Policy Opportunity

Africa's social policy development occurs within the framework of continental integration initiatives. The AfCFTA, for example, as indicated earlier provides unprecedented opportunities for policy coordination and learning (Yaya et al., 2025). This continental dimension, combined with advances in digital technology and innovative financing mechanisms, creates possibilities for social policy models that transcend traditional categories identified in comparative literature. Unlike the fragmented integration processes that characterised Latin American development or the bilateral arrangements common in Asian cooperation, continental advances in Africa offer a comprehensive framework that provides unique opportunities for social policy harmonisation and learning.

1.5 The Legacy of External Dependency

The post-independence trajectory of African governments reveals a complex pattern of social policy development heavily influenced by donor agendas and external policy recommendations (see Chipenda & Tom; Ndhlovu & Mokone in this volume). This external influence has often resulted in fragmented approaches that prioritise donor preferences over locally defined priorities, creating what Mkandawire (2001) identified as states more accountable to external forces than to their citizens. This contrasts with the more autonomous policy development processes observed in successful Asian developmental states, where domestic priorities largely shaped social policy trajectories, even when learning from international experiences.

1.6 Analytical Framework: Four Pillars of African Social Policy Transformation

This volume contributes to evolving theoretical debates by addressing four interconnected questions that capture the dynamics of contemporary African social policy. Each is informed by a social policy framework that emphasises transformation, integration, and contextual adaptation.

1.6.1 The Transformative Potential of Social Policy

The first pillar examines whether social policy can serve as a transformative force in addressing structural inequalities and promoting inclusive development. This involves moving beyond residual approaches that merely provide safety nets to embrace social policy as a productive factor that can drive economic growth, social cohesion, and political stability. The transformative potential of social policy lies in its capacity to redistribute resources, create human capital, and generate social solidarity that underpins democratic governance. Drawing on TSP principles, this approach recognises that social policy interventions must simultaneously address immediate needs while transforming the structural constraints that perpetuate inequality. Unlike the stratification traps that characterised Latin American welfare development, African social policy can be designed from the outset to promote inclusion and equality through integrated approaches that combine economic and social objectives.

1.6.2 Envisioning Transformative Social Policy

The second pillar explores what transformative social policy could look like in African contexts, building on the TSP framework's multifunctional understanding of social policy. This involves designing policy frameworks that can address the continent's challenges while harnessing emerging opportunities. Key considerations include developing universal social protection systems that can accommodate informal economies; creating integrated approaches that address food security, health, education, and social protection simultaneously; and leveraging digital technologies to enhance service delivery and expand coverage. Unlike the employment-based insurance models that dominated both Latin American and Asian social protection development, African models must innovate around the realities of informal economies and extensive rural-urban linkages while maintaining the TSP commitment to equality and solidarity.

1.6.3 Global Political Constellations and Policy Space

The third pillar examines how African social policy development is embedded within broader global political constellations while seeking to expand policy space for autonomous development. This involves understanding how international trade rules, aid architectures, and global governance frameworks constrain or enable policy innovation. The TSP framework provides a foundation for challenging external dependency by emphasising locally defined priorities and indigenous knowledge systems. The challenge involves learning from Asian experiences of maintaining policy autonomy while engaging with global markets while avoiding the external dependency that has characterised much of Africa's post-independence development trajectory.

1.6.4 Decolonising Social Policy Discourse

The fourth pillar addresses the decolonisation of social policy discourse through the assertion of African epistemologies and value systems, directly aligned with TSP's emphasis on contextual adaptation and local ownership. This involves challenging the dominance of Western theoretical frameworks and developing indigenous approaches that reflect African values, institutions, and development priorities. The decolonisation project requires both critique of existing approaches and construction of alternatives grounded in African experiences and aspirations. This goes beyond the adaptation of existing models to embrace truly indigenous approaches that can draw selectively from global experiences while remaining rooted in African contexts and the TSP commitment to transformation through local agency.

1.7 Structure and Organisation of the Volume

This volume assembles voices from across the African continent to provide comprehensive coverage of social policy themes while maintaining coherence around the four analytical pillars outlined above. The contributions are organised to progress from foundational theoretical discussions through sectoral applications and cross-cutting themes to vulnerable populations and future directions. This structure culminates in forward-looking analyses of emerging trends and continental integration possibilities.

Part I: Theoretical Foundations
The opening section establishes theoretical foundations by critically engaging with existing social policy literature while proposing new frameworks suited to African contexts. Chapter 2 by Clement Chipenda and Tom Tom engages directly with the intellectual contributions of Thandika Mkandawire and Jimi Adesina, demonstrating how their vision for post-crisis African development challenges external paradigms and provides theoretical foundations for TSP. Their examination reveals how these scholars' combined ideas around the concept of TSP challenge dominant external paradigms while providing vital tools for reconceptualising social policy in ways that centre African epistemologies and developmental aspirations. Chapter 3 by Newman Tekwa extends this analysis by examining gender as an important dimension of TSP, moving beyond mainstreaming approaches to embrace structural transformation. Tekwa demonstrates how transformative approaches must inherently address gender inequities to achieve meaningful social change. Chapter 4 by Emmanuel Ndhlovu and Neo Mokone explores how social policy frameworks in Africa are significantly influenced by colonial history and continued relationships with former colonisers. It examines social policy's potential for inclusive development in Africa through reviewing existing debates on African social policies. This chapter argues that African social policy needs reimagining as part of a decolonial agenda for inclusive development, engaging directly with social policy's role as an instrument for change rather than merely a welfare mechanism.

Part II: Sectoral Innovations
Contemporary challenges facing African social policy are examined through multiple lenses in this section. It critically examines specific policy domains through both comparative and country-specific lenses, demonstrating how social policy operates across different sectors. Chapter 5 by Ambrose T. Kessy examines Tanzania's COVID-19 response, demonstrating how political leadership transitions can significantly reshape health governance outcomes and showing how health emergencies can trigger social policy realignment and innovation.

Chapter 6 by Emmanuel Ndhlovu and Daniel F Meyer examines social protection measures in Southern Africa, while Chap. 7 by Jeremiah M. Muneeni studies Kenya's healthcare policy realignment. These chapters reveal the dynamics between crisis response and systemic transformation, bringing the dynamics of COVID-19 into the spotlight. Chapter 8 by Theshaya Naidoo addresses ethical and technological innovations in social protection, revealing both the potential and risks of incorporating digital technologies into social protection systems, especially in contexts where data privacy frameworks remain underdeveloped. In addition, Chap. 9 by Alex Nadège Ouedraogo provides historical analysis of food and nutrition policy actors in Senegal (1918–2021), illuminating the importance of understanding policy trajectories and figurations in effective intervention design. Chapter 10 by Clement

Chipenda and Jimi Adesina examines agricultural market participation among resettled farmers in Zimbabwe, substantiating the argument that land reform functions as a vital social policy instrument with significant welfare implications.

Each domain is analysed through the imperative that social policy has to be transformative. The contributions explore how different sectors from a social policy perspective can contribute to the multifunctional objectives of production enhancement, social protection, redistribution, reproduction, and social cohesion.

Part III: Cross-Cutting Themes
This section addresses cross-cutting themes that span multiple sectors and demonstrate the integrated nature of social policy in the African context. The role of civil society organisations in shaping and implementing social policy emerges as a crucial theme through Chap. 11 by Otshepeng Mazibuko's examination of South African civil society and Chap. 12 by Takudzwa Musekiwa's broader continental analysis. These chapters demonstrate how non-state actors serve as essential catalysts for inclusive governance and policy innovation while highlighting the challenges they face in navigating complex political landscapes. Chapter 13 by Neo Mokone and Emmanuel Ndhlovu bridges gaps between informal economies and transformative social policy, highlighting how rigid regulatory frameworks often fail to accommodate the characteristics of informal work. Chapter 14 by Deborah Andrew Ngusa examines Village Community Banks (VICOBAs) in Tanzania, revealing both the potential and limitations of informal financial mechanisms in addressing women's economic empowerment and demonstrating how community-based financial mechanisms can serve as social policy tools for economic empowerment. Chapter 15 by Winnet Hokoza and Clement Chipenda advances gender transformation perspectives through investigation of masculinity and mental health in a comparative analysis of Zimbabwe and South Africa. It reveals how restrictive gender norms impact help-seeking behaviours and mental wellbeing. Chapter 16 by Raymond Chirowamhangu analyses inclusive education policy implementation in South Africa, offering vital perspectives on practical interventions for learners with disabilities, highlighting both policy achievements and implementation gaps.

Part IV: Vulnerable Populations and Social Justice
The fourth section focuses on transformative approaches to addressing structural inequalities and supporting marginalised populations. Regional perspectives on youth development challenges are examined in Chap. 17 by Edward Muzondo et al. Their analysis focuses on drug abuse among youth in Southern Africa, advocating for rehabilitative rather than punitive approaches. The chapter emphasises social development principles through the ARISE Cycle as a decolonised rehabilitation approach. Chapter 18 by Primrose Hove and Clement Chipenda explores children's rights in rural Zimbabwe through a comparative analysis with South Africa, using an integrated framework that combines TSP and human rights approaches. The chapter reveals how this integrated theoretical framework can illuminate both structural barriers and opportunities for transformative change. Their analysis of institutional arrangements and power relations offers important perspectives for rights-based transformative interventions in rural contexts.

Part V: Synthesis and Future Directions
The concluding chapter synthesises lessons from diverse country experiences while exploring the implications of continental integration for social policy coordination. Chapter 19 examines regional perspectives and policy harmonisation opportunities, drawing on the volume's diverse contributions to identify patterns and possibilities for South-South learning. It provides synthesis and recommendations for the future of social policy in Africa, addressing post-2030 development frameworks and the continent's evolving social policy landscape.

1.7.1 Analytical Coherence and Contributions

Each contribution engages with the volume's overarching themes while providing important and unique perspectives based on authors' expertise and research. The diversity of perspectives ranging from macro-level policy analysis to micro-level implementation studies reflects the complex dynamics of social policy challenges while maintaining analytical coherence. The progression from theoretical foundations through sectoral applications to cross-cutting themes and future directions demonstrates how social policy operates across multiple levels and dimensions.

The volume's organisation around the four pillars of transformation, policy design, global context, and decolonisation ensures that each section contributes to the broader analytical framework while addressing specific dimensions of African social policy development. The concept of social policy unique to the African context provides the unifying theoretical foundation that allows diverse contributions to speak to common themes of transformation, integration, and contextual adaptation.

1.7.2 Contribution to Global Social Policy Debates

This volume makes several important contributions to global social policy debates. First, it provides comprehensive examination of the social policy discourse across the African continent, addressing a significant gap in comparative social policy literature through concrete country experiences and sectoral analyses. Second, it centres African voices and perspectives in social policy analysis, challenging the dominance of Northern theoretical frameworks through the intellectual contributions of scholars like Mkandawire and Adesina while demonstrating indigenous approaches to policy innovation. Third, it offers new analytical tools for understanding social policy in contexts characterised by informal economies, persistent rural-urban linkages, and continental integration processes. The volume's attention to village community banks, informal economy integration, and civil society innovations provides models for policy design that transcend traditional categories identified in comparative literature. Fourth, it advances the TSP framework as an important African contribution to global social policy theory, demonstrating how transformative approaches can address the interconnected challenges of development, democracy, and social cohesion while contributing to SDG achievement and post-2030 development preparation.

The volume's emphasis on understanding social policy as a transformative instrument that can simultaneously address immediate needs while transforming structural constraints provides valuable insights. This multifunctional approach offers valuable perspectives for other regions confronted with similar challenges of inequality, informality, and weak state capacity while demonstrating the potential for social policy to accelerate progress towards global development goals. The comparative approach, drawing lessons from Latin American and Asian experiences while remaining grounded in African realities, provides a model for South-South learning and knowledge exchange.

By addressing the four key pillars of transformation, policy design, global context, and decolonisation through the TSP lens, this volume contributes to broader debates about the future of social policy in an increasingly interconnected but unequal world. The African experience, understood through social policy, offers valuable lessons for other regions about the possibilities for social policy to serve as an instrument for inclusive development, democratic governance, and social cohesion. It also provides useful perspectives for the design of post-2030 development frameworks that can address the systemic challenges of the twenty-first century. As the global community prepares for the next phase of international development cooperation, Africa's social policy innovations and theoretical contributions provide essential foundations for more inclusive, sustainable, and transformative approaches to global development.

1.8 Conclusion: Charting New Pathways and the Imperative for Decolonising Social Policy for African Futures

Through these diverse chapters, this volume makes a significant contribution to ongoing efforts to decolonise social policy thinking and practice in Africa. It advocates for approaches that emphasise policy sovereignty, indigenous knowledge systems, and structural transformation while maintaining African agency in policy formulation and implementation. The chapters collectively challenge conventional wisdom by demonstrating how colonial legacies continue to influence contemporary frameworks across multiple countries. The volume establishes robust theoretical foundations that centre African epistemologies, moving beyond external paradigms to embrace social policy approaches that address the continent's unique historical context, cultural diversity, and developmental aspirations. Critical perspectives into how external pressures create policy circumnavigation reveal the ongoing challenges to local policy autonomy.

An important aspect of this volume is that it advances discourse beyond critique alone, offering constructive alternatives rooted in African realities. Case studies from COVID-19 responses, healthcare reforms, and social protection systems highlight instances where locally grounded policies have delivered meaningful outcomes. Analyses of informal economies, Village Community Banks (VICOBAs), and agricultural market participation reveal both institutional barriers and innovative pathways forward. The exploration of ethical and technological innovations demonstrates how digital technologies can enhance inclusion while addressing critical concerns about privacy and equity in contexts where data privacy frameworks remain underdeveloped.

The collection shows that decolonising social policy requires addressing multiple dimensions simultaneously. Gender transformative approaches examining masculinity and mental health demonstrate how social policy must confront embedded power relations. The importance of contextual understanding and participatory processes emerges through analyses of healthcare reforms, food and nutrition policy trajectories, and resource governance in mineral-rich communities. Civil society organisations emerge as essential catalysts for inclusive governance. The critical challenge of translating inclusive education policy into practice for learners with disabilities highlights the gap between policy formulation and implementation.

The volume addresses emerging challenges facing African youth through innovative approaches that emphasise rehabilitative rather than punitive measures. The ARISE Cycle emerges as a decolonised rehabilitation approach, while TSP principles demonstrate potential for addressing unemployment, strengthening social safety nets, rebuilding community structures, and enhancing institutional coordination. The integration of TSP with human rights-based approaches highlights both structural barriers and opportunities for transformative change in rural contexts. The volume's comprehensive scope demonstrates the multifaceted nature of social policy in the African context through diverse methodological approaches spanning historical analysis, empirical case studies, and policy implementation research. This methodological diversity reflects the dynamics of social policy challenges while maintaining analytical coherence through the shared TSP framework.

Ultimately, this volume provides both theoretical frameworks and practical insights for policymakers, scholars, and practitioners committed to reimagining social policy in Africa. By bringing together voices from across the continent, the collection demonstrates the diversity of African experiences while identifying common patterns and possibilities for South-South learning. It serves as an invitation to collaborative action towards creating more effective, equitable, and culturally appropriate social policy solutions that honour the continent's rich diversity while addressing shared aspirations for inclusive and sustainable African futures. The volume's emphasis on social policy as a multifunctional instrument that can simultaneously address immediate needs while transforming structural constraints offers valuable perspectives not only for Africa but for other regions confronted with similar challenges of inequality, informality, and the imperative for decolonised approaches to development.

References

Adesina, J. (2011). Beyond the social protection paradigm: Social policy in Africa's development. *Canadian Journal of Development Studies, 32*(4), 454–470.

Adesina, J. (2021). Introduction: Social policy in the African context. In J. Adesina (Ed.), *Social policy in the African context* (pp. 1–12). CODESRIA.

Adesina, J. (2025). *Meeting the Cesare challenge: Locally grounded, globally relevant scholarship (in postgraduate training).* Paper presented at the UNISA College of Graduate Studies Research and Innovation Week, 19 May.

AfDB. (2025). *African Economic Outlook 2025*. African Development Bank.

Atukunda, P., Eide, W. B., Kardel, K. R., Iversen, P. O., & Westerberg, A. C. (2021). Unlocking the potential for achievement of the UN Sustainable Development Goal 2—'Zero hunger'—In Africa: Targets, strategies, synergies and challenges. *Food & Nutrition Research, 65*, 7686. https://doi.org/10.29219/fnr.v65.7686

Barrientos, A. (2024). *Social protection in Latin America: Causality, stratification and outcomes*. Palgrave Macmillan.

Chipenda, C. (2022). Critical insight on joint venture partnerships in post-land reform Zimbabwe. Progress or retrogression? In G. Mkodzongi (Ed.), *The future of Zimbabwe's agrarian sector. Land issues in a time of political transition* (pp. 146–167). Routledge.

Chipenda, C. (2023). Rethinking the role of social policy in responding to the Covid-19 pandemic in South Africa: Critical reflections using the intellectual contribution of Thandika Mkandawire. In V. Gumede & E. Ndhlovu (Eds.), *Development in Africa: Remembering Thandika Mkandawire* (pp. 197–231). Ace Press.

Chipenda, C. (2024). A transformative social policy perspective on land and agrarian reform in Zimbabwe. *Africa Spectrum, 59*(1), 89–118.

Chipenda, C., & Adesina, J. (2025). The shifting role of small-scale farms in agriculture and food systems transformation in a post-COVID-19 context: Critical lessons from Zimbabwe. In E. Ndhlovu (Ed.), *African food systems*. Palgrave Macmillan. https://doi.org/10.1007/978-3-031-90823-1_4

Commission for Gender Equality. (2024). *Exploring barriers to women's access to communal land in selected provinces of South Africa*. Commission for Gender Equality.

Cortés, R. (2008). The contemporary social policy debate in Latin America. *Global Social Policy, 8*(1), 109–114. https://doi.org/10.1177/1468018107086091

Croissant, A. (2004). Changing welfare regimes in East and Southeast Asia: Crisis, change and challenge. *Social Policy & Administration, 38*(5), 504–524. https://doi.org/10.1111/J.1467-9515.2004.00404.X

Cruz-Martínez, G., Vargas Faulbaum, L., & Velázquez Leyer, R. (2021). *Social security and pension systems: The deep stratification of Latin American societies* (pp. 163–194). Palgrave Macmillan. https://doi.org/10.1007/978-3-030-61270-2_6

Danboyi, J. C. (2025). *ECOWAS and authoritarian governments*. ODUMUNC 2025 Issue Brief. Old Dominion University. https://www.odu.edu/sites/default/files/2025/documents/ECOWAS%20authoritarian%20governments.pdf

De Haan, A. (2011). Social policies in Asia and Europe in the twenty-first century: Convergence or path dependence? *International Journal of Sociology, 41*(3), 46–67. https://doi.org/10.2753/IJS0020-7659410302

Diyakova, L. (2022). Social policy in the period of COVID-19: Latin American approaches. *Latinskaia Amerika, 5*, 28–44. https://doi.org/10.31857/S0044748X0019914-4

Filgueira, F., Arza, C., & Blofield, M. (2023). The slow and reluctant development of social citizenship in Latin America. In K. J. Baehler (Ed.), *The Oxford handbook of governance and public management for social policy*. Oxford Academic. Retrieved July 11, 2025, from https://doi.org/10.1093/oxfordhb/9780190916329.013.57

Gumede, V., & Ndhlovu, E. (2023). Introduction: Honouring Thandika Mkandawire. In V. Gumede & E. Ndhlovu (Eds.), *Development in Africa: Remembering Thandika Mkandawire* (pp. 1–18). Ace Press.

Haggard, S., & Kaufman, R. R. (2008). *Development, democracy and welfare states*. Princeton University Press.

Huber, E.., & Ponce de León, Z. (2019). The changing shapes of Latin American welfare states. In *Oxford research encyclopedia of politics*. Retrieved July 11, 2025, from https://oxfordre.com/politics/view/10.1093/acrefore/9780190228637.001.0001/acrefore-9780190228637-e-1656

Huber, E., & Stephens, J. D. (2012). *Democracy and the left in Latin America. Social policy and inequality in Latin America*. University of Chicago Press.

Jones, R. S. (2012). Meeting the social policy challenges facing Korea. *Asian Economic Policy Review, 7*(1), 91–108. https://doi.org/10.1111/J.1748-3131.2012.01221.X

Kebede, D., Tesfay, G., & Emana, B. (2021). Impact of land acquisition for large-scale agricultural investments on income and asset possession of displaced households in Ethiopia. *Heliyon, 7*(12), e08557. https://doi.org/10.1016/j.heliyon.2021.e08557

Le Roux, A., & Cilliers, J. (2025). *Agriculture*. Published online at futures.issafrica.org. Retrieved July 11, 2025, from https://futures.issafrica.org/thematic/04-agriculture/

Lin, K., & Chan, R. K. H. (2015). Repositioning three models of social policy with reference to East Asian welfare systems. *International Social Work, 58*(6), 831–839. https://doi.org/10.1177/0020872813503857

Martin, M. (2022). *Asia's extreme inequality crisis: Building back fairer after COVID-19*. Retrieved July 11, 2025, from https://doi.org/10.21201/2022.9547

Mesa-Lago, C. (1978). *Social security in Latin America: Pressure groups, stratification, and inequality*. University of Pittsburgh Press.

Mesa-Lago, C. (2004). Models of development, social policy and reform in Latin America. In T. Mkandawire (Ed.), *Social policy in a development context. Social policy in a development context* (pp. 175–202). Palgrave Macmillan. https://doi.org/10.1057/9780230523975_8

Mkandawire, T. (2001). Thinking about development states in Africa. *Cambridge Journal of Economics, 25*(3), 289–313.

Mkandawire, T. (2004). Introduction: Social policy in a development context. In T. Mkandawire (Ed.), *Social policy in a development context* (pp. 1–33). UNRISD/Palgrave Macmillan.

Mkandawire T (2007) Transformative social policy and innovation in developing countries. *The European Journal of Development Research 19*(1), 13–29.

Mkandawire, T. (2009). Institutional monocropping and monotasking in Africa. In N. Akbah, B. Kwesi, & S. Howard (Eds.), *Good growth and governance in Africa: Rethinking development strategies* (pp. 1–28). UNRISD.

Mkandawire, T. (2010). How the new poverty agenda neglected social and employment policies in Africa. *Journal of Human Development and Capabilities, 11*(1), 37–55.

Mkodzongi, G. (2022). Zimbabwe's changing agrarian sector in perspective. In G. Mkodzongi (Ed.), *The future of Zimbabwe's agrarian sector: Land issues in a time of political transition* (pp. 1–13). Routledge.

Mok, K. H., Kühner, S., & Yeates, N. (2017). Introduction—Managing welfare expectations and social change: Policy responses in Asia. *Social Policy & Administration, 51*(6), 845–856. https://doi.org/10.1111/SPOL.12335

Mtero, F., Gumede, N., & Ramantsima, K. (2023). Elite capture in South Africa's land redistribution: The convergence of policy bias, corrupt practices and class dynamics. *Journal of Southern African Studies, 49*(1), 5–24. https://doi.org/10.1080/03057070.2023.2187969

Odipo, E., et al. (2024). The path to universal health coverage in five African and Asian countries: Examining the association between insurance status and health-care use. *The Lancet Global Health, 12*(1), e123–e133.

Rakšnys, A. V., Svirbutaitė-Krutkienė, G., & Guogis, A. (2022). The specificity and development issues of welfare state models in East and Southeast Asia. *Tiltai, 87*(2), 53–67. https://doi.org/10.15181/tbb.v87i2.2317

Saunders, P., & He, A. J. (2017). Social protection in East Asian Chinese societies: Challenges, responses and impacts. *Journal of Asian Public Policy, 10*(2), 121–127. https://doi.org/10.1080/17516234.2017.1306611

Segura-Ubiergo, A. (2007). *The political economy of the welfare state in Latin America*. Cambridge University Press.

Tiwari, I. P. (2013). State welfarism and social welfare in Asia. *Journal of Arts and Humanities, 2*(2), 64–78. https://doi.org/10.18533/JOURNAL.V2I2.70

Tom, T. (2024). Social policy, development and democracy in Africa: Reconnoitering Thandika Mkandawire's philosophy. *Journal of Asian and African Studies*, 1–19. https://doi.org/10.1177/00219096241270705

United Nations. (2024a). *The sustainable development goals report 2024*. UN.

United Nations. (2024b). *Promotion of durable peace through sustainable development in Africa*. Report of the Secretary-General 23 July 2024. UN.

UNRISD. (2023). *Crises of inequality: Shifting power for a new eco-social contract. Research and Policy Brief 39*. UNRISD.

Wong, J. (2025). *The welfare state in East Asia*. Cambridge University Press. https://doi.org/10.1017/9781108887120

World Bank. (2020). *The African Continental Free Trade Area: Economic and distributional effects*. World Bank.

Yaya, I. M., Ndiaye, C. T., Asongu, S., & Diop, S. (2025). Analysis of the effects of the African Continental Free Trade Area (AfCFTA) on the labor market in the Central African Economic and Monetary Community (CEMAC). *International Journal of Manpower, 46*(10), 77–102.

Zhuang, J., Kanbur, R., & Rhee, C. (2014). Rising inequality in Asia and policy implications. *Research Papers in Economics*. Retrieved July 11, 2025, from https://www.econstor.eu/bitstream/10419/101277/1/779008596.pdf

Part I
Theoretical Foundations

Chapter 2
Beyond External Paradigms: Theorising Transformative Social Policy Through Thandika Mkandawire and Jimi Adesina's Vision for Postcrisis African Development

Clement Chipenda and Tom Tom

2.1 Introduction

In the aftermath of the COVID-19 pandemic, Africa finds itself at a critical juncture where the inadequacy of externally imposed development paradigms has become starkly apparent. Global supply chains remain disrupted, inflation continues to surge, and geopolitical tensions are reshaping international relations, creating what the UNRISD (2022, 2023) describes as 'a global state of fracture'. This era is characterised by multiple intersecting crises that include violent conflict, environmental destruction, political and social polarisation, the lingering consequences of the COVID-19 pandemic, a cost-of-living crisis, and ever-rising inequalities. The continent's experience during the pandemic, characterised by limited fiscal space, over-reliance on external aid, and the fragility of health systems designed according to donor priorities, has exposed the weaknesses of development models that prioritise external validation over local ownership. Africa continues to experience multiple socio-economic and political challenges with both endogenous and exogenous roots. This is significant in a context of serpentine neoliberalism, covert and overt neocolonialism, and an obstinately unequal global order (Gumede, 2015; Nhemachena, 2023). In addition, the emergence of new global power configurations, including the growing influence of China, Brazil, and India in African development, alongside the resurgence of protectionist policies in traditional donor countries, has created both opportunities and challenges that demand fresh theoretical frameworks for understanding social policy in the African context. As the UNRISD (2023:3) notes, 'inequality, environmental degradation and lack of resilience associated with the current global economic system is not an unfortunate

C. Chipenda · T. Tom (✉)
South African Research Chair in Social Policy, College of Graduate Studies, University of South Africa, Pretoria, South Africa
e-mail: chipec@unisa.ac.za

E. Ndhlovu, C. Chipenda (eds.), *Redefining the Social Policy Discourse*,
https://doi.org/10.1007/978-3-032-24067-5_2

by-product but rather built in by design' (see also Kempf et al., 2023). These multiple intersecting crises are exacerbated by persistent North-South power dynamics in social policy discourse, where African development agendas remain subordinated to external paradigms (Adesina, 2020). The urgency of these challenges is compounded by the accelerating effects of climate change, which is negatively affecting African nations while simultaneously exposing the inadequacy of existing social protection systems to address climate-induced vulnerabilities. This power asymmetry manifests in policy prescriptions that often disregard local contexts and undermine national sovereignty, creating a problem on how social policy is conceptualised and implemented across the continent (Tom, 2024a, 2024b; Chipenda, 2024, 2025).

What makes this historical moment significant is the convergence of multiple transitions. These include the digital revolution transforming how social services are delivered, the youth demographic dividend offering unprecedented opportunities for development, and the growing assertiveness of African institutions like the African Union in articulating continental development agendas. The contemporary significance of theoretical contributions by African scholars to African development is evident as the continent engages with these entrenched challenges. These transformations occur alongside the increasing recognition that traditional development metrics, focused primarily on GDP growth and poverty reduction, fail to capture the multidimensional nature of human wellbeing and social transformation that contemporary African societies seek. While knowledge generation on African affairs to improve the continent's development is ascending (Adesina, 2022; Hormeku-Ajei et al., 2022), the adoption of transformative philosophies that link and advance social policy and development remains largely under-explored (Tom, 2024b). The emergence of homegrown solutions from Rwanda's community-based health insurance schemes to Ethiopia's Productive Safety Net Programme demonstrates that African countries have the potential and are increasingly willing to experiment with alternatives to donor-driven models, yet these innovations remain insufficiently theorised within broader frameworks of transformative social policy (TSP).

This chapter makes a contribution to knowledge by examining how preeminent African intellectuals - Thandika Mkandawire and Jimi Adesina - whose scholarship has significantly reshaped social policy discourse on the continent, offer a theoretical framework that not only challenges dominant external paradigms but also provides vital tools for reconceptualising social policy in ways that centre African epistemologies, lived experiences, and developmental aspirations. Their work gains relevance in the present context where African countries are challenged with the legacies of structural adjustment, the imperatives of sustainable development goals, and the need to build resilient systems capable of withstanding future shocks. This comes in a context where the quest for improving Africa's wellbeing in its diversity has been explored through various lenses, from African renaissance to changing development discourse (Nyerere, 1968; Shivji, 2006). However, postcolonial socialist ideologies and expansionist approaches, despite being noble, were derailed by several factors, chiefly neoliberalism and the hegemony of the Global North in Africa's development, governance, and social policy. Today, this hegemony manifests in new forms, through the conditionalities attached to climate finance, the digital divide that reinforces technological dependence, and the continued extraction of natural resources under the guise of 'green transitions' that serve primarily to meet Northern consumption demands. Africa has become 'a laboratory of development models and donor practices' (Hujo, 2021:33), especially since the debt crisis of the early 1980s paved the way for stronger integration of African economies into global markets.

The timeliness of revisiting Mkandawire and Adesina's theoretical contributions is in the context of present global debates about 'building back better' post-COVID-19, which often reproduce the same externally driven approaches that have historically failed to deliver transformative outcomes for African societies (see Chipenda, Introduction in this volume). For the purposes of this chapter, the research objectives are systematically analysing the theoretical contributions of Mkandawire and Adesina, articulating how their ideas provide alternatives to externally imposed paradigms, while demonstrating the practical applications of TSP in addressing contemporary African challenges. These objectives are urgent given the present moment where African countries face the dual challenge of recovering from the pandemic while simultaneously preparing for future crises, including climate change impacts, demographic pressures, and potential future pandemics. Through an interpretive approach that examines their scholarly works, the chapter seeks to highlight pathways towards policy sovereignty. The relevance of their framework extends beyond academic discourse to practical policymaking, as evidenced by the growing number of African countries that are beginning to assert greater control over their development agendas. The urgency of their ideas is highlighted by how neoliberal economics have permeated all areas of Africa's social policy and development, resulting in policy regimes that fail to support production, redistribution, reproduction, and social compact of the present and future generations (Adesina, 2015; Mkandawire, 2006). This failure has been evident in the context of the pandemic, where countries with stronger social protection systems and more robust public health infrastructure, characteristics associated with TSP approaches, demonstrated greater resilience and more effective responses to the crisis. This is critical in Africa, where policy merchandising by the Global North is rife and social policy has been downgraded to charity rather than an investment in human capabilities and development (Adesina, 2020). The contemporary manifestation of this 'policy merchandising' includes the proliferation of fintech solutions for social protection that prioritise technological innovation over social transformation and the emphasis on 'evidence-based policy' that often privileges randomised controlled trials over contextual understanding and local knowledge systems. Social policy in the African context is seen as having been highly limited in addressing intersecting inequalities of class, gender, generation, disability, and space, while the synergy between social and economic policies necessary for sustainable improvement in people's wellbeing remains elusive (Mkandawire, 2010).

What distinguishes this analysis from previous examinations of African social policy is its focus on how Mkandawire and Adesina's theoretical framework provides tools for navigating the dynamic terrain of the twenty-first-century development challenges, from the fourth industrial revolution to climate adaptation while maintaining African agency and sovereignty. This chapter's unique contribution lies in its systematic analysis of the contributions by Mkandawire and Adesina combined through the transformative social policy theoretical framework that provides tools for transcending these limitations. Their work offers a prescient analysis of how African countries can leverage present global transitions, including the shift towards renewable energy, the growing importance of the care economy, and the increasing recognition of indigenous knowledge systems to advance TSP agendas. Mkandawire's vision, rooted in radical nationalism, focuses on dismantling racialised neoliberal policies through transformative approaches that prioritise emancipation and social justice (Phiri, 2023). His work critiques the dominance of externally imposed neoliberal frameworks that perpetuate inequality, advocating instead for policies that reflect local contexts, promote African agency, and prioritise human welfare (Cheru, 2022; Nagar, 2020). In the contemporary context, this vision takes on new dimensions as it addresses how African countries can harness technological innovations, demographic advantages, and natural resource endowments to build more equitable and sustainable societies. Complementing this vision, Adesina's contributions emphasise the need for inclusive policies that address historical injustices and the importance of tailoring social policies to local realities (Adesina, 2007). His framework becomes relevant in addressing contemporary challenges such as youth unemployment, climate-induced migration, and the need for social protection systems that can adapt to changing labour markets and emerging forms of work. For Adesina (2021a: 2–3), '...progressive social policy is fundamentally about ensuring human flourishing' through enhancing productive capacity, protecting people throughout the life cycle, attending to distributive outcomes, and advancing social cohesion.

The contemporary relevance of their combined scholarship is further evidenced by how their theoretical framework anticipates and addresses many of the challenges that have become prominent in post-pandemic development discourse (as indicated by Chipenda, see Introduction in this volume). These challenges include the need for universal social protection, the importance of public investment in health systems, the recognition of care work as essential labour, and the imperative of building resilient societies capable of withstanding multiple shocks. Their combined scholarship thus advances beyond the traditional focus on colonial arrangements and postcolonial political economy (Onimode, 1998; Rodney, 1981) construction of alternative pathways through which African states can assert sovereignty over their development agendas. Through an interpretive analysis of their ideas, this chapter demonstrates how their theorisation of TSP, with its emphasis on policy sovereignty, indigenous knowledge systems, and transformative approaches, provides an important foundation for addressing contemporary challenges while asserting African agency in policy formulation. This analysis is timely as African countries navigate the dynamic landscape of achieving the Sustainable Development Goals (SDGs) while simultaneously addressing the specific challenges of post-pandemic recovery, climate adaptation, and the need to harness the demographic dividend. The chapter proceeds in four main sections. First is the methodology section, which highlights the employment of a semi-systematic review of literature from Web of Science, Google Scholar, Research Gate, and other scholarly databases. Second is the theoretical framework which highlights Mkandawire and Adesina's theorising of TSP. The TSP analytical framework, which was initiated by Mkandawire through the UNRISD's Social Policy in a Development Context project with attention to policy sovereignty, the developmental state, and Adesina's theoretical extensions which reconceptualise social policy beyond welfare, its basis for inclusive development, and the incorporation of indigenous knowledge systems, is presented in this section. Third, the discussion section addresses contemporary challenges in decolonising social policy discourse and tackling power asymmetries in policy formation, with particular attention to post-pandemic recovery and national sovereignty implications. Finally, the conclusion synthesises their theoretical contributions to African social policy while outlining future research directions.

2.2 Research Methods and Materials

This study employed an interpretive research paradigm to analyse the theoretical contributions of Mkandawire and Adesina to African social policy discourse. As Gadamer (2004) argues, interpretivism offers valuable insights into social phenomena by emphasising the subjective meaning-making processes that shape social reality. This approach is appropriate for examining scholarly works that challenge dominant paradigms and propose alternative frameworks (see Bhattacherjee, 2012; Pervin & Mokhtar, 2022), as is the case in this chapter. The research methodology centred a semi-systematic literature review, utilising content and discourse analysis techniques. We prioritised reputable academic databases including Scopus, Web of Science, EBSCO, Google Scholar, and JSTOR, among others, selected for their international credibility, impact factor, and accessibility (Krippendorff, 2019). Initial keyword searches using terms 'Thandika Mkandawire', 'Jimi Adesina', 'development', 'social policy', and others yielded 196 potential sources. Due to the wide breadth of these keywords, the literature had to be trimmed based on inclusion and exclusion criteria established prior to initiating the study. To enhance the analysis, we implemented structured inclusion criteria. Articles had to (1) engage with Mkandawire and Adesina's perspectives on social policy in the African context, (2) demonstrate clear connections to social policy and development frameworks, (3) have undergone peer review, and (4) be available in English. This filtering process resulted in a final corpus of 50 core articles, supplemented by contextual literature examining socio-economic and political dynamics in Africa.

Our analytical framework draws primarily from the TSP approach. Content analysis enabled systematic examination of recurring themes, theoretical constructs, and policy implications across the selected literature. Scholars explored the significance of content analysis (Krippendorff, 2019). This involved identifying key concepts, examining contextual relationships, and mapping theoretical interconnections between social policy and development paradigms. Navigating the challenges of interoperability and fragmentation across diverse source materials, we sought common threads and patterns across disparate materials to establish coherent theoretical frameworks. To mitigate potential biases, we implemented methodological safeguards including cross-verification across multiple sources, critical examination of ideological underpinnings, and contextual interpretation within broader sociopolitical landscapes. As emphasised by scholars (see Bhattacherjee, 2012), these measures enhanced the analytical integrity and interpretive validity of our findings. The analysis focused specifically on extracting Mkandawire and Adesina's theoretical frameworks regarding critiques of externally imposed policy paradigms, conceptualisations of TSP, approaches to policy sovereignty, perspectives on the developmental state, and the integration of indigenous knowledge systems. These analytical categories facilitated systematic examination of their intellectual contributions while providing structured comparative frameworks.

2.3 Conceptualising Social Policy: Definitions, Contestations, and Transformative Approaches

2.3.1 Defining Social Policy in the African Context

The conceptualisation of social policy in Africa reveals significant tensions arising from competing theoretical traditions and political ideologies. As Aina (2021) notes, dominant definitions of social policy often stem from Western thought and practices, leading to conceptual limitations when applied to African contexts. This definitional challenge can be seen as being further complicated by what Hujo (2021:33) describes as '…an increasing hybridisation of models and fragmentation of approaches rather than a consolidation of a social model…' across the African continent. Traditional definitions have characterised social policy as encompassing '…social insurance schemes, social assistance programmes, and labour market policies… which are either directly financed and provisioned through the state or at least regulated by the state…' (Hujo, 2021:34). This conventional understanding, however, becomes problematic when transplanted into African societies, where colonial legacies, structural adjustment programmes, and external donor influences have fostered diverse institutional environments.

Mkandawire's work has been very useful in offering important perspectives into social policy within the developmental state. His emphasis has been that approaches to social policy should be holistic, normative, deliberate, and coherent (Mkandawire, 2004, 2006). His early definition framed social policy as 'collective interventions directly affecting transformations in social welfare, social institutions, and social relations… [and] access to adequate and secure livelihoods and income' (Mkandawire, 2001:1). This evolved into a more expansive conceptualisation, where he can be seen describing social policy as '…a collective intervention in the economy to influence access to and the incidence of adequate and secure livelihoods and income' (Mkandawire, 2004:1). Adesina (2007:1–2; 2009) expanded this view, stating that social policy involves:

> ...the collective public efforts aimed at effecting and protecting the social well-being of the people within a given territory. Beyond immediate protection from social destitution, social policy might cover education and healthcare provision, habitat, food security, sanitation, and guarantee some measure of labour market protection. The idea of a tolerable, minimum level of livelihood and decency is intuitive and socially constructed; and normative (ideological) rather than technical.

These definitions highlight social policy's multidimensional character, functioning simultaneously as an intellectual framework and a practical intervention strategy (Noyoo & Boon, 2021). Aina (1997: 17) notes that social policy represents '...systematic and deliberate interventions in social and economic life for the purpose of ensuring the basic needs and well-being of the peoples of a society'. This understanding is crucial for achieving comprehensive and transformative development, as Mkandawire (2007, 2009: 141) has emphasised that social policy is important in the nation-building project. Considering these explanations, social policy, according to Adesina (2021b), encompasses five main concerns—redistribution, protection, production, reproduction, and social cohesion. In addressing these concerns, policymakers and practitioners utilise various tools, including education, health, land and agrarian policies, affirmative action, income support, and fiscal policy. The success of social policy is reliant on a mix of these different instruments within a shared normative context. This is as important, if not more so, than financial resources, and it becomes critical for policymakers and implementers to be cautious so as not to equate high expenditure on a policy instrument with its outcomes (Adesina, 2021b). Adesina (2021a, 2021b) further argues that social policy should be viewed through an endogenous lens, based on norms of mutuality and generosity, where everyone contributes. He provides an example of the Swahili proverb, '...treat your guest as a guest for two days; on the third day, give him a hoe!' Such norms emphasise collective planting and harvesting, shared financial obligations among family members, micro-savings schemes, and extensive family support systems (Adesina, 2021b).

2.3.2 *Contestations in Social Policy Discourse*

Given the definitions of social policy given in the preceding section, it is not surprising that the social policy landscape in Africa is characterised by notable contestations. Over the past three decades, Adesina (2021a: 2) observed the predominance of a more residual approach to social policy, with influential external and local actors leveraging the continent as a testing ground for various social experiments. This trend, which is rooted in anti-developmental perspectives, is seen reducing poverty alleviation to simplistic measures such as targeted cash transfers while overlooking the deeper structural issues that contribute to entitlement failures and inequality (Ouma & Adesina, 2019). For Adesina (2011), what is being sold to developing countries (especially African countries) is the more regressive version of liberal social policy regime, within a methodological framework (see Myrdal, 1978). Adesina believes that it isolates the problem of vulnerability from the broader, systemic processes that in a 'cumulative causative' process, created what is being treated, which is extreme poverty and vulnerability. The dualist social policy regime allocates social risks differently with 'good risks' being allocated to the market and 'bad risks' to the public domain, and much of the latter is based on Poor Law norms (Adesina, 2011:466).

A significant objection to this prevailing 'social protection paradigm' in the African context is its tendency to reduce social policy to social protection, which in turn further reduces the goal of social protection to simply cash transfers (Adesina, 2009, 2011; Mkandawire, 2004). This approach stands in stark contrast to successful economic and social development efforts, which derive from visionary agenda setting that embraces a wide vision of society with a long-term perspective (Adesina, 2011). Adesina (2020) has highlighted that instead of utilising proactive social policy tools designed to enhance productive capacity, employment opportunities, and collective social provisioning, the present framework prioritises market mechanisms for resource allocation. A defining characteristic of this social protection paradigm is its refusal to challenge the economic instruments and philosophy that generated the crisis it purports to address. The narrowing of ideational vision (Adesina, 2010) stems from this aversion, as evidenced in the policy instruments often recommended and the focus on 'extreme' vulnerability or the poor, targeting, and the disconnecting of social policy and economic policy (Adesina, 2010, 2011). The contemporary SPP therefore reflects an ideational crisis at the heart of the dominant international, multilateral, bilateral, or 'development community' in the wake of neoliberal orthodoxy's hegemonic power. This raises critical questions about how to respond to the existential crisis it has created (Adesina, 2010). For Adesina (2011), the paradigm is characterised by three main features, namely, excessive focus on the ultra-poor, the preference for means testing and targeting, and the disconnection between the 'social' and the broader economic aspects of development policymaking. Equally important is that the above has happened in a context of reversal of the mode of engagement between the Global North and the Global South (Adesina, 2011). The contemporary 'new poverty agenda' thus adopts a safety net strategy, operating within a stratified social provisioning system that offers distinct public services specifically for the 'poor' (Adesina, 2020; see also Fischer, 2018).

The allure of conditional cash transfers in the African context is considered as particularly significant. It lies in their presentation as development panaceas or interventions that are purported to achieve transformative outcomes with minimal investment. This framing appeals to international donors and helps persuade governments to adopt such programmes. However, as Adesina (2011) critically observes, evaluation methodologies for these interventions typically emphasise micro-level impacts while systematically neglecting the broader structural dynamics that perpetuate vulnerability. In addition, the focus on chronic poverty has often involved a superficial consideration, if not a benign neglect, of inequality. These approaches are seen to echo classical Poor Law norms and their resurrection in neoliberal thinking. High levels of inequality, such as those witnessed globally and within sub-Saharan Africa, are not only detrimental to building an inclusive society, but they also violate the ethical basis of successful social policy regimes (Adesina, 2011).

While addressing entitlement failures and poverty reduction demands urgent attention, such efforts, according to Adesina (2011), must be integrated within comprehensive development frameworks rather than treated as stand-alone interventions. Significantly, it is the neoliberal policy regime that is seen as having exacerbated national and global inequality and the vulnerability that the social protection paradigm claims it wishes to fix that has 'maladjusted' African economies (Mkandawire, 2005). In addition, Adesina (2020) emphasises the necessity of contextualising rising poverty and inequality within the very neoliberal paradigms that generate these conditions. This critical perspective is especially relevant in the African context, where, as Ouma (2020) note, external agencies aggressively market narrowly conceived policy solutions that can undermine holistic development approaches. African development trajectories have been severely constrained by what are identified as unjust trade engagement structures, conditional aid arrangements, and oppressive debt mechanisms imposed by Global North countries seeking to maintain their exploitative accumulation patterns (Gumede, 2022; Gumede & Ndhlovu, 2023). This neoliberal perspective on social policy creates a disconnection between economic and social policies. It separates anti-poverty measures from wider developmental goals and responds to life's uncertainties and challenges through a segmented system (Adesina, 2007, 2011). The consequences extend beyond fragmented policy approaches, resulting in African nations being unable to conceptualise sovereign development pathways outside frameworks dictated by Northern powers, ultimately intensifying poverty and social marginalisation (Gumede & Ndhlovu, 2023; Chipenda, 2023). Consequently, access to public social services often requires proof of an inability to self-sustain, perpetuating stigma around seeking assistance (Adesina, 2020).

The reductionist approach to social policy is further evidenced in Aina's (2021) observation of fragmented, residual, and sectoralised policy frameworks. These frameworks treat areas like housing, education, public health, and cash transfers as isolated components rather than interconnected parts of a comprehensive development strategy. What has emerged, as Mkandawire (2001) and Adesina (2021a, 2021b) articulate, are functionally compromised nation-states that demonstrate greater accountability to external actors than to their own citizenry. Aina notes that the growing focus on social protection, as opposed to holistic social policy, has contributed to the emergence of '…a whole research and management industry… around social protection rather than social policy' (Aina, 2021). This narrowed focus aligns with Gumede and Ndhlovu's (2023) observation that such externally driven development paradigms have systematically undermined Africa's capacity to foster authentic, contextually appropriate social policy frameworks.

These contemporary challenges in the social policy discourse in relation to the African context can be seen as having deep historical roots. Hujo (2021) highlights that present social policy approaches in Africa reflect both colonial legacies and neoliberal influences. During the colonial period, colonial powers pursued an extractive agenda with minimal social investment in their colonies. The few social services that existed were primarily run by religious groups and societies, resulting in stratified, segregated, and segmented social provisioning that prioritised colonials and settlers over indigenous peoples (Adesina, 2021b). This has, in different ways, had a lasting legacy on the social policy discourse and provision in contemporary Africa. The postcolonial era was progressive, and it initially saw promising developments, with massive expansion in education and significant investment in public health. Social policy was deliberately employed to foster social cohesion and nation-building, with deliberate attention to addressing gender equity gaps in most policies. However, this progressive trajectory was interrupted when expenditure on social policy stagnated due to the fiscal and debt crises of the 1970s and 1980s (Adesina, 2021b), creating serious challenges for sustainable social development across the continent. These multiple developmental challenges, coupled with growing recognition of social policy's potential progressive role in addressing structural inequalities, have stimulated a rethinking of social policy frameworks and their transformative possibilities. It has led to the development of the TSP concept which is examined in greater detail in the next section.

2.3.3 The Concept of Transformative Social Policy

The concept of transformative social policy (TSP) was developed by Thandika Mkandawire during his tenure as director of the United Nations Research Institute for Social Development (UNRISD), specifically under the UNRISD Social Policy in a Development Context Research Project. This framework, in line with discussions in the preceding sections, can be seen as reconceptualising social policy beyond residual welfare approaches to address complex societal challenges. Premised on the ideas of Mkandawire and later Adesina, the principles of TSP emphasise a broader understanding of social policy that integrates economic objectives. While Mkandawire acknowledges that addressing poverty is '...a welcome antidote to policy-making that had expunged poverty from the central agenda', he cautions that focusing exclusively on immediate causes risks 'narrowing the overall development agenda' (Mkandawire, 2010: 37). This perspective has been of importance in highlighting TSP's dual purpose which is to alleviate poverty while simultaneously aligning social policy with economic policy to foster comprehensive socio-economic development (Mkandawire, 2009: 22).

To operationalise TSP, Mkandawire (2007:10) identified four instrumental values that clarify social policy objectives. These include legitimation and embedding of the development project, accumulation and investment, human resource development, and labour market relations. These values contribute to TSP's overarching goal of transforming societies rather than merely addressing poverty symptoms. He emphasises that '...social policy consists of major transformative instruments, many of which are simply unavoidable for any meaningful policy of "catch-up" and development' (Mkandawire, 2007:10). It is in line with this line of thought that Mkandawire, as has been alluded to earlier, further defined social policy as 'collective interventions to directly affect social welfare, social institutions, and social relations…concerned with the redistributive effects of economic policy, the protection of people from the vagaries of the market and the changing circumstances of age, the enhancement of the productive potential of members of society, and the reconciliation of the burden of reproduction with that of other social tasks' (Mkandawire, 2010:149). This comprehensive definition has been of critical importance as it has been the bedrock of the TSP concept and it aligns with its transformative aspirations by highlighting social policy's multifaceted role in shaping societal outcomes.

Building on Mkandawire's foundation, Adesina (2010:10) has expanded the definition and understanding of TSP. He emphasises that it pertains '…not only to the transformation of an economy or protection from destitution, but also to the transformation of social relations'. This holistic perspective has reinforced the idea that TSP is '…based on norms of equality and solidarity, grounded in a specific national vision and embodying the functions of production, protection, reproduction, redistribution, social cohesion, and nation-building' (Adesina, 2010:10). By incorporating social relations, TSP articulates the interconnectedness between economic and social objectives, highlighting the necessity for policies that reshape both structures. TSP thus, transcends the conventional narrow approaches to social policy by embracing a comprehensive vision that involves multitasking for multiple objectives. Beyond mere safety nets, it is argued that it encompasses the aforementioned multiple functions which are vital in the African context (Adesina, 2007, 2009, 2011; Mkandawire, 2006). In addition, Yi (2015:1) has further strengthened the idea around this integrated approach. He asserts that TSP requires examination of '…economic, social, and political relations, policy linkages, and the comprehensiveness of social policy interventions to transform existing unequal and unjust social, economic, and political relationships'. Given this, TSP extends beyond mere poverty alleviation to encompass broader societal transformation. As Mkandawire (2007) notes, while reproduction, redistribution, and social protection remain vital, development contexts also demand attention to 'the vital issue of production'.

Central to the TSP framework is the assertion that 'progressive social policy is fundamentally about ensuring human flourishing' (Adesina, 2021a:2–3). This objective is achieved through various interconnected means, which include '…enhancing the productive capacity of citizens through public investment in education, healthcare, housing, etc.; reconciling 'the burden of reproduction with that of other social tasks' (Mkandawire, 2011); protecting people from the vagaries of life throughout the life cycle; attending to the distributive outcome of economic performance; advancing social cohesion; and achieving the nation-building objectives which are vital in the African context (Adesina, 2021a). For African nations facing significant development challenges, TSP represents a significant departure from prevailing limited social protection paradigms, advocating instead for a return to what Adesina (2009) terms a wider vision of social policy. This approach embeds a wider social policy and societal vision within broader development strategies firmly grounded in principles of equality and solidarity. Rather than focusing exclusively on destitution, vulnerability, and short-term risk management, TSP offers a holistic framework that recognises that social policy extends beyond minimum safety nets to encompass comprehensive public provisioning of education, healthcare, employment, and housing (Adesina, 2009; Chipenda, 2022, 2023). Adesina (2015:99) has further clarified that TSP 'stretches from the economy to social relations and social institutions... [and] involves a wide range of instruments to raise human well-being, transform social institutions, social relations, and the economy'. This comprehensive understanding sharply contrasts with more limited visions of social policy often promoted by external donors, which Hujo (2021) argues are '...not in line with the TSP concept or with the UNRISD definition of transformative change as attacking the root causes of poverty, inequality, and unsustainability'.

Through linking the role of social policy to national and social questions, Mkandawire provides important perspectives on addressing intersecting inequalities of gender, race, and class (Tom, 2024a). These questions, which are national and social, are important in all African countries especially in a context of the increasing search for effective and inclusive development. Mkandawire (2009:130) explains that 'the intersection between race and class as well as horizontal and vertical inequalities make attention on social and national questions in Africa critical'. This concern can be viewed from Mkandawire's (Mkandawire, 2009:141) argument for the nation-building role of social policy. The eminent view is that both the social question and the national question can be addressed through nation-building if social policy is robust and transformative. Mkandawire's philosophy was also crystallised around the importance of addressing gender inequality within the broader remit of TSP and development. This is notable in some of his most impactful scholarly contributions. Mkandawire (2006) reiterated that social policies with a reproduction function aim to reconcile the burden of family and childcare with other social tasks and share this burden of responsibility. Social policy in this context can transform gender, social, and other relations through diverse instruments including affirmative action, anti-discrimination legislation, and laws that relate to marriage and the family. This is linked to the search for gender equity and investment in human capital development (Adesina, 2015). An analysis of most African countries and broadly the Global South shows that they are experiencing multiple development, social policy, and governance crises. Application of the ideas of Mkandawire (see Mkandawire, 2005, 2009) reveals that at the heart of the crisis are ideological and epistemological muddles, external determination, and super-imposition by the global economic powers (see Tom, 2024a).

What is of significance is that the TSP framework emphasises the critical integration of social and economic policies. Adesina (2021a:3) argues that 'successful advancement in human well-being has always involved the integration of social and economic policies and constructing social policy regimes focused on its multiple tasks'. This integration is seen creating a virtuous cycle where 'public provisioning of education, healthcare, and housing, as social investment based on solidarity and advancing equality, supports economic development' while simultaneously ensuring that 'economic development grounded in the same norms of solidarity and advancing equality ensures the resources necessary for the extension of social policy' (Adesina, 2021a:3). While distinct in their focus, social and economic policies are inherently interconnected and interdependent in the TSP framework. Economic policy, while it is concerned with fiscal and monetary instruments affecting economic functioning, must work in concert with social policy to achieve transformative outcomes. The TSP framework illustrates this relationship, highlighting how shared normative commitments to equality and solidarity bind together social and economic realms, creating a model that has parallels in various successful social policy regimes worldwide, from the Bismarckian and Beveridgean to the Nordic and Nationalist models (Adesina, 2009). The resulting development outcomes feed into social, economic, and political development, which further reinforce the normative foundations of the social policy regime.

From these insights, it can be argued that the TSP conceptual framework offers a crucial theoretical foundation for decolonising social policy in Africa, one that challenges North-South power dynamics and transcends externally imposed neoliberal paradigms. The scholarship of Mkandawire and Adesina emphasises policy sovereignty, indigenous knowledge systems, and transformative approaches that address structural inequalities while promoting human flourishing. The framework provides vital tools for reconceptualising social policy in the African context through an African-centred lens, asserting policy autonomy in response to contemporary challenges. While implementation challenges exist, it can be argued, as shall be demonstrated in later sections, that the theoretical insights from Mkandawire and Adesina remain essential for developing context-specific, sovereignty-conscious social policies that can drive inclusive development in contemporary Africa. This reinforces their central argument that effective, locally determined social policies are important for achieving genuine development and human wellbeing in the African context.

2.4 Theoretical Contributions to Transformative Social Policy

The intellectual contributions of Mkandawire and Adesina to the framework of TSP represent what can be seen as a significant reconceptualisation of social policy in the African context. Their work not only largely challenges dominant paradigms but also provides theoretical foundations for a more comprehensive understanding of development that emphasises policy sovereignty, indigenous knowledge systems, and structural transformation.

As we engage with their theoretical contributions, it is important to start by briefly analysing the key theoretical innovations in the TSP framework. The ideas of Mkandawire and Adesina, through the TSP framework, can be seen redefining social policy's scope and purpose through several key innovations. At its core, their work establishes social policy as encompassing 'collective interventions to directly affect social welfare, social institutions, and social relations' (Mkandawire,

2010:149). This theoretical positioning can be seen transcending traditional welfare approaches by integrating four instrumental values, namely, legitimation and embedding of the development project, accumulation and investment, human resource development, and labour market relations (Mkandawire, 2007:10). The framework emphasises the critical importance of physical social infrastructure and welfare state provisions as essential components of transformative social policy (see also Neuhuber, 2025). This aligns with Adesina's (Adesina, 2021a:2–3) assertion that 'progressive social policy is fundamentally about ensuring human flourishing' through enhancing productive capacity, protecting people throughout the life cycle, and attending to distributive outcomes. The theoretical framework operates across multiple levels which are inclusive of micro (policy shape), meso (policy context), and landscape (policy universe), emphasising the need for systemic reconfiguration and new social competencies as has been argued elsewhere by Sousa (2024). Hirvilammi et al. (2023) further elaborate on these levels by identifying five key interrelated factors that enhance transformative capacity. These include social networks, collaboration and participation, knowledge and learning, shared policy frameworks and visions, and financial resources. These elements support sustainable social and economic development within eco-welfare states, providing concrete mechanisms through which TSP can be operationalised.

The second aspect worth looking at is in relation to policy sovereignty and indigenous knowledge systems. A cornerstone of the TSP theoretical framework is the emphasis on policy sovereignty and indigenous knowledge systems. Mkandawire's work particularly challenges what Adesina (2021a:2) describes as '…the predominance of a more residual approach to social policy, with influential external and local actors leveraging the continent as a testing ground for various social experiments'. The framework advocates for policy autonomy that allows African nations to develop context-specific solutions grounded in local realities. This theoretical position is strengthened by contemporary emphasis on intersectoral coordination and inclusive resource allocation (see Rustamova et al., 2025), which aligns with Adesina's (2021b) argument for viewing social policy through an endogenous lens based on norms of mutuality, solidarity, and generosity. The framework advocates for adaptive social planning, collaboration, and inclusivity in policy formation as argued by Tiamiyu and Odendaal (2023) and as exemplified in Adesina's reference to traditional African proverbs and practices that emphasise collective responsibility. Yi et al. (2024) reinforce this perspective by highlighting key components of a TSP framework including a rights-based approach to social welfare, integration with public policies, active civil society participation, supranational social policy emergence, recognition of informal workers, and national ownership in development cooperation contexts. These elements ensure that TSP remains grounded in local contexts while engaging with global frameworks.

The third aspect relates to the reconceptualisation of social policy beyond welfare. The TSP framework goes a long way in reconceptualising social policy beyond residual welfare approaches. As Adesina (2020) has emphasised, rather than utilising proactive social policy tools designed to enhance productive capacity, employment opportunities, and collective social provisioning, present frameworks often prioritise market mechanisms for resource allocation. The work of Mkandawire and Adesina advocates for equal opportunities in employment, addressing labour market fragmentation, and promoting evidence-informed policymaking, issues which have been flagged as critical by Papadakis and Tzagkarakis (2025). This has been in a context where there is an urgent contemporary need for public-community partnership governance and innovative strategies (see Chaves Avila & Gallego-Bono, 2020). This reconceptualisation is significant in challenging what Adesina terms the 'social protection paradigm' (SPP), which reflects an ideational crisis within the international development community. Their TSP framework, instead, advocates for comprehensive public provisioning that creates what Adesina (Adesina, 2021a:3) describes as a virtuous cycle where the public provisioning of services based on solidarity can advance equality while concurrently supporting economic development. Yi and Kim (2015) further expand this reconceptualisation by identifying five essential dimensions of TSP, namely, production, redistribution, protection, reproduction, and social cohesion. They emphasise solidarity-based redistributive programmes working alongside other policies to achieve socially, economically, and politically sustainable and equitable development, particularly in the context of the SDGs. This multidimensional approach helps operationalise TSP in ways that address both immediate welfare needs and long-term structural transformation.

The last aspect is in relation to the theoretical framework being a tool for challenging external paradigms. Mkandawire and Adesina view TSP as offering robust tools for challenging external paradigms. They critique what Adesina (2011) describes as 'the more regressive version of liberal social policy regime' being sold to developing countries, which isolates vulnerability from broader systemic processes. In line with this, TSP can be interpreted as emphasising what is presented in contemporary society as the need for transformative resilience and just transitions as has been argued by Neuhuber (2025). At the core of this debate are arguments for the promotion of sustainable social and economic development through the integration of social, economic, and environmental characteristics. This theoretical positioning directly confronts what Mkandawire identifies as the 'maladjustment' of African economies through neoliberal policy regimes. The TSP framework instead, advocates what Adesina (2009) terms a 'wider vision of social policy' that is grounded in the principles of equality and solidarity. Emphasised in this context is the necessity of contextualising rising poverty and inequality within the very neoliberal paradigms that generate these conditions (Adesina, 2020). Koehler (2017) reinforces the challenge to external paradigms by asserting that TSP frameworks must be rights-based, universal, and address the root causes of inequalities. It requires political contestation and active civil society engagement to sustain progress, as rights can be dismantled amidst political backlashes, necessitating ongoing defence and advocacy. This perspective situates TSP within broader political struggles for social justice and equity.

Given the issues raised in the preceding sections, the theoretical contributions of Mkandawire and Adesina provide a comprehensive framework for reconceptualising social policy in ways that centre African epistemologies and developmental aspirations while addressing contemporary challenges of sustainability, resilience, and social inclusion. Their work continues to influence present theoretical developments in TSP, particularly in understanding how policy frameworks can effectively bridge transformative resilience and just transitions while promoting sustainable development. Kempf and Dutta (2021) highlight the importance of integrating TSPs to reach vulnerable groups, enhancing targeting within universalism, and linking TSPs to crisis management. These components can ensure long-term development and support the implementation of the SDGs. In addition, the need and necessity of revitalising the relationship between nature and humans, promoting alternative resource governance, and aligning policies with SDGs to achieve sustainable transformative pathways that balance human welfare and environmental health across various policy levels is of critical importance. These recent contributions demonstrate the continuing evolution and relevance of the TSP framework in addressing contemporary global challenges.

2.5 Implementation Challenges and Contemporary Relevance

The theoretical foundations of the TSP framework as conceptualised by Mkandawire and Adesina face significant implementation challenges in the contemporary global context while maintaining critical relevance for addressing the dynamic socio-economic issues facing Africa today. These challenges and opportunities are interconnected across multiple dimensions, from post-pandemic recovery to confronting persistent global power imbalances that continue to shape development discourse.

2.5.1 Application to Post-COVID Recovery

The COVID-19 pandemic created an unprecedented moment of collective human vulnerability, with WHO data revealing 7,074,400 recorded deaths and 776,798,873 confirmed positive cases as of 3 November 2024 (WHO, 2024). Africa emerged as the least affected WHO region, reporting a comparatively modest 9,584,113 confirmed cases and 175,531 deaths. This outcome challenged prevailing narratives of continental healthcare vulnerability and demonstrated remarkable systemic resilience despite resource constraints. This resilience, however, coexists with the pandemic's disproportionate impact on vulnerable populations, particularly those in the informal sector who were largely excluded from assistance efforts.

As Adesina (2021b) observes, 'both the current pandemic and the history of policymaking on the continent suggest that in a post-COVID Africa, policy makers need to transcend the stratified, segregated, and segmented social policy systems prevailing across the continent and to return to a broader vision of social policy'. This observation highlights how the pandemic exposed existing gaps in social protection systems. It created an urgent need for comprehensive frameworks that can address structural vulnerabilities while building resilient societies capable of withstanding future shocks. The TSP framework offers a comprehensive approach to post-pandemic recovery by addressing these systemic gaps. Yunus et al. (2021) emphasise the role of social policies in promoting sustainable human development through strengthening educational systems and environmental awareness. Building on this foundation, effective recovery requires integrating existing welfare systems, constructing robust safety nets, and investing in human capital (Schady et al., 2023; Gilligan, 2020). These arguments align closely with TSP's multidimensional framework. This framework connects immediate recovery needs with long-term structural transformation objectives. This comprehensive approach connects directly to Kempf and Dutta's (2021) observation that TSPs are essential for reaching vulnerable groups during crisis periods. When linked to long-term development strategies, they ensure inclusive progress towards the SDGs. This creates pathways from immediate recovery to sustainable transformation that centre African agency and sovereignty.

2.5.2 *Addressing Climate Change and Environmental Challenges*

The environmental dimensions of development have become increasingly central to TSP implementation. Climate change presents both challenges and opportunities for applying the framework's principles in new domains. Bostrom (2021) argues that social policies can promote sustainable lifestyles and reduced consumption patterns that directly address environmental degradation. This environmental focus extends TSP beyond its traditional socio-economic dimensions. It creates opportunities for progressive social policies to simultaneously address wealth inequality, foster community engagement, and tackle unsustainable resource utilisation while ensuring just transitions to sustainable economies. Environmental degradation, including deforestation, soil erosion, and biodiversity loss, compounds climate challenges and undermines development gains. This multidimensional approach resonates with Mkandawire's (2007) vision of integrating social, economic, and political aspects of development. The integration of environmental concerns with social policy reflects growing recognition of their interdependence. This recognition highlights the need for holistic approaches that address both human welfare and ecological sustainability.

Climate change significantly disrupts agricultural productivity through altered weather patterns, threatening food security and farmers' incomes (Mirzabaev et al., 2023), while unsustainable resource extraction further undermines ecosystems that are critical for rural livelihoods. McClelland et al. (2022) demonstrate how comprehensive social policies address interconnected issues across communities, health systems, and governance structures. These frameworks are increasingly incorporating principles of ecological sustainability and resource conservation. This aligns with TSP's emphasis on structural transformation and policy integration. Climate change also exacerbates health risks, including the spread of vector-borne diseases and mental health issues stemming from climate anxiety (Dye, 2024; Isham et al., 2003). The interconnectedness of environmental and human health necessitates integrated adaptation strategies that promote both sustainability and resilience. The overall quality of life is threatened by climate-induced changes such as rising sea levels, pollution, extreme weather events, and habitat destruction, which disproportionately impact poorer nations while limiting their capacity to implement sustainability measures (Shukla et al., 2024).

While climate change poses significant threats, it also presents opportunities for innovation in sustainable practices, circular economy approaches, and community-led conservation efforts. This potential for improved wellbeing through adaptive strategies demonstrates how environmental challenges can catalyse TSPs that integrate economic, social, and ecological dimensions of development, making sustainability a central pillar of contemporary frameworks.

2.5.3 Confronting Persistent North-South Power Dynamics

Perhaps the most significant challenge to TSP implementation lies in the persistent power imbalances that continue to shape global development discourse and practice. Tom (2024a, 2024b) analyses how most African countries and broadly the Global South are experiencing multiple development, social policy, and governance crises resulting from ideological and epistemological muddles, external determination, and super-imposition by global economic powers. This analysis connects directly to Mkandawire and Adesina's critique of externally imposed development paradigms (see Ndhlovu & Mokone; Tekwa in this volume). The work by all these scholars emphasises the need for policy sovereignty and indigenous knowledge systems.

African leaders need to develop a culture of consciousness and interrogation of the roots and routes of Western manipulation and domination in governance, development, and social policy, which often operates through multilateral institutions, international nongovernmental organisations, and policy merchandising (Adesina, 2020; Mkandawire, 2010). The historical record reinforces this concern, with Euro-generated strategies implemented in Africa including structural adjustment programmes, poverty reduction strategy papers, and various social protection programmes proving 'manipulative and palliative' rather than transformative. Contemporary manifestations of this power asymmetry include the proliferation of fintech solutions for social protection that prioritise technological innovation over social transformation. There is also the emphasis on 'evidence-based policy' that often privileges randomised controlled trials over contextual understanding and local knowledge systems. This experience demonstrates that 'democratic and inclusive development and social policy for Africa and by Africa are better than imported variants' (Tom, 2024a).

Confronting these power dynamics requires strategic approaches. Buscher et al. (2020) advocate transformative policies like universal basic income and progressive taxation that promote economic redistribution and recognise the value of essential services. These approaches can challenge neoliberal hegemony while building community resilience. They also assert African agency in development processes. Similarly, Cheng et al. (2021) highlight how innovative social policies can strengthen multilateral cooperation while maintaining indigenous priorities and perspectives. This strategic engagement reflects what Gumede (2018:10) describes as approaches that 'reject, oppose and pursue an alternative path to complete liberation of the continent and reclaim its dignity'. Such approaches invoke the need to adopt what Tom (2024a) calls the Bandung approach to development, emphasising policy sovereignty and indigenous knowledge systems as significant to meaningful transformation.

2.5.4 Relevance to Present African Development Challenges

The TSP framework maintains significant relevance for addressing contemporary African development challenges. This is precisely because it provides tools for navigating complex power dynamics while fostering inclusive growth. This growth centres African agency and sovereignty. Mkandawire (2010:37) articulated this ongoing relevance when he argued that 'the new challenge in Africa is to bring back development, but now one that is democratically anchored and socially inclusive'. This vision aligns with Sen's (1999) conception of development as freedom. In this conception, economic development, equity, and democracy become mutually reinforcing elements of transformative change.

Tom (2024a) emphasises how TSPs enhance autonomy and development by addressing structural poverty and inequality through redistributive, productive, protective, reproductive, and social compact functions. These functions operate through diverse instruments that reflect the TSP framework's recognition that development challenges vary within and across countries. This requires context-specific applications while maintaining core transformative principles. Implementation approaches include investments and reforms in education, health, human settlement, agrarian structures and relations, labour, child and family care, old age, social insurance, and fiscal systems (Tom, 2024a). This multifaceted approach enables context-specific application while maintaining core transformative principles. These principles prioritise African epistemologies, lived experiences, and developmental aspirations.

The framework's relevance extends to addressing intersecting inequalities. Mkandawire (2009: 141) argued that 'the intersection between race and class as well as horizontal and vertical inequalities make attention on social and national questions in Africa critical'. This perspective recognises how multiple forms of marginalisation interact. It requires comprehensive policy responses that address both immediate welfare needs and long-term structural transformation. Gender equality represents a central dimension of this intersectional approach. Mkandawire (2006:1) emphasised that 'social policies with a reproduction function aim to reconcile the burden of family and childcare with other social tasks and sharing this burden of responsibility'. This focus on gender relations connects with broader development objectives. As Adesina (2015:17) argues, 'public investment in education is a critical element in redressing gender inequality and the transformation of gender relations'.

Yi and Kim (2015) highlight how TSPs contribute to social cohesion through solidarity-based redistributive policies that promote sustainable development and political stability. Rather than functioning merely as safety nets, they foster equitable development through production, redistribution, protection, and social cohesion functions. These functions address both immediate welfare needs and long-term structural transformation. Chipenda (2024, 2025) emphasises how TSPs prioritise local needs and perspectives, enhancing welfare and resilience especially in rural contexts. By addressing poverty and inequality through equitable access to land and agrarian reforms, TSP connects immediate welfare concerns with structural transformation objectives. This approach recognises that social protection programmes can synergistically enhance economic and social outcomes by integrating redistributive and productive efforts while maintaining focus on African sovereignty and self-determination.

While TSP implementation faces significant challenges in contemporary contexts, these challenges highlight the framework's ongoing relevance and urgency. By integrating concerns around post-pandemic recovery, climate change, global power dynamics, and intersecting inequalities, TSP provides comprehensive guidance for African development strategies in the twenty-first century. At its core, the framework emphasises autonomy, inclusivity, and equity as important principles for transformative change. It offers alternatives to externally imposed paradigms while recognising that achieving such alternatives requires Africa to be agentive both by individuals and the state in asserting policy sovereignty.

2.6 Conclusion: Theorising African Social Policy beyond External Paradigms

This chapter examined the theoretical contributions of Thandika Mkandawire and Jimi Adesina to TSP discourse, demonstrating their significance for reconceptualising social policy in the African context. Through systematic analysis of their combined scholarship, we have shown how their theoretical framework provides vital tools for transcending externally imposed neoliberal paradigms while asserting African agency in policy formulation. Their work gains urgency in the contemporary context of global fragmentation, where multiple intersecting crises including the lingering effects of the COVID-19 pandemic, climate change, and persistent

North-South power imbalances expose the inadequacy of externally imposed development models.

The TSP framework, as articulated by Mkandawire and Adesina, offers a comprehensive alternative to residual approaches that have dominated contemporary social policy discourse. Their theoretical innovations, which emphasise policy sovereignty, the developmental state, and indigenous knowledge systems, provide the conceptual architecture necessary for addressing Africa's complex development challenges in ways that centre African epistemologies, lived experiences, and developmental aspirations. By reconceptualising social policy beyond welfare to encompass production, protection, reproduction, redistribution, and social cohesion, they have established a foundation for holistic development approaches that challenge North-South power dynamics.

The analysis reveals that Africa's relatively successful pandemic response, despite resource constraints, demonstrates the potential of approaches that prioritise local knowledge and contextual understanding over externally imposed frameworks. This experience validates Mkandawire and Adesina's argument that development strategies must be grounded in African realities rather than donor-driven models. Their framework becomes relevant as contemporary manifestations of external control including policy merchandising through fintech solutions and evidence-based approaches that privilege external validation over local knowledge continue to undermine African sovereignty in development processes.

Our analysis reveals that the implementation of TSP faces significant challenges in contemporary contexts, especially regarding post-pandemic recovery, climate adaptation, and persistent global power imbalances. However, these challenges highlight rather than diminish the framework's relevance, as they demonstrate the continued need for approaches that can address multiple, interconnected crises while maintaining focus on structural transformation and human flourishing. The theoretical contributions of Mkandawire and Adesina provide essential guidance for navigating these territories while resisting the temptation to rely on externally imposed solutions.

The environmental dimensions of contemporary development challenges further highlight the framework's relevance, as climate change creates opportunities for TSP approaches that integrate ecological sustainability with social transformation. Rather than viewing environmental and social policies as separate domains, the TSP framework enables African countries to develop integrated approaches that address climate adaptation while advancing social justice and equity objectives.

What emerges most clearly from this analysis is that effective social policy must be grounded in African epistemologies and contextual realities rather than externally imposed frameworks. As African nations confront the challenge of building back better from the pandemic while simultaneously preparing for future crises, the principles of policy sovereignty and indigenous knowledge systems become increasingly vital for ensuring that recovery strategies serve African interests rather than external agendas. The TSP framework offers not merely a critique of existing paradigms but a constructive alternative that centres African agency and developmental aspirations. The framework's emphasis on confronting North-South power dynamics becomes urgent in the contemporary context, where new forms of dependency emerge through digital technologies, climate finance conditionalities, and green transition initiatives that primarily serve Northern consumption patterns. Mkandawire and Adesina's theoretical contributions provide tools for recognising and resisting these forms of contemporary colonialism while building alternatives that prioritise African sovereignty and self-determination.

Future research directions should explore practical applications of the TSP framework across diverse African contexts, examining how policy sovereignty can be operationalised within existing global power structures. In addition, an investigation into how indigenous knowledge systems can inform climate adaptation strategies would strengthen the framework's contemporary relevance. Comparative analyses of TSP implementation across different African states could yield valuable insights into contextual adaptations and implementation challenges, especially in post-pandemic recovery contexts.

The theoretical contributions of Mkandawire and Adesina provide an essential foundation for decolonising social policy discourse in Africa precisely because they offer practical alternatives to external paradigms while maintaining focus on structural transformation. Their TSP framework, with its emphasis on comprehensive approaches to human flourishing, offers a path towards more equitable, sustainable, and sovereignty-conscious development. In an era marked by what UNRISD describes as 'a global state of fracture', their vision of social policy as a transformative force remains not just relevant but essential for addressing Africa's contemporary development challenges while asserting continental agency and dignity. The urgency of their theoretical contributions lies not only in their critique of existing paradigms but in their provision of constructive alternatives that can guide African countries through the challenges of the twenty-first century. As global power configurations continue to shift and new forms of dependency emerge, the TSP framework provides a foundation for building resilient, equitable, and truly African approaches to development that prioritise human flourishing over external validation.

References

Adesina, J. O. (2007). Social policy in the sub-Saharan African context: In search of inclusive development. In *Social policy in the sub-Saharan African context: In search of inclusive development: Introduction*. Palgrave Macmillan.

Adesina, J. (2009). Social policy in sub-Saharan Africa: A glance in the rear-view mirror. *International Journal of Social Welfare, 18*(1), 37–51.

Adesina, J. (2010). *Return to a wider vision of development: Social policy in reframing a new agenda. Keynote address delivered at the 48th session of the UN Commission for Social Development*. UN Commission on Social Development.

Adesina, J. (2011). Beyond the social protection paradigm: Social policy in Africa's development. *Canadian Journal of Development Studies, 32*(4), 454–470.

Adesina, J. (2015). Return to a wider vision of social policy: Rereading theory and history. *South African Review of Sociology, 46*(3), 99–119.

Adesina, J. (2020). Policy merchandising and social assistance in Africa: Don't call dog monkey for me. *Development and Change, 51*(2), 561–582.

Adesina, J. (2021a). Introduction: Social policy in the African context. In J. Adesina (Ed.), *Social policy in the African context* (pp. 1–12). CODESRIA.

Adesina, J. (2021b). *The evolution of social policy: Up to and beyond COVID-19. CSPS Social Policy Brief 2021/01*. Centre for Social Policy Studies, University of Ghana.

Adesina, J. (2022). Variations in postcolonial imagination: Reflection on Senghor, Nyerere and Nkrumah. *Africa Development, XLVII*(1), 31–58.

Aina, T. (1997). Globalization and social policy in Africa: Issues and research directions. CODESRIA.

Aina, T. A. (2021). Reclaiming transformative social policy for inclusive development. In J. Adesina (Ed.), *Social policy in the African context* (pp. 13–28). CODESRIA.

Bhattacherjee, A. (2012). *Social science research: Principles, methods and practices* (2nd ed.). Global Text Project.

Bostrom, M. (2021). Take the opportunity afforded by the COVID-19 experiences: Progressive non-growth policies for sustainable lifestyles. *Frontiers in Sustainability, 2*(2021).

Buscher, B., Koster, M., Spierenburg, M. & Fletcher, R. (2020). *Planning for post-corona: Five proposals to craft a radically more sustainable and equal world.* Retrieved February 16, 2025, from https://repository.ubn.ru.nl/handle/2066/218075

Chaves Avila, R. & Gallego-Bono, J. R. (2020). Transformative policies for the social and solidarity economy: The new generation of public policies fostering the social economy in order to achieve sustainable development goals. *The European and Spanish cases. Sustainability, 12*(10), 4059.

Cheng, Y., Liu, H., Wang, S., & Li, Q. (2021). Global action on SDGs: Policy review and outlook in a post-pandemic era. *Sustainability, 13*(11), 6461.

Cheru, F. (2022). On resuscitating the aborted national project: A restrospective and prospective view (Notes from my last conversation with Thandika Mkandawire). *Africa Development, 47*(2), 1–14.

Chipenda, C. (2022). Comparative perspectives on Zimbabwe and South Africa's social policy response to Covid-19. *Africa Today, 35*, 235–269.

Chipenda, C. (2023). Rethinking the role of social policy in responding to the Covid-19 pandemic in South Africa: Critical reflections using the intellectual contribution of Thandika Mkandawire. In V. Gumede & E. Ndhlovu (Eds.), *Development in Africa: Remembering Thandika Mkandawire* (pp. 197–231). Ace Press.

Chipenda, C. (2024). A transformative social policy perspective on land and agrarian reform in Zimbabwe. *Africa Spectrum, 59*(1), 89–118.

Chipenda, C. (2025). Transformative social policy through land reform: Women's economic agency and rural social change in Zimbabwe. *Journal of Asian and African Studies.*

Dye, C. (2024). Climate change: A healthy price for carbon. In C. Dye (Ed.), *Investing in health and wellbeing: When prevention is better than cure* (pp. 220–247). Oxford Academic.

Fischer, A. (2018). *Poverty as ideology: Rescuing social justice from global development agendas.* Zed Books.

Gadamer, H. G. (2004). *Truth and method.* Continuum Publishing Group.

Gilligan, D. (2020). Social safety nets are crucial to the COVID-19 response: Some lessons to boost their effectiveness. In COVID-19 and global food security. In J. Swinnen & J. McDermott (Eds.), *COVID-19 and global food security* (pp. 102–105). International Food Policy Research Institute (IFPRI).

Gumede, V. (2015). *Political economy of post-apartheid South Africa.* CODESRIA.

Gumede, V. (2018). Social policy for inclusive development in Africa. *Third World Quarterly, 39*, 122–139.

Gumede, V. (2022). Editorial: Rethinking and reclaiming development in development. In S. Ndlovu-Gatsheni & B. Mpofu (Eds.), *Rethinking and unthinking development: Perspectives on inequality and poverty in South Africa and Zimbabwe.* Berghahn Books.

Gumede, V., & Ndhlovu, E. (2023). Introduction: Honouring Thandika Mkandawire. In V. Gumede & E. Ndhlovu (Eds.), *Development in Africa: Remembering Thandika Mkandawire* (pp. 1–18). Ace Press.

Hirvilammi, T., Peltomaa, J., Pihlajamaa, M., & Tiilikainen, S. (2023). Towards an eco-welfare state: Enabling factors for transformative eco-social initiatives. *European Journal of Social Security, 25*(4), 445–463.
Hormeku-Ajei, T., Balaji, A., Olukoshi, A., et al. (2022). Introduction: Early post-independence progressive policies—Insights for our times. *Africa Development, 47*(1), 1–30.
Hujo, K. (2021). Rethinking social policy in Africa—A transformative approach. In J. Adesina (Ed.), *Social policy in the African context* (pp. 29–44). CODESRIA.
Isham, J., Woolcock, M., Pritchett, L., and Busby, G. (2003). *The varieties of resource experience: How natural resource export structures affect the political economy of economic growth.* Middlebury College Discussion Paper No. 2003–08. Middlebury College.
Kempf, I., & Dutta, P. (2021). Transformative social policies as an essential buffer during socio-economic crises. *Sustainable Development, 29*(3), 517–527.
Kempf, I., Hujo, K., & Ponte, R. (Eds.). (2023). *Global study on new eco-social contracts.* UNRISD.
Koehler, G. (2017). The politics of rights-based, transformative social policy in South and Southeast Asia. *International Social Security Review, 70*(4), 105–126.
Krippendorff, K. (2019). *Content analysis: An introduction to its methodology* (4th ed.). SAGE Publications Inc.
McClelland, A. G., Shaw, D., O'Grady, N., & Fattoum, A. (2022). Recovery for development: A multi-dimensional, practice-oriented framework for transformative change post-disaster. *Journal of Development Studies, 59*(1), 1–20.
Mirzabaev, A., Bezner Kerr, R., Hasegawa, T., Pradhan, P., Wreford, A., Tirado von der Pahlen, M. C., & Gurney-Smith, H. (2023). Severe climate change risks to food security and nutrition. *Climate Risk Management, 39*, 100473.
Mkandawire, T. (2001). *Social policy in a development context. Paper no. 25.* UNRISD.
Mkandawire, T. (2004). Introduction: Social policy in a development context. In T. Mkandawire (Ed.), *Social policy in a development context* (pp. 1–33). UNRISD/Palgrave Macmillan.
Mkandawire, T. (2005). Maladjusted African economies and globalization. *Africa Development, 30*(1), 1–33.
Mkandawire, T. (2006). Transformative social policy: Lessons from UNRISD research. *UNRISD Research and Policy Brief 5*, 4–5.
Mkandawire, T. (2007). Transformative social policy and innovation in developing countries. *The European Journal of Development Research, 19*(1), 13–29.
Mkandawire, T. (2009). Institutional monocropping and monotasking in Africa. In N. Akbah, B. Kwesi, & S. Howard (Eds.), *Good growth and governance in Africa: Rethinking development strategies* (pp. 1–28). UNRISD.
Mkandawire, T. (2010). How the new poverty agenda neglected social and employment policies in Africa. *Journal of Human Development and Capabilities, 11*(1), 37–55.
Mkandawire, T. (2011). Running while others walk: Knowledge and the challenge of Africa's development. Africa Development 36(2), 1–36.
Myrdal, G. (1978). Institutional economics. *Journal of Economic Issues, 12*(4), 771–783.
Nagar, M. (2020). A historical political economy approach to Africa's economic development: A critique of Thandika Mkandawire's interests and incentives, ideas and institutions. In S. O. Oloruntoba & T. Fayola (Eds.), *The Palgrave handbook of African political economy. Palgrave handbooks in IPE* (pp. 221–243). Palgrave MacMillan.
Neuhuber, T. (2025). One and the same or worlds apart? Linking transformative regional resilience and just transitions through welfare state policies. *Sustainability, 17*(2), 637.
Nhemachena, A. (2023). Even chameleons stalk their prey: Rinonyenga rinohwarara rinosumudza musoro rawana, twenty-first century imperial sleights of hand and the recolonisation of Africa. *The African Review, 51*(5), 496–524.
Noyoo, N., & Boon, E. (2021). Nation-building and the nationalist discourse: Revisiting social policy in Ghana and Zambia in the first decade of independence. In J. Adesina (Ed.), *Social policy in the African context* (pp. 45–90). CODESRIA.
Nyerere, J. K. (1968). *Ujamaa: Essays on socialism.* Oxford University Press.

Onimode, B. (1998). *A political economy of the African crisis*. Zed Books.
Ouma, M. (2020). Social protection policymaking in Kenya. *Africa Development/Afrique et Développement, 45*(2), 125–142.
Ouma, M., & Adesina, J. (2019). Solutions, exclusion and influence: Exploring power relations in the adoption of social protection policies in Kenya. *Critical Social Theory, 39*(3), 376–395.
Papadakis, N., & Tzagkarakis, S. I. (2025). Welfare state, social policy and social sustainability, within the context of the permacrisis. *Frontiers in Political Science, 6.*
Pervin, N., & Mokhtar, M. (2022). The interpretivist research paradigm: A subjective notion of a social context. *International Journal of Academic Research in Progressive Education and Development, 11*(2), 419–428.
Phiri, M. (2023). Against imperial social policy in Africa: Recasting Mkandawire's transformative ideas for Africa's liberation. *Critical Sociology, 49*(3), 437–455.
Rodney, W. (1981). *How Europe underdeveloped Africa*. Howard University Press.
Rustamova, N., Sharifzoda, S., Burxanxodjaeva, X., Rahimqulova, L., Turdialiev, M., Nurullaev, F., & Eshchanova, D. (2025). Social protection in developing countries: Legal, economic and social trends. *Qubahan Academic Journal, 5*(1), 118–149.
Schady, N., Holla, A., Sabarwal, S., Silva, J., & Yi Chang, A. (2023). *Collapse and recovery: How the COVID-19 pandemic eroded human capital and what to do about it*. World Bank.
Sen, A. (1999). Development as freedom. Oxford University Press.
Shivji, J. (2006). *The changing development discourse in Africa*. Global Policy Forum. Retrieved February 18, 2025, from https://archive.globalpolicy.org/socecon/develop/africa/2005/1007discourse.htm
Shukla, K., Patil, Y., Estoque, R. C., & López de Haro, P. A. (2024). *Global perspectives on the impact of climate change on quality of life*. Retrieved February 11, 2025, from https://www.igi-global.com/chapter/global-perspectives-on-the-impact-of-climate-change-on-quality-of-life/341608
Sousa, J. (2024). A policy and evaluation framework for sustainable transitions—An energy policy approach. *Advances in Environmental and Engineering Research, 5*(1).
Tekwa, N. (2020). *Gender, land reform and welfare outcomes. A case study of Chiredzi district, Zimbabwe*. PhD thesis. UNISA.
Tiamiyu, B., & Odendaal, N. (2023). Conceptual framework of adaptive social planning for sustainable urban transformation. In A. Almusaed & A. Almssad (Eds.), *Sustainable regional planning* (pp. 1–12). IntechOpen.
Tom, T. (2024a). Decades of land and agrarian reform in Zimbabwe. A review of transformative social policy lessons for other African countries. *The African Review*.
Tom, T. (2024b). Social policy, development and democracy in Africa: Reconnoitering Thandika Mkandawire's philosophy. *Journal of Asian and African Studies*, 1–19.
UNRISD. (2022). *Crises of inequality: Shifting power of a new eco-social contract*. UNRISD.
UNRISD. (2023). *Crises of inequality: Shifting power for a new eco-social contract. Research and Policy Brief 39*. UNRISD.
WHO. (2024). *World Health Organisation*. Retrieved February 13, 2025, from https://data.who.int/dashboards/covid19/deaths?m49=716&n=o
Woolcock, M., Pritchett, L., and Busby, G. (2003). The varieties of resource experience: How natural resource export structures affect the political economy of economic growth. Middlebury College Discussion Paper No. 2003-08. Middlebury College.
Yi, I. (2015). *New challenges for and new directions in social policy*. UNRISD.
Yi, I., & Kim, T. (2015). Post 2015 development goals (SDGs) and transformative social policy. *Oughtopia, 30*(1), 307–335.
Yi, I., Kaasch, A., & Stetter, K. (2024). Conclusion: Toward new social policies for transformation. In *Emerging trends in social policy from the South: Challenges and innovations in emerging economies* (pp. 268–274). Bristol University Press.
Yunus, M., Biggeri, M., & Testi, E. (2021). Social economy and social business supporting policies for sustainable human development in a post-COVID-19 world. *Sustainability, 13*(21), 12155.

Chapter 3
Rethinking Gender, Social Policy and Development with Thandika: A Transformative Social Policy Perspective

Newman Tekwa

3.1 Introduction

While Thandika Mkandawire's work persistently engaged with social policy and development challenges in postcolonial Africa, the two literatures, on the one hand social policy research and development studies on the other hand, are yet to converge in mainstream scholarly writing on Africa. This dates back to the formative stages of these academic disciplines at the beginning of the twentieth century. While they all have their origins in universities of western and northern Europe, North America and other Western countries, their geographical focus was quite distinct. Midgley (2019) tracing the origins and subsequent development of these two disciplines of academic inquiry argues that social policy, geographically, concerned itself with analyses of the provision of social services by national government of the Western countries. Whereas, development studies, while emerging as a discipline of academic inquiry in western and northern universities at the same time with social policy, geographically, was more concerned with economic policies and programmes of newly independent developing nations in the Global South and how these governments can transform their economies from predominantly agrarian to modern industrial states (see Midgley, 2019; Lateef, 1995, p. 14). The academic disciplines while emerging at the same place and time, their geographical focus was different but will significant implications for theory, policy and practice in the Global South. Midgley (2019, p. 14) is quick to highlight the extent to which the work of academics in social policy and development studies began to influence governments and the work of international organisations not only in the Global North but the Global South as well. With the development economics taking the lead in influencing finance for development in international organisations,

N. Tekwa (✉)
South African Research Chair in Social Policy, College of Graduate Studies, University of South Africa, Pretoria, South Africa

E. Ndhlovu, C. Chipenda (eds.), *Redefining the Social Policy Discourse*,
https://doi.org/10.1007/978-3-032-24067-5_3

particularly the World Bank and International Monetary Fund (IMF) in the dealings with newly independent states, it is the latter institutions' development programmes which form the point of departure for this paper with a particular focus on the African continent.

In the chapter, I ask the following research questions:

- How does Thandika Mkandawire's approach to social policy challenge traditional gender norms and contribute to the development of more inclusive policies in African contexts?
- What role does social policy play in addressing intersectional inequalities related to gender, class and ethnicity in development processes, according to Thandika Mkandawire's transformative perspective?
- In what ways can a transformative social policy perspective informed by Thandika Mkandawire's lens assist in rethinking the relationship between gender, social policy and development for gender-equitable African futures?

The rest of the chapter is structured as follows. The next section provides the methodological approach adopted in the chapter. This is followed by a review of relevant literature on gender, social policy and development in Africa. The discussion section interwoven with the literature provides a context and support for the analysis and arguments made. The chapter concludes providing actionable policy recommendations to centre issues of gender in development theory, policy and practice in Africa.

3.2 Methods

A qualitative literature analysis constituted the research method adopted for the chapter to answer the research questions that guided the research study. According to Onwuegbuzie et al. (2012, p. 3), the approach involves the identification of a research problem, in this case, gender, social policy and development in Africa, and the search for relevant sources of secondary literature. These first two steps are followed by the formulation of search terms, keywords and phrases pertinent to the research problem or question in interest (Onwuegbuzie et al., 2012, p. 2). As the research focus used Mkandawire's lens to understand gender, social policy and development in Africa, by implications key search words included phrases that included social policy, Africa, development and Mkandawire, to which a vast number of search outputs comprising peer-reviewed articles, book chapter, conference proceedings and memorial lectures written by and on Mkandawire appeared. In addition to the literature by and on Mkandawire, published and unpublished secondary literatures on and by the World Bank on development in Africa were also included. The search produced vast amounts of literature from which the study could be based. While not all Mkandawire's literature had a focus on gender, it remained critical as it sets the background to discuss trajectories of development in Africa, particularly Mkandawire's critique on the involvement of the latter on development in Africa. Thus, all the literature was downloaded in a file saved on the computer with the majority covering scholarly works by Mkandawire and from and on the World Bank in addition to a few others. In total, 22 scholarly works by Mkandawire were reviewed though not all cited in the chapter. The subsequent step after downloading the research articles involved the reading of the relevant secondary sources, note taking and summarising the key points and main themes as they emerged from the sources. A perfect fit was discernible between the research questions and the collected secondary sources as social policy issues and development in Africa were Mkandawire's preoccupation in all his writing and in several of his publications analysed including the extent to which gender provides a 'micro-level' analysis of development in Africa, from which the chapter got its pitch. Based on the notes, themes that emerged from the qualitative analysis were then merged and used to weave a narrative as presented on the chapter. No ethical issues were applicable in this research as it involved no human participants. While the next sections provide a review of literature, the chapter begins to synthesise and analyse existing research with the literature interwoven throughout the discussion providing context and support for the analysis and arguments made.

3.3 Literature Review

In the post-independence period, most vital sectors in Africa needed immediate state attention and heavy investment including education, agriculture, industrialisation, transport and energy generation. Colonial governments had severely neglected the development of not only rural areas but also servicing those areas within urban areas that were resided by indigenous black Africans (Nalwimba, 2023, p. 4; Gulali, 2018, p. 1). For instance, in 1935, there was not a single secondary school in Zambia with colonialists spending all the resources on the needs of white settlers arguing that the country is too poor to afford one (Andreasson, 2001, p. 190). This meant that many postcolonial governments, including Zambia, had to embark on a massive public investment programme to correct the imbalances in service provision between rural and urban areas including those urban spaces reserved for indigenous Africans (Nalwimba, 2023, p. 4). With many postcolonial leaders holding a suspicious view of the private sector which they associated with the former colonisers, they embarked on state-led development in which the state intervened in the economy to mobilise resource, establish industries and accelerate public investment (Husain, 1995, p. 178; Brunswijck, 2019, p. 2). Economic planning became the key strategy with the state as an instrument of such planning and intervening in marketing, agriculture, labour markets, external trade and public enterprise (Husain, 1995, p. 178). Many of these public investments by African independent government, especially in infrastructure, were supported by the commodity prices of the 1950s and the early 1960s (Lateef, 1995, p. 14). It appeared to many that the dividend of political independence was paying off as many parts of the continent recorded positive economic growth and the desire to do more even grew stronger in many of the newly independent African nations.

One and half to two decades preceding the 1960s political developments in Africa, seemingly unrelated were the post-war establishment of twin Bretton Woods Institutions (BWIs) of the International Bank for Reconstruction and Development (IBRD) commonly referred to as the World Bank (WB) and International Monetary Fund (IMF) in 1944 (Bretton Woods Project, 2019, p. 2; MENAFEM, 2024, p. 2). These institutions had an original mandate of providing capital for the re-building of Europe and Japan that were devastated by the two world wars (Bretton Woods Project, 2019, p. 2). Opening its doors for business on 25 June 1946, the World Bank provided direct budget support for infrastructure development through direct lending to European governments and Japan in addition to providing support to the private sector (Brunswijck, 2019, p. 2; Bretton Woods Project, 2019, p. 2; Lateef, 1995, p. 6; MENAFEM, 2024, p. 2; Gulali, 2018, p. 2). Africa and the other countries of the developing world were not a priority of the Bretton Woods Institutions (Lateef, 1995; Bretton Woods Project, 2019, p. 2; MENAFEM, 2024, p. 2). Through the provision of loan programmes to states with balance of payment problems as well as policy advice through either technical assistance or bilateral and multilateral macroeconomic surveillance, the IMF focused on promoting international fiscal and monetary cooperation, securing international financial stability, facilitating international trade and promoting high employment and sustainable economic growth (Bretton Woods Project, 2019, p. 2; Lateef, 1995, p. 56). These institutional features were critical and had significant implications in understanding the institutions' later dealing with development issues in Africa. Additionally, the institutions are structured to rely on private resources to fund their operations and *vigorously seek* to promote private foreign investment (Lateef, 1995, p. 7; emphasis is the author's).

To cater to the needs of African and other developing countries that are not creditworthy, a corollary subsidiary, the International Development Association (IDA), was established in 1960. The newly established IDA had a mandate to provide finance on concessional terms to low-income countries that lack creditworthiness for IBRD borrowing and grants funded by donors (MENAFEM, 2024, p. 2). With an institution strategically established specifically to serve loan requirements for developing countries, developments at this time are well captured by Trevor Manuel, the former South African Minister of Finance when he delivered the annual Francis Wilson Memorial Lecture on 16 April 2024 at the Southern Africa Labour and Development Research Unit (SALDRU), University of Cape Town:

> *Our friends in Washington could tell us how it was easy to borrow and there was never a limitation on how much money you can borrow. (Trevor Manuel 16 April 2024)*

To drive a state-led development strategy rather than relying on domestic resource mobilisation, many post-independence governments turned to external sources, particularly the World Bank and IMF loans (Gulali, 2018, p. 2). While there might be nothing wrong with borrowing to finance development, Lateef (1995, p. 11) gives a strong caution regarding the BWIs, particularly their intensive policy advice and dealings with its borrowers. In attempts to influence the outcomes of investment they financed, their advice on macroeconomics and public expenditure policies is usually to the detriment of borrower countries. The institutions would seek to shape its borrower's policies rather than the finance it provides compromising the latter's macroeconomic policy sovereignty (Lateef, 1995, p. 11; Mkandawire, 2014). It was particularly on the last point that Thandika Mkandawire had a contention with the influence of institution, particularly the World Bank, on development policy in Africa and its implications for gender.

3.3.1 The World Bank and Thandika's Critique of Mainstream Development Policy in Africa

The role of the World Bank in shaping development policy in Africa and its implications on gender has been a subject of extensive debate and critique. Thandika Mkandawire, a prominent scholar in African development and economics, provides a compelling analysis of the World Bank's policies and their implications for the continent's socio-economic landscape. Before delving into the gender implications, the proceeding section provides Mkandawire's critiques examining the complexities of structural adjustment programmes (SAPs), governance issues, the impact of globalisation and the potential for developmental states in Africa. By delving into Mkandawire's work, this review aims to illuminate the shortcomings of mainstream development policies in Africa. The latter is particularly important as the World Bank assumed the moniker 'knowledge bank' positioning itself as a repository of development expertise on the continent (Bretton Woods Project, 2019, p. 2). With the establishment of the International Development Association (IDA) and the short-lived commodity boom of the 1960s and early 1970s, many newly independent African countries borrowed extensively from the World Bank and the IMF to finance their state-led development strategies. This was partly informed by modernisation theories of the 1950s and basic needs approaches (BNAs) of the 1970s that placed emphasis on infrastructure development (Midgley, 2019, p. 21, 24; Lateef, 1995, p. 14). However, the two oil shocks first in the 1970s and the second in the 1980s impacted the balance of payment of many oil-importing developing nations including those in Africa, many of whom were owing to the World Bank. With inadequate attention paid to the potential dangers of debt resulting from underwriting poor investment and policies, many African countries found themselves in very precarious financial positions. The second oil shock in 1980 worsened the situation, and the World Bank and the IMF in their meeting the same year agreed that structural adjustments were necessary in many borrower countries amid falling commodity prices and rising interest rates (Lateef, 1995, p. 2). The Zimbabwean case given as a recount by Trevor Manuel in his memorial lecture captures the experience more vividly:

> *...for me personally, the most important experience was what happened in Zimbabwe. You remember from. 1980 to 1990 the tremendous in education, health especially building infrastructure sort of tightly held together but borrowed extensively from the World Bank and IMF. And after a decade the chicks were called in that was when structural adjustments arrived in Zimbabwe. (Trevor Manuel 16 April 2024)*

As illustrated in the Zimbabwean case above, the World Bank's structural adjustment programmes (SAPs) emerged in the 1980s as a response to the economic crises that many African nations faced. These programmes were designed to stabilise economies through a series of neoliberal reforms, including austerity measures, deregulation and privatisation (Brunswijck, 2019, p. 1; Lateef, 1995, p. 14; Bretton Woods Project, 2019, p. 2). Mkandawire and Soludo (1999) in their seminal work, *African Voices on Structural Adjustment*, argue that these policies often had detrimental effects on the very populations they aimed to help (Mkandawire & Soludo, 1999, p. 15). The authors contend that SAPs were imposed without adequate consideration of local contexts, leading to increased poverty and social dislocation on the African continent (see also Mkandawire, 2014). One of the central critiques articulated by Mkandawire is that SAPs tended to prioritise macroeconomic stability over social welfare with far-reaching effects on gender inequality. The emphasis on reducing public spending resulted in significant cuts to essential services such as health and education, disproportionately affecting the most vulnerable groups in society, particularly women (Mkandawire & Soludo, 1999, p. 22). For instance, in countries like Zambia and Tanzania, the implementation of SAPs led to a decline in health-care access and educational opportunities, exacerbating existing inequalities. In Malawi, the removal of agricultural subsidies as part of SAP conditions exacerbated food insecurity and malnutrition among smallholder farmers, particularly women.

The Bretton Woods Project (2019, p. 2), in its submission on the main criticism of the World Bank and the IMF on countries of the Global South, links the introduction of structural adjustment programmes with the decimation of the health systems—devastating impacts that persist to date. Moreover, Mkandawire highlights that the conditionalities associated with World Bank loans often undermined local governance structures. The requirement for governments to adopt specific neoliberal policies curtailed their ability to respond effectively to the needs of their populations, particularly women. This lack of local ownership and accountability not only diminished the efficacy of these programmes but also fostered a sense of disillusionment among citizens regarding their governments (Mkandawire & Soludo, 1999, p. 30). On the social front, the ramifications of SAPs have been profound. As Mkandawire and Soludo (1999) point out, the reduction of social spending and the prioritisation of market mechanisms often led to increased inequality and social unrest. In many cases, the most marginalised populations bore the brunt of these reforms, as access to public goods became increasingly limited (p. 35). This critique underscores a significant flaw in the mainstream development narrative that equates economic growth with social progress, revealing how the prioritisation of market forces can lead to social fragmentation (Adelman, 2000, p. 1). The emphasis on macroeconomic stability often overshadowed the need for social investment resulting in a disconnection between economic policy and social reality (Mkandawire, 2014).

A concomitant effect of the structural adjustment programmes was the rolling out of targeted social protection programmes across the African countries and other developing world. Thandika Mkandawire critically examined the World Bank's targeted social protection programmes in Africa, arguing that these initiatives often failed to address the structural causes of poverty and inequality in developing countries (Mkandawire, 2014). By design, targeted social protection programmes are designed to assist specific groups identified as vulnerable or poor. While promoted as efficient means to allocate resources, Mkandawire (2005) argues that the approaches often overlook the broader socio-economic context of poverty, leading to several critical shortcomings (Mkandawire, 2005, 2014). Many deserving individuals particularly women are excluded from benefits, while those who do not need assistance may receive support (Mkandawire, 2010). By focusing on specific groups, targeted programmes tend to operate in silos, failing to address the interconnected nature of poverty, health, education and employment. This lack of integration can hinder the development of comprehensive social protection systems that are essential for sustainable poverty alleviation (Mkandawire, 2009). In contrast, universal approaches can foster synergies across sectors, leading to more holistic solutions to poverty. With many beneficiaries of targeted social protection programmes mainly female, another critical issue raised by Mkandawire is the potential for stigmatisation associated with targeted social protection programmes. Beneficiaries often face social stigma, which can deter individuals from seeking assistance or participating in broader community activities (Mkandawire, 2005). Furthermore, the conditionality often attached to assistance can create dependency, where individuals become reliant on aid rather than developing sustainable livelihoods (Mkandawire, 2010). This dependency undermines the very objectives of social protection, which should aim to empower individuals to escape poverty.

In addition to the above and forming a background to the paper is Thandika Mkandawire's critical analysis of the World Bank's involvement in Africa, particularly during the debt crisis that erupted in the 1980s and its contemporary implications on gender and African women in particular. Mkandawire's work reveals the extent to which the contemporary debt crisis in Africa is inseparable with structural adjustment programmes. Mkandawire (2001) contends that the World Bank's approach to the debt crisis has been characterised by a one-size-fits-all mentality. The diverse economic and political contexts of African countries were often ignored in favour of uniform prescriptions that did not account for local conditions. This lack of contextual understanding has led to the failure of many SAPs and has perpetuated cycles of dependency. In response to growing criticism of its policies, the World Bank launched initiatives such as the Heavily Indebted Poor Countries (HIPC) initiative in the late 1990s. This programme aimed to provide debt relief to eligible countries, contingent upon the implementation of poverty reduction strategies. However, Mkandawire critiques the HIPC initiative for its slow implementation and bureaucratic hurdles, which hindered timely debt relief for many countries. Additionally, the focus on macroeconomic stability within the HIPC framework often overshadowed the need for comprehensive social policies aimed at poverty alleviation. Mkandawire (2001) argues that while debt relief is essential, it must be accompanied by meaningful investments in social infrastructure to ensure sustainable development. The limited scope of the HIPC initiative raises questions about the commitment of the World Bank to address the root causes of poverty and inequality. It is that interference and the implications of the latter on gender equality which, using Thandika Mkandawire's transformative social policy lens, forms the basis of the ensuing discussion of results section.

3.4 Discussion

3.4.1 Development in Africa: A Critical Gender and Social Policy Analysis

Attempts by newly independent African government to modernise their economies through import substitution industrialisation with financial support from the World Bank and the IMF, particularly the greater emphasis on economic growth, were highly gendered (Midgley, 2019, p. 21; Bretton Woods Project, 2019, p. 2; Adelman, 2000, p. 1; Lateef, 1995, p. 14). Mkandawire provided a gender critique of the import substitution industrialisation policies and strategies, particularly the extent to which it marginalised smallholder farmers and rural economies where African women were concentrated due to its urban bias resulting in a disproportionate allocation of industrial resources to the latter at the expense of the countryside (Mkandawire & Soludo, 1999, p. 91). In her classic book *Women's Role in Economic Development*, Ester Boserup (1970) analysed how women's lives in developing economies were affected by modernisation's conflation of development with economic growth. The author provided initial critiques of the extent to which import substitution industrialisation policy and strategies fostered male wage employment and perpetuated traditional gender roles (Boserup, 1970). At international level, the author highlighted gender biases in development, particularly the way in which women's economic roles and contribution to development either working in agriculture, crafts, trade or other productive capacities were not recognised and ignored.

This misalignment not only stunted agricultural development but also exacerbated social inequalities including gender, ultimately undermining the broader developmental goals of African nations. In response, widespread international feminist movement emerged in the early 1970s including women in development (WID) which campaigned not only for the recognition of women's economic contribution but also the integration of women into global processes of economic, political and social growth and change (Rathgeber, 1990, p. 2). The potential gender progressiveness of the early 1970s which saw the coining of WID was a direct critique to modernisation theories of the mainstream development. With the second oil shock in 1980 amid deepening economic crisis, falling commodity prices, rising interest rate and worsening balance of payment challenges in many developing countries including those on the African continent, the implications for gender cannot be overemphasised. Irrespective of context, the World Bank pushed a particular set of macroeconomic policy prescriptions across all borrower countries (Bretton Woods Project, 2019, p. 6). Implemented through stabilisation structural adjustment programmes (SAPs), these World Bank macroeconomic developmental initiatives had far-reaching implications on gender equality that persists today (Mkandawire, 1999, p. 321; Bretton Woods Project, 2019, p. 2). In his critique on the effects of structural adjustment programmes (SAPs) on gender in Africa, Thandika Mkandawire emphasised that these programmes have often exacerbated existing gender inequalities. He argues that the fiscal consolidation measures mandated by SAPs led to significant cuts in public spending on essential services such as health and education, which disproportionately affected African women. As women typically bear the primary responsibility for caregiving and household management, the reduction in social services by many African governments shifted much of the care burden to family, particularly women, resulting in increased burdens and further entrenching traditional gender roles and limiting their opportunities for empowerment and economic participation (Mkandawire & Soludo, 1999; Mkandawire, 2001).

Unlike the developmental approach in the early post-independence period whereby newly independent African governments sought to ensure universal access to publicly provided services in health and education where women could potentially benefit, the targeted social protection programmes that accompanied SAPS were deleterious to women and progress towards gender equality. Embedded within the macroeconomic strategies were targeted social protection policies which Mkandawire heavily criticised, particularly due to their impact on gender equality. The bank's continued emphasis on poverty targeting in social protection which has proven over and over to be ineffective more acutely on transforming structural inequalities of gender and reaching the most vulnerable remains problematic (Mkandawire, 2005, p. 9). This selective approach tends to perpetuate the marginalisation of women, as it assumes a homogeneity among the poor that fails to account for the diverse experiences and needs of different gender groups. Moreover, Mkandawire highlights that targeting undermines the broader goals of social transformation by framing social policy merely as a tool for poverty alleviation rather than a means for achieving social justice and equality (Mkandawire, 2005). He emphasised that universal policies, which provide benefits to all, are more likely to foster gender equity by ensuring that women have access to the same resources and opportunities as men. In contrast, the targeting paradigm can lead to a dual structure of service provision that segregates benefits, ultimately reinforcing patriarchal norms and limiting women's empowerment in society.

The current debt restructuring across many borrower developing countries had even deepened issues of gender inequalities. Debt servicing curtails the ability of African governments to finance infrastructure and social services (Bretton Woods Project, 2019, p. 4; Brunswijck, 2019, p. 2). World Bank's advice to use public resources to leverage or subsidise private sector through its cascading approach, particularly the Maximising Finance for Development (MFD) programme which reframes the state's role from the earlier social provisioning to essentially creating an enabling environment for the so-called private sector investment geared towards facilitating the private sector, has far-reaching implications on gender equality (Bretton Woods Project, 2019, p. 1, 4; Brunswijck, 2019, p. 2). Despite the gendered impacts of private sector-driven models of development that had failed in several if not all contexts, Mkandawire persistently critiqued the bank's insistence to 'crowd in' private sector investment in agricultural, energy and water sectors in many countries of the Global South (Mkandawire, 2005, p. 4; see also MENAFEM, 2024, p. 4; Brunswijck, 2019, p. 2). In the current debt restructuring, borrower countries are now expected to adopt required 'prior actions' to receive funding from the Bretton Woods Institutions, and in 2017 the World Bank issued a total of 434 prior actions. Brunswijck (2019, p. 1) argues that 27% of the prior actions are intended to enhance the role of the private sector by improving the business climate and privatisations. The bank's new gender strategy (2024–2030), which is part of the Evolution Roadmap, has been criticised for its narrow gender equality metrics that are mainly focused on economic empowerment which many women in developing countries have found ambivalent if not accompanied by measures that address care (MENAFEM, 2024, p. 3). These metrics fail to address systemic gender inequality in a comprehensive manner, particularly in the African context.

3.4.2 Towards a Gender Transformative Social Policy and Development with Thandika

Contributing to the gendered analysis of development, with a particular emphasis on Africa's development, Thandika Mkandawire (2004, p. 6) reminds that gender studies in Africa provide an important micro-level argument for rethinking gender and social policy in the context of development, particularly the need to pay attention to social reproduction. To this realisation, Mkandawire proffered a context-specific approach to social policy and development in Africa through the formulation of the transformative social policy (TSPA) framework that defines social policy as collective interventions directly affecting transformation in social welfare, social institutions and social relations (Mkandawire, 2001, p. 1; Mkandawire, 2006). The utility of the transformative social policy perspective in Africa's development lies in its prioritisation of gender equity as a fundamental component of sustainable development. The latter is particularly important as it brings Mkandawire's perspective on the gender-social policy-development nexus in Africa that will be given greater attention subsequently. Citing North (1990), Thandika defines social institutions as the humanly devised constraints that shape human interaction or the rules of the game in a society. The above three concepts, Thandika Mkandawire contends, are now widely recognised as important determinants of economic development, which, in turn, facilitates achievements in these areas (Mkandawire, 2004, p. 1; Mkandawire, 2007, p. 15).

Important to highlight within the transformative social policy framework, and in contrast to the 'social protection framework' pushed by the World Bank, is the role of the African state to drive economic growth and social progress (Mkandawire, 2001, p. 291; Mkandawire, 2006). Critiquing the World Bank's approach which undermines state capacity, Mkandawire argues that successful developmental states are characterised by state ability to articulate and implement a coherent development agenda, including that on gender (Mkandawire, 2001, p. 295). He argued that the African state should play a proactive role in steering economic development, rather than adopting a laissez-faire approach that cedes control to market forces. With the state more amenable to the needs of women relative to the market, it plays a critical role for gender progressive development. Thus, Mkandawire called for a paradigm shift in development policy advocating for a re-evaluation of the role of the state and policies that prioritise social equity and sustainable development (Mkandawire, 2001, p. 305). Furthermore, rather than relying on externally generated knowledge for Africa's development, particularly from the World Bank, Mkandawire emphasises the importance of local ownership in development initiatives. By empowering local communities and incorporating their voices into policy formulation, African nations can enhance the effectiveness and legitimacy of development strategies (Mkandawire, 2001, p. 310). This participatory approach not only fosters accountability but also ensures that policies are context-specific and responsive to the unique needs of different regions and social groups including women.

Thinking beyond World Bank' targeted safety nets in Africa with their limitations not only in merely counteracting policy failures and developmental disasters but also gender insensitivity, Thandika Mkandawire makes a strong case for universal social protection. Highlighting the limitations and unintended consequences of targeting, particularly on gender, Mkandawire advocated for a more universal approach to social protection in Africa that fosters inclusivity and social cohesion. The utility of universalism and its implications for social policy and development in Africa lies in the approach's effectiveness in addressing structural dimensions of poverty and inequality including that of gender often overlooked by targeted social protection programmes (Mkandawire, 2015). Such a comprehensive approach to social policy and development is vital in Africa for tackling the systemic issues that perpetuate poverty and inequality and are crucial for sustainable development in Africa (see Tekwa & Adesina, 2018). He contends that the one-size-fits-all approach to social protection advocated by the World Bank does not adequately address the diverse realities of poverty in Africa and other developing regions. This calls African policymakers to be attuned to local conditions and the intersections of cultural practices and historical contexts and gender when designing social protection programmes.

Zeroing in on gender, social policy and development, Mkandawire lamented the extent to which research on gender and development in Africa is yet to comprehensively engage with social reproduction issues, particularly 'labour power' as a produced means of production and the central role of women in this work—a persistently ignored but critical source of gender inequality in development contexts like Africa (Mkandawire, 2004, p. 6). Citing several feminist political economy social reproduction scholars, Mkandawire highlighted the extent to which in the African context, contrasted to the Global North, significant investments in human capital take place in the household sector, in the form of women investing long hours of unpaid work in the care and socialisation of children who will become the next generation of adults/workers (Mkandawire, 2004, p. 6; see also Tekwa, 2024). African researchers on gender, social policy and development are yet to comprehensively examine how social reproduction—activities and relations that sustain daily life—intersects with localised economic processes and gender relations and the crucial role of unpaid labour, often performed by women, in sustaining both the economy and society and to highlight the systemic inequalities within these frameworks. Thandika argued that such an attention to human capital offers greater opportunities than older theories to integrate gender relations into both social analysis and development policies. Such a lens as promulgated by Thandika Mkandawire, unlike the present approaches to the study of gender on the continent, highlights the significance of integrating gender at the outset rather than treating it as an afterthought (Mkandawire, 2004, p. 10). This calls for gender to be given its full social meaning when making policy and the need to re-analyse and change the underlying assumptions about labour input to recognise some constraint arising from the reproductive sector (Mkandawire, 2004, p. 10). Nowhere is this particularly important than the African continent with its largely untheorised agrarian and informal sectors.

3.5 Conclusion

Contextualised within the political economy of Africa's development, this chapter historicised the development trajectory of the continent from the early independence period but nested within the hegemonic economic and social policy influences of the Bretton Woods Institution of the World Bank and the International Monetary Fund. The institutions' establishment, not originally focused on African development, quickly re-oriented themselves as sources not only for development finance but also a repository of 'development knowledge' with far-reaching implications for gender, social policy and development on the continent. Based on a qualitative literature review of Thandika Mkandawire's critical analysis of the World Bank's involvement in shaping Africa's development trajectory, the method involved the collation of Thandika Mkandawire's work on Africa's development and then the application of a gender lens into his critique of mainstream development in Africa.

Three distinct phases are identifiable in the World Bank involvement in shaping the trajectories of development in Africa, each with its own gendered implications with the latter deepening overtime with the growing institutions' influences on the domestic economic and social policies of borrower countries. While the institutions' development finance support for modernisation and import substitution industrialisation produced modest economic growth in many newly independent

African countries, a gender critique reveals not only an urban bias with industrial resources disproportionately channelled in the latter at the expense of rural areas where the majority of African women resided; the approach has tended to reinforce male breadwinner models creating wage work for men at the expense of African women who remained excluded from formal participation in the economy. While this phase did not last long, the economic crisis of the 1980s saw the World Bank introducing structural adjustment programmes in many of the borrower countries to curtail ballooning debt as part of the mainstream development strategy for African countries. Through his extensive body of work, Mkandawire highlighted the gendered detrimental effects of structural adjustment programmes, particularly the substantial cut in social expenditure and privatisation of social services which exacerbated gender inequalities as the burden off-loaded by the state was transferred to family, particularly on women. The attendant targeted social protection programmes rolled out across all the developing world including Africa underscore their insensitivity to local conditions and shortcomings in addressing structural complexities of poverty and inequality in Africa including those of gender. With the deepening debt crisis and many borrower countries undergoing debt restructuring, the accompanying austerity measure had exacerbated rather than ameliorated existing inequalities and undermined women's rights and economic opportunities. The persistent private sector-led and market-oriented policies are not amenable to objective of gender equality and transformation. As African nations continue to navigate the challenges of the contemporary debt crisis and seek sustainable development pathways, it is crucial that policymakers heed Mkandawire's insights. Under the prevailing polycrisis, it remains imperative for African government to move towards inclusive, universal gender transformative social policies that not only empower individuals including women to promote sustainable development but also make gender an integral part of the policymaking process. This is in addition to prioritising social equity and local ownership for a more gender inclusive, effective and sustainable future for Africa. Ultimately, Mkandawire reminds us that the continent's path to development must be shaped less by external influences but the voices and experiences of African people, ensuring that policies are not only effective but also reflective of the continent's diverse realities.

References

Adelman, I. (2000). Fifty years of economic development: What have we learned?. *Keynote address at the Annual Bank Conference on Development Economics, June 26–28, Paris.*

Andreasson, S. (2001). Divergent paths of development: The modern world-system and democratization in South Africa and Zambia. *Journal of World-Systems Research, 7*(2), 175–223.

Boserup, E. (1970). *Woman's role in economic development.* St. Martin's Press.

Bretton Woods Project. (2019). What are the main criticisms of the World Bank and the IMF? *Critical Voices of the World Bank and the IMF.* Retrieved December 16, 2024, from https://www.brettonwoodsproject.org/2019/06/what-are-the-main-criticisms-of-the-world-bank-and-the-imf/

Brunswijck, G. (2019). *Flawed conditions: The impact of the World Bank's conditionality on developing countries.* European Network on Debt and Development (EURODAD) Briefing Paper, April 2019. Retrieved April 27, 2020, from https://eurodad.org/files/pdf/5cac92cd557d8.pdf

Gulali, G. (2018). *The World Bank in Africa: An analysis of World Bank aid and programmes in Africa and their impact*. ITRD Consulting Group and the Lending for Education in Africa Partnership Programme, a pilot programme of Volta Capital. https://www.realityofaid.org/wp-content/uploads/2018/12/1-The-World-Bank-in-Africa-An-analysis-of-World-Bank-aid-and-programmes-in-Africa-and-their-impact.pdf

Husain, I. (1995). The challenge of Africa. In K. S. Lateef (Ed.), *The evolving role of the World Bank: Helping meet the challenge of development* (pp. 177–208). World Bank.

Lateef. (1995). *The evolving role of the World Bank: Helping meet the challenge of development*. World Bank.

MENAFEM. (2024). *IDA21: A missed opportunity for transformative development in the Global South, December 2nd, 2024*. Retrieved December 16, 2024, from https://menafemmovement.org/ida21-a-missed-opportunity-for-transformative-development-in-the-global-south/

Midgley, J. (2019). Social policy and development: An overview. In J. Midgley, R. Surender, & L. Alfers (Eds.), *Handbook of social policy and development* (pp. 14–34). Edward Elgar.

Mkandawire, T. (1999). The political economy of financial reform in Africa. J. Int. Dev., 11: 321–342. https://doi.org/10.1002/(SICI)1099-1328(199905/06)11:3321::AID-JID5943.0.CO;2-V.

Mkandawire, T. (2001). Thinking about developmental states in Africa. *Cambridge Journal of Economics, 25*, 289–313. https://doi.org/10.1093/cje/25.3.289

Mkandawire, T. (2004). Introduction. Social policy in a development context. In T. Mkandawire (Ed.), *Social policy in a development context* (pp. 1–33). UNRISD/Palgrave Macmillan.

Mkandawire, T. (2005). Maladjusted African economies and globalisation. *Africa Development, 30*(1), 1–33. ISSN 0850-3907.

Mkandawire, T. (2006) *Transformative social policy: Lessons from UNRISD research*. UNRISD Research and Policy Brief 5. UNRISD.

Mkandawire, T. (2007). Transformative social policy and innovation in developing countries. *European Journal of Development Research, 19*(2), 13–29.

Mkandawire, T. (2009). Institutional monocropping and monotasking in Africa. In N. Akbar, B. Kwesi, S. Howard, et al. (Eds.), *Good growth and governance in Africa: Rethinking development strategies* (pp. 1–28). United Nations Research Institute for Social Development.

Mkandawire, T. (2010). Running while others walk: Knowledge and the challenge of Africa's development. Revised version of the paper delivered as inaugural lecture for chair, African Development at the London school of economics on 27 April 2010. *Africa Development, 36*(2), 1–36.

Mkandawire, T. (2014). The spread of economic doctrines and policymaking in postcolonial Africa. *African Studies Review, 57*(1), 171–198.

Mkandawire, T. (2015). *Africa: Beyond recovery*. Sub-Saharan Publishers. http://www.jstor.org/stable/j.ctvgc60z7

Mkandawire, T., & Soludo, C. C. (1999). *Our continent, our future: African perspective on structural adjustment*. Africa World Press.

Nalwimba, N. (2023). *Africa's social policy trajectories since the colonial period: Towards an understanding of social policies and gender in Zambia*. GETSPA Working Paper, GETSPA WPE 1 2023, University of Ghana.

North, D. C. (1990). Institutions, Institutional Change and Economic Performance. Cambridge University Press.

Onwuegbuzie, A. J., Leech, N. L., & Collins, K. M. (2012). Qualitative analysis techniques for the review of the literature. *Qualitative Report, 17*, 1–28.

Rathgeber, E. M. (1990). WID, WAD, GAD: Trends in research and practice. *The Journal of Developing Areas, 24*(4), 489–502. http://www.jstor.org/stable/4191904

Tekwa, N. (2024). Women's work and land reform in Zimbabwe: A feminist political economy of social reproduction. *Cross-Sections Social Science Journal, 13*(2), 59–80. https://doi.org/10.18392/metsz/2024/2/6

Tekwa, N., & Adesina, J. (2018). Gender, poverty and inequality in the aftermath of Zimbabwe's land reform: A transformative social policy perspective. *Journal of International Women's Studies, 19*(5), 4. https://vc.bridgew.edu/jiws/vol19/iss5/4

Chapter 4
Rethinking and Unthinking Social Policy in Africa: Towards Decolonisation and Prospects for Inclusive Development

Emmanuel Ndhlovu and Neo Mokone

4.1 Introduction

Africa is presently experiencing several challenges. In addition to a widespread leadership crisis, the region struggles with the vagaries of climate change and associated natural hazards, war and conflict, hyperinflation, and incessant energy crises. The result is the worsening of numerous challenges, including food insecurity, inequality, and retarded development. According to the Food and Agriculture Organization (FAO), the United Nations Economic Commission for Africa (ECA), African Union Commission (AUC), or the World Food Programme (WFP) (2023), around 282 million people in Africa (20% of the population) were undernourished in 2022. Approximately 868 million people were moderately or severely food-insecure, and 342 million people were severely food-insecure. The region is not on track to meet the food security and nutrition targets of the Sustainable Development Goal (SDG) 2—and the Malabo goals of ending hunger and all forms of malnutrition by 2025. Hunger has worsened significantly. Unlike in the Euro-North American countries where the welfare service state is well developed and improving, African welfare regimes must contend with the impact of colonial and postcolonial policies and transitioning democracy (Surender, 2023).

The present situation in Africa is that most people are not in formal employment. Approximately 70% of the people rely on agriculture for their livelihoods and income (Ndhlovu, 2025). A significant percentage is also located in the informal

E. Ndhlovu (✉)
School of Public Management, Governance and Public Policy, College of Business and Economics, University of Johannesburg, Auckland Park, South Africa

N. Mokone
Centre on African Philanthropy and Social Investment (CAPSI), Wits Business School, University of the Witwatersrand, Johannesburg, South Africa
e-mail: neo.mokone@wits.ac.za

E. Ndhlovu, C. Chipenda (eds.), *Redefining the Social Policy Discourse*,
https://doi.org/10.1007/978-3-032-24067-5_4

sector, particularly in West and East Africa. Only a small number of African citizens have access to employment with good salaries, good education, good healthcare, or good social services (Conradie, 2018). A large section of the population is, therefore, left out in the modern economy and associated levels of social services, and poverty and inequality are thus high (International Labour Organisation (ILO), 2023). Presently, there are two main strands of social protection systems in Africa: those for middle-income countries such as Botswana, Mauritius, South Africa, Rwanda, and Morocco, dominated by age-based (the elderly and children) social transfers which are rights-based and tax-funded, and those in low-income countries such as Angola, Malawi, Ethiopia, Somalia, and Zimbabwe, often funded with development aid, and thus less sustainable.

With poverty so high in Africa, it is crucial that social policies be redesigned in a way that can enable the people to emerge from poverty and begin to contribute towards the region's commitment to development. Many studies have detailed the reasons for Africa's failure to attain development since independence. Among them has been the pursuit of inappropriate policies (Ayittey, 2005; Ndhlovu & Mhlanga, 2023). This view is supported by Gumede (2015), who problematised the African responses to the impact of the skewed socio-economic and political order and the historical experience of various forms of enslavement. Cheru (2009) argued that the post-1945 global order was the fundamental constraint to Africa's development. This chapter argues that African countries can benefit by rethinking their social policy systems and aligning them with local needs and commitment. This chapter (i) explores the potential of social policy to improve human welfare in Africa and (ii) examines the possibility of social policy as the basis for inclusive development.

Since Africa is basically an agrarian community, with a significant percentage engaged in farming, countries could tilt social policies to strengthen the sector as the basis for regional development. The focus can be on helping the people graduate from subsistence farming, which is vulnerable and unprofitable. While age-based social protection, mainly the social assistance approach of food aid and cash transfers to the elderly and children, is vital, it does not generate prospects for development as it is mainly to meet immediate consumption needs. Targeting various population categories and activities in social protection programmes could actually provide the means by which the African population can begin to actively participate in contributing to the broader economy.

This chapter contributes to the ongoing debate on African development. It seeks to highlight the diagnostic potential of social protection as an intervention towards inclusive development in Africa. Combining both development and social policy debates, the chapter could be useful for policymakers, academics, students of development, donor actors, and politicians. The review nature of the chapter allows it to provide a comprehensive summary of the key debates around these topics and thus propose context-specific interventions. This chapter makes significant contributions to contemporary social policy debates in African development processes against the onslaught of neoliberal policies that either disregard or adopt a narrower view of social policy's transformative potential. By focusing on social protection, the chapter argues that in a development context, social protection does not necessarily require adopting a one-size-fits-all approach. Instead, each policy area must be creatively designed to serve intrinsic and instrumental resolves (Kpessa-Whyte, 2021).

The chapter is organised as follows: After the present introduction, the next section provides a background to the origin of the social policy, followed by a conceptualisation of this concept. This is followed by a review of development debates in Africa, focusing on the possibilities of attaining inclusive development. After that, the research methodology for the study is presented. This is followed by a discussion of the major debates around social policy and inclusive development in Africa. Lastly, conclusions are drawn from the discussion.

4.2 Origin of Social Policy

The advent of formal social policy systems is traced to Europe, where they were used as response strategies to the disruptions of the 'great transformation' (Polanyi, 1944) of the eighteenth and nineteenth centuries. In the contemporary world, these systems have again been called upon to respond to rapid economic transitions and social dislocation (Devereux, 2013). The Industrial Revolution of the eighteenth and nineteenth centuries triggered a mass rural-urban migration as people took up jobs in factories and offices. The labour which was traditionally used for household food production in the countryside was not commodified to produce profits for employers. This rendered both workers and rural dwellers vulnerable to the impact of market forces. The social policy then emerged as an 'intervention in the market economy in order to secure the cultural and social integrity of humankind and/or society (e.g., welfare-state redistribution and the regulation of the economy)' (Drahokoupil, 2004, p. 837). Social policy was, therefore, meant to protect society against self-regulating markets.

With the First Industrial Revolution, people who were once farmers lost their access and capacity to their own production and to sell their labour for survival. They now relied on regular wages and salaries from employers. Before the Industrial Revolution, informal social policy was organised according to the principles of reciprocity, redistribution, and householding. However, with the advent of the Revolution, these principles were eliminated by the economic rules of the market (Devereux, 2013). The traditional informal social policy system was damaged when community bonds were disrupted, leaving communities more vulnerable. The state and other well-wishers then stepped in to provide assistance to individuals and households that could not support themselves. This saw the introduction of an array of formal social policy instruments to ensure the reproduction of the labour force (Awortwi, 2018). The introduced social policy interventions were not meant to emancipate the affected households but rather to enable them to survive so that, where possible, their labour could be called upon to contribute to the market. Therefore, contributory social security was introduced for adults with labour capacity for periods when they would not be working—unemployment insurance, retirement pensions, and paid maternity and sick leave. Non-contributory social welfare was also introduced for dependants who were not working and whose private support systems were not sufficient—children, older persons, and people living with disabilities. This social protection model was first introduced in Germany in the 1880s. It was then institutionalised in the United Kingdom in the wake of the recommendations of the 'Beveridge Report' of 1942, which proposed that a social insurance system would guarantee a minimum standard of living. With time, the model was widely adopted across Europe (Esping-Andersen, 1990).

The model was also exported to Africa, Asia, and Latin America by European administrators during the colonial era. Devereux (2013) reveals that, in Africa, the earliest social security systems were first introduced in South African and North African countries. Although the earliest social security systems were based on employee contributions, farmers and those working in the informal sector were at times encouraged to join, particularly in countries such as Egypt and Tunisia, where such would be provided with incentives such as lower contribution rates (Nasri et al., 2022). In the French-speaking countries of West Africa, retirement plans for civil servants, which were once voluntary during the colonial period, became compulsory schemes after independence. In the former British colonies of Ghana, Nigeria, Kenya, Zimbabwe, Zambia, and Malawi, provident funds for public sector employees were introduced after independence.

The European social protection system, however, did not align with the needs of Africans and the conditions of their existence. Unlike in Europe, most of the African population is still located in rural areas where farming is the main livelihood activity. In industrialised economies, the majority of the people are located in urban areas where they have access to formal jobs, with contracts that entitle them to a regular income from which contributions to social security can be automatically deducted (Devereux, 2013). In these economies, very few people live in poverty and have no access to regular paid work and rely on social welfare. For instance, the European Commission reveals that, in October 2023, the Euro area seasonally adjusted unemployment rate was 6.5% (European Commission, 2023), while in Africa unemployment in most of the countries was over 50% and reached as high as 80% in countries such as Zimbabwe (Mhlanga & Ndhlovu, 2021). Therefore, most Africans live outside the formal job market, with no job security, regular income, and no opportunity to contribute to social security. Many employed Africans are also located in the informal sector, with employment mostly seasonal, casual, and low-paid.

The rural folks also face different challenges and opportunities compared to their urban colleagues, most of whom have paid formal jobs. The everyday challenges of rural folks include food crises, often caused by unfavourable weather. They often lack adequate resources to respond positively to the challenges that disturb their livelihood activities. In responding to these challenges, many African governments and donor communities have often introduced cash transfers to rural households instead of providing them food aid because, with adequate resources and a favourable climate, most African peasant farmers can increase productivity and remain with a surplus, thus decreasing food insecurity chances. In Southeast Zimbabwe, Ndhlovu (2024a) found that smallholder farmers who were beneficiaries of social assistance, mainly cash transfers, whether from the state, the private sector, or donor actors, used it to purchase inputs and acquire information to improve their farming activities.

4.3 Conceptualisation of Social Policy

Social policy is conceptualised as public actions or interventions that can either protect or harm people in certain areas (Adesina, 2007). It refers to 'collective interventions directly affecting transformations in social welfare, social institutions and social relations... [as well as] access to adequate and secure livelihoods and income' (Mkandawire, 2001, p. 1). In the context of national development, Mkandawire (2001, p. 12) also describes the social policy as 'an instrument for ensuring a sense of citizenship [and] an important instrument for conflict management, which is in turn a prerequisite for sustained economic development...'. At the centre of social policy is ensuring that all people live a life of dignity irrespective of their status, culture, age, gender, or any other distinctive feature. Social policy is located within the public policy system of countries. Therefore, it can also be understood as referring to comprehensive programmes and actions that governments adopt to protect and promote social security and the welfare of citizens (Mkandawire, 2011a). Social security is one of the key features of social policy, and it is generally used to refer to a system through the state or community that provides support to people in distress or need due to old age, physical handicap or disability, unemployment, and long-term infirmity, among others, to enable them to maintain or enhance their socio-economic welfare.

Social policy entails the adequate rolling out of social services—defined as facilities and services provided by the state to promote and sustain the security and welfare of citizens (Ndhlovu, 2024c). Social services include public healthcare, housing, education, and counselling, particularly for vulnerable members of society (Taylor, 2021). Social policy aims to provide social security and social services and thus safeguard human dignity, promote social justice, and facilitate inclusive development in society. The notion of social justice is essential in the conceptualisation of social policy since justice is the goal of social policy. Nielson (1996) argues that the question of justice is the question of what a suitable social order can guarantee human welfare. For instance, in an agrarian community such as Africa, social justice will ensure that people have equal and adequate access to farmland for crop production and pastures for their livestock. The recent land grabs that have been unfolding in Africa as monopoly-finance capital invested in the agricultural sector in Africa shows African governments' preference for economic growth ahead of human welfare and social justice. In Zimbabwe, the government threatened that where peasants are not productive enough, land acquired under the country's Fast-Track Land Reform Programme of the early 2000s would be repossessed and redistributed to potential investors under the 'Zimbabwe Is Open for Business' development approach (Nussbaum, 2004). In South Africa, land redistribution has also been slow due to the prioritisation of monopoly-finance capital ahead of citizens. Rethinking social policy would reroute the African government's priorities to promote human welfare rather than ultimately focus on economic growth.

In unthinking colonial versions of social policy, Africa needs to recentre the justice that was denied its people through colonial subjugation. Justice is the main principle of the existence and coexistence of societies and peoples and, therefore, should inform any development policy. Social policy is largely pro-poor and, therefore, resonates with development pursuits that are both sustainable and also satisfactory to the African population that is large. This chapter examines how African governments could deploy social policy to boost production, improve social protection, and ensure social reproduction, social cohesion, and redistribution as part of their campaign on development.

In view of the above differences between African countries and their former colonisers, it is argued the copy-and-paste strategy of policies is unlikely to generate any meaningful results. Therefore, there have been increasing calls to rethink and reframe social policy in Africa as the basis for the attainment of inclusive development.

4.4 Inclusive Development in Africa

There is no clear definition of what 'inclusive development' entails. However, the concept generally entails growing an economy to ensure wealth distribution, reduce inequality, and improve well-being rather than just focusing on economic growth. Ndhlovu (2025) defines it as a type of development that delivers a progressive, sustained, and sustainable transformation of people's lives and societies. It entails a fair distribution and preferable development returns (Dörffel & Schuhmann, 2022). It is a development that places humanity at the centre. Anand and Sen (2000) argue that development should be human-centred because humans are both the end and the means of development (Anand & Sen, 2000). Therefore, using gross domestic product (GDP) to measure growth is inadequate (Jansen et al., 2024). Liu et al. (2024) also argue that examining income levels and stocks of (material) wealth does not allow adequate extrapolations about people's preferences and the extent to which they are satisfied. So, they can only be transitional goals with instrumental value (Dörffel & Schuhmann, 2022). Talmage (2020) describes inclusiveness as a shared outcome emanating from systems of inclusion that employ diversity as a resource. Inclusivity denotes that all people have equal access to social, political, and economic opportunities.

Many African scholars support a type of development that transcends GDP limits. Thandika Mkandawire, one of Africa's most prolific intellectuals on social policy, always associated development with human freedom and aspiration (Mkandawire, 2011b). Ndlovu-Gatsheni (2012, p. 2) describes the development as a 'liberatory human aspiration to attain freedom from political, economic, ideological, epistemological, and social domination…'. Mensah (2019, p. 4) defines development as 'an evolutionary process in which the human capacity increases in terms of initiating new structures, coping with problems, adapting to continuous change, and striving purposefully and creatively to attain new goals'. Gumede (2019, p. 51) defines development as '…improvements in well-being, involving socio-economic progress', while Brobbey (2010, p. 1) describes it as 'the capacity of a state with the aim of achieving higher outcome of production for the satisfaction of citizens and empowering them to make demands'. Reyes (2001) avers that development entails a social condition whereby existing natural resources and methods satisfy people's needs. It is also understood as a practice that involves improving people's conditions (Todaro & Smith, 2006). These definitions of 'development' are a clear rebuff of the capitalistic features of W.W. Rostov's definition, which considered development as an evolutionary process through which the West evolved and through which the non-West still in the process of evolving (Ndhlovu & Mhlanga, 2023). African definitions of development revolve around the need for the well-being of all people—inclusive development. Possibilities of attainment development in Africa necessitate sovereignty commitments as well as decolonisation and decoloniality moves to enable countries on the continent to pursue development strategies that are context- and cultural-specific. African countries need to wean themselves from their former colonisers.

Most African countries continue to depend on their former colonisers for donations, loans, and development models. This enabled the former colonisers to control such countries many years after direct colonialism. As a result, many African countries are unable to improve the conditions of their people (Ndhlovu, 2024b). For instance, from 1980 to 2019, the human development index (HDI) of sub-Saharan Africa (SSA) was the lowest compared to other regions (see Table 4.1). Although the HDI for the Caribbean, Latin America, Europe, Central Asia, and the Arab states showed no significant improvement in the 2011–2015 period, their HDI levels were still better than for SSA.

In the same period, Africa also underachieved in terms of life expectancy, expected years of schooling, mean years of schooling, gross national income (GNI)

Table 4.1 Human development by region

	1980	1990	2000	2010	2015	2017	2018	2019
Very high human development	0.766	0.791	0.858	0.888	0.892	0.894	0.896	0.898
High human development	0.614	0.791	0.687	0.739	0.746	0.744	0.748	0.753
Medium human development	0.420	0.465	0.548	0.625	0.631	0.624	0.627	0.631
Low human development	0.316	0.356	0.383	0.453	0.497	0.507	0.509	0.513
Arab states	0.444	0.556	0.578	0.639	0.687	0.699	0.702	0.705
East Asia and the Pacific	0.428	0.516	0.581	0.666	0.720	0.735	0.740	0.747
Europe and Central Asia	0.644	0.652	0.68	0.744	0.756	0.785	0.787	0.791
Latin America and the Caribbean	0.582	0.626	0.68	0.728	0.751	0.762	0.764	0.766
South Asia	0.356	0.438	0.468	0.545	0.621	0.635	0.637	0.641
Sub-Saharan Africa	0.365	0.399	0.401	0.460	0.523	0.542	0.544	0.547

Source: United Nations Development Programme (UNDP) (2020)

per capita, gender equality, and women empowerment compared to other global regions (UNDP, 2020). In a study that covered 28 African countries for the period 1960–1987 and using panel analysis, Savvides (1995) observed positive economic performance correlated with trade openness, investment, initial income, schooling, and growth of government. Ojo and Oshikoya (1995) examined the economic performance of 17 African countries using ordinary least squares and generalised least squares techniques and found that investment, external debt, population growth, human capital, and proxies for the macro-economic environment substantially determined the long-run development in Africa. Gyimah-Brempong and Traynor (1999) examined the link between political instability, investment, and economic growth in the SSA region. Through a dynamic panel approach, these scholars confirmed the inverse relationship between political instability and economic growth. Bertocchi and Canova (2002) state that factors such as the degree of economic penetration by former colonial powers explain the different growth performances in African countries.

The central thread running through the reviewed literature is that the development challenges that African countries are presently experiencing are closely linked to their colonial encounter with Europe and the failure of Africa to operate independently after gaining independence. In early post-independence Africa (1960–1980), most countries performed well (Ndhlovu & Mhlanga, 2023). The HDI improved from 0.081 in 1950 to 0.173 in 1980. Considering that HDI grows very slowly, this was a significant improvement. Although the growth was not consistent, both GDP and GDP per capita (GDP/C), life expectancy, and education improved between 1960 and 1980 (see Table 4.2).

The development trajectory of African countries was disrupted by the advent of neoliberalism in the late 1970s and the structural adjustment programmes of the 1980s (Ndhlovu, 2025). Most countries fell for these neoliberal and capitalist development policies, hoping to steer their development agendas. By their nature, neoliberal policies enforce the privatisation and individualisation of public institutions. It is believed that institutions are more effective and efficient in providing social services when they operate as private entities. However, in Africa, the adoption of these policies has increased poverty, job loss, and the weakening of income distribution (Ndhlovu, 2025). From the mid-1980s, many African countries saw the breakdown of public institutions due to underfunding. With the majority of Africans depending on the state and public institutions for healthcare, education,

Table 4.2 Sub-Saharan Africa development indicators

Year	1950	1960	1970	1980
HDI	0.081	0.108	0.139	0.173
GDP	–	255,6131	402,2042	581,0636
GDP/capita	–	1124	1384	1516
Life expectancy	0.076	0.098	0.123	0.146
Education	0.030	0.049	0.072	0.117

Source: Prados de la Escosura (2015)

and other welfare needs, the breakdown of public institutions revived the misery that commenced with the advent of colonialism. To rescue themselves from the clutches of their former colonisers, African intellectuals have recommended delinking, disengaging, and disconnecting from the Euro-American centric development framework and pursuing home-grown versions that are cognisant of African social and political realities (Amin, 1990; Gumede, 2019). Social policy emerged as part of the collective effort to pursue a development that will benefit all citizens of the African continent.

4.5 Materials and Methods

This chapter is based on a review of existing texts on social policy in Africa, focusing on social security and social assistance as the most common social protection interventions in Africa (Ndhlovu, 2024c). Social policy is one of the areas that have attracted heated academic and policy attention. Thus, a review approach enabled the authors to succinctly summarise and critically analyse existing debates and research on the topic (Snyder, 2019). The review method allowed the authors to review texts in both grey and academic formats to provide a comprehensive overview of the present understanding of the issue, identify knowledge gaps within the social policy literature, and highlight conceptual limitations. The authors reviewed government documents, reports by policy and research institutions, peer-reviewed texts, and non-peer-reviewed documents. The aim was to provide a comprehensive summary of existing research on social policy and development to generate a broad understanding of these subjects. Reviewed texts were accessed on the websites of institutions, the Internet, and institutional repositories.

Content analysis—the study of what is contained in the text—was used to examine current views on the role of social policy in African countries. Latent content analysis—an analysis of the meaning of texts—was then used to establish the major issues raised on the reviewed texts. The use of this type of review necessitated the need to establish the themes that emerged from the review; thus, thematic data analysis was adopted.

The advantage of the review method was that the chapter managed to synthesise and integrate findings from several studies, allowing the authors to easily identify patterns, gaps, and contradictions in the texts. This helped the authors to structure the argument in support of social protection as a potential intervention for the attainment of inclusive development, particularly in an agrarian society such as SSA.

The disadvantages of this method, however, relate to originality concerns (Albadarin et al., 2024). The chapter relied on existing research, which may have limited its novelty and originality compared to original research articles. The other disadvantage was related to the scope of the study. Review articles often focus on a particular subject (Palmatier et al., 2018). This chapter explored development and social policy, focusing on social security and social assistance, and therefore might have failed to provide a comprehensive understanding of other areas of social protection. Lastly, some of the texts reviewed were outdated. The lack of recent information forced the authors to rely on these outdated texts, thus compromising the quality of the conclusions reached.

4.6 Findings and Discussion: Social Policy and Prospects for Inclusive Development

Five themes, based on the components of the social policy approach as discussed by Mkandawire (2011a) and Adesina (2007), were identified in the review. These themes were production, distribution, social protection, social cohesion, and nation-building. These themes are discussed in the following sub-sections.

4.6.1 *Production as an Engine for Development*

The review problematised the tendency to limit social protection as a mere reactionary strategy in African countries, a tendency used by the colonial administration and sustained by the new government at independence. Humanitarian relief, particularly food aid and sometimes cash transfers, is vital in the context of providing temporary relief. It provides temporary support to individuals and households whose livelihoods have been disrupted and destroyed, for instance, following natural hazards such as drought, storms, and floods that destroy harvests or pandemics such as COVID-19 that result in loss of employment (Beierl & Dodlova, 2022; Government of Lesotho (GoL), 2020; Ndhlovu, 2024c). Unlike social security, which is rules-based (Government of Angola (GoA), 2018; Sakala et al., 2022), humanitarian relief is completely non-contributory. Emergency food aid is widely criticised because while it eases food insecurity for a particular period, it does not address its underlying causes (Devereux, 2013; White & Devereux, 2011; Seepamore, 2018). From the onset, humanitarian food aid is not meant to be a solution but a reaction. Considering that most of the food insecurity-inducing conditions are predictable or are already known, interventions that shield production are a better strategy. The social policy proposed by African scholars such as Mkandawire (2011a) and Adesina (2007) emphasises the need to support the 'production' capacity of people as part of completing the decolonisation process and help them '… to attain freedom from political, economic, ideological, epistemological, and social domination…' (Ndlovu-Gatsheni, 2012, p. 2).

Development is attainable when all citizens have the opportunity and means to participate in the processes that can generate it. In agrarian Africa, particularly in SSA, most of the countries suffered colonial rule by European countries such as Britain, Germany, Portugal, and Belgium, for instance. At independence, most of them were faced with deep-rooted poverty among their vulnerable populace. Most of them, therefore, considered their development agendas in the agricultural sector. This was because these countries lacked the resources to commence their development agendas in other economic sectors. In addition, most had adequate farmland to provide for their people to practise farming. Furthermore, the African population possessed indigenous knowledge in farming than in other sectors (Ndhlovu, 2025). Therefore, social policies that support 'production' in the agricultural sector are more suitable in African agrarian communities. Instead of rolling out food aid in moments of disaster, It is crucial to strengthen food production systems and practices through promoting and facilitating easier farm equipment distribution and input access through subsidised arrangements, improving education and training programmes to improve awareness of new farming challenges and rapidly changing weather patterns, securing markets for farmers, providing subsidised farm produce transportation, and investment in agro-processing plants closer to farmers to boost production and reduce food waste. This form of support differs from what could be conceptualised for communities in developed countries where the majority of the populace is located in towns and cities where they do not necessarily engage in farming but depend on employment for livelihoods and income. There is, therefore, the need to rethink social policies in Africa so that they do not merely mirror policies being utilised by developed countries but rather address local needs.

The review acknowledges that food aid, in all its types, including food-for-work arrangements, supplementary feeding, and school feeding schemes rolled out in many SSA countries to address food insecurity challenges, wields the potential to reduce food insecurity through providing additional nutrients to people who lack adequate access to nutritious food (Beyene, 2023; Devereux, 2013; Van Ittersum et al., 2016). However, these studies also found that such interventions have little or no apparent impact on chronic food insecurity. Scholars such as Devereux (2013) and Barrett and Maxwell (2005) argue that many years of food aid in Africa have instead sustained the demand for food aid because they somehow weakened farmers' incentives to grow food, inhibited market development, distorted consumer preferences, and allowed donors to institutionalise surplus disposal. In addition, this has caused many African governments to be aid-dependent, thus neglecting investment in agriculture and rural livelihoods (Ndhlovu, 2025). Devereux (2013), however, recognises the broader developmental objectives of food-for-work and school feeding schemes but argues that they are not a pathway to inclusive development.

4.6.2 Resource Distribution for Inclusive Development

The attainment of inclusive development also requires that social policies ensure equitable and broad-based access to resources to allow all citizens to participate in development activities. Distribution is one of the components of social policy identified by African scholars (Adesina, 2007; Mkandawire, 2011b). In many African countries, natural resource distribution programmes have been implemented or promised. In Southern Africa, Zimbabwe implemented a broad-based land reform programme in which farmland was acquired from the whites and distributed to blacks. At independence in 1980, whites, who constituted only 5% of the population, owned 51% of all the farmland, while blacks, who constituted 72% of the population, scrambled for 42% of the land, most of which was in ecologically dry areas. The remaining 7% was classified as state land. The land reform of the 2000s distributed the land to about 1.3 million peasant households, 30,000 small-to-medium capitalist farmers, and 250 large-to-corporate capitalist farmers (Moyo, 2005). In South Africa and Namibia, land reform programmes are also underway, highlighting the acknowledgement of the importance of broad-based resource distribution as a pathway towards inclusive development (Ndhlovu & Majova, 2023). Land reforms are not common in developed economies where the agrarian question has long been addressed (Bernstein, 2010), but it is still a burning issue in the Global South, including Africa (Moyo et al., 2012). The attainment of inclusive development in Africa, therefore, could be based on a social policy approach that promotes equitable and broad-based natural resource distribution. This could also enable citizens to increase their income-generation capacity and contribute to many countries and national economies, thus growing economic growth.

4.6.3 Social Protection and Inclusive Development

Inclusive development is more attainable when all citizens are protected to pursue their capabilities at full potential. Social protection is one of the key features of transformative social policy. Adesina (2015, p. 4) avers that the '… transformative role of social policy stretches from the economy to social relations and social institutions… [and] involves a wide range of instruments to raise human well-being, transform social institutions, social relations and the economy…'. This approach goes beyond the mere distribution of food and cash transfers because these interventions do not hold development potential but only help people to survive. Food aid responds directly to real food security needs but offers no sustainable solution to food insecurity (Martins, 2011). Therefore, it is not adequate as a social protection intervention. Devereux (2013) also argues that whether people will use cash transfers to buy food in a community that normally depends on agrarian activities depends on the capacity and functioning of local food markets. If food is available at the local or national level, cash transfers can improve access. Even when they have no access to aid, people can find means to access food when it is available.

Social protection as a development approach, therefore, transcends the provision of social assistance programmes to include the provision of infrastructure that can enable people to meet other social security needs such as health, education, and resource distribution to enable them to operate at their full potential. Infrastructure such as transport networks, health and educational facilities, markets, and water supplies is vital in transforming the lives of people in farming communities. When this infrastructure is available, people can likely meet their food needs without relying on social assistance. Infrastructure such as markets can also enable people to sustain their food reserves (Uchendu, 2018). For instance, Devereux (2013) posits that where drought has collapsed food production activities, cash transfers can be effective in ensuring food security but only where markets are available so that people can buy food. This will also allow people to bounce back from the shock. Therefore, social protection should not only focus on ensuring that people do not sleep with empty stomachs but also that they are provided with the necessary infrastructure that can enable them to be food-sufficient. This not only increases self-esteem and welfare but also allows them to put their labour and passion to productive use and contribute to local and national development.

4.6.4 Social Cohesion and Cooperation

Development should not be merely reduced to GDP growth. GDP is an estimate of market throughput, a combination of the value of all final goods and services produced and traded for money within a specific period (Stiglitz et al., 2018). GDP statistics measure current economic activity but disregard inequalities in income and wealth, destruction of the natural environment, longevity, or the quality of social relations (Ravetti & Cambini, 2021; Simonova, 2019). It is not a comprehensive measure of well-being or even economic welfare (Giannetti et al., 2015). GDP is a flawed measurement of welfare (Germain, 2023). Therefore, when analysing development, it is vital to go beyond material things that accompany production and protection features and also to consider the possibility of social cohesion and cooperation, which are more important in achieving development and inclusive development.

Social cohesion is a component of the transformative social policy framework (Adesina, 2021). Even though it is multi-faceted, it can be described in four main classes, namely, social, task relations, perceived utility, and emotion (Ndhlovu, 2022). It is defined by Burns (2018, p. 10) as '… the extent to which people are cooperative, within and across group boundaries, without coercion or purely self-interested motivation'. It involves 'understanding the social infrastructure, institutions, customs and material and non-material relations that either constrain or enable the individual in whatever pursuit they are engaged' (Murisa, 2007, p. 2). It denotes the capacity for people to build a shared identity and a sense of belonging. In agrarian communities, social cohesion is vital as people draw on it to engage in collaborative farming activities such as land clearing, planting, weeding, and harvesting. Social cohesion is possible in rural communities where households know each other, share some interests and identities, and may even be related. Unlike in developed countries, a bigger percentage of the African population lives in rural areas where farming activities are being practised (Ndhlovu, 2025). Therefore, the pursuit of social policy in Africa needs to take into consideration the importance of social cohesion as integral to development interventions. Adesina (2021) links social cohesion to the concept of nation-building.

4.6.5 Social Policy and Nation-Building

The other aspect of social policy introduced by Adesina (2015) under the transformative social policy is the aspect of 'nation-building'. This aspect refers to the use of government policies meant to promote social inclusion, equality, and collective ideals within a diverse society, contributing to the development of an integrated national identity and a sense of belonging among citizens, principally in circumstances where historical divisions or inequalities exist (Chikwendu et al., 2024; Kpessa-Whyte, 2021). Governments can achieve this by using social programmes to actively build a nation, bridge societal gaps, and foster unity across different groups (Béland et al., 2018).

The reconfiguration of the territorial boundaries of multi-ethnic African nations during colonialism forced these ethnic nations to also reconfigure their identities and communities in response to colonial modernity (Khapoya, 2013). Colonial administrations were poised to thwart any prospects for nation-building and, therefore, disrupted national boundaries to ensure that Africans would not live as communities. Living as communities would facilitate the 'nation-building' process, which would represent a challenge to the colonial project. After gaining independence from colonial rule, the former colonisers restructured the economic system to brood dependence in the newly independent states. The social policies recommended and funded by these former colonisers were meant to sustain the influence of European countries in African affairs.

At the height of political transformation in the 1960s and 1970s, the first generation of African leaders commenced to restructure the colonial institutions of modernity to construct a new Africa (Ndhlovu & Mhlanga, 2023). Drawing from the nationalist ideology, African leaders pursued wide-ranging, all-inclusive nation-building policies and implemented several national projects to lift their people out of poverty and foster inclusive development (Brobbey, 2010). Various national projects implemented signified African leaders' positive vision of nation-building. Contrary to the dominant European ideologies of Marxism and socialism, African leaders pursued socialist thought with an emphasis on African values (Ndhlovu, 2025). In Tanzania, Julius Nyerere pursued familyhood or 'Ujama' (Nyerere, 1969), while, in Ghana, Kwame Nkrumah reconceptualised Marxism to construct the ideology of conscience (Nkrumah, 1964). In Zambia, Kenneth Kaunda's 'humanism' is an ideology that involves an active concern for human welfare as the central object of development policy. Damachi (1976, p. 7) avers that: 'Humanism is supposed to be derived from the traditional African culture…, the African society was progressive and human, and the present generation of Zambians must bear this in mind in all its political, economic, social, and cultural activities'. In Botswana,

Seretse Khama pursued *Kagisano* to teach people the importance of social virtues: unity, peace, harmony, and a sense of community (Ikechukwu, 2016). Khama believed that socio-economic and political development could be achieved if people only embraced these values. Therefore, guided by nationalism, the first generation of African leaders sought to achieve nation-building by integrating African ethnic groups into the new African states. Social policies for inclusive development could, therefore, build on this foundation.

The development approaches pursued by African leaders differed from those used by Europeans, which pursued Marxism and socialism. The experiences of Africa and Europe are also different, and thus African countries need to pursue social policies that speak to their contexts of circumstances. This will enable countries to mobilise their people to work together for the collective good and build their nations, enhance human welfare, and increase prospects for the achievement of inclusive development.

4.7 Conclusion

This chapter explored the potential of social protection to improve human welfare in Africa. It also examined the possibility of social protection as the basis for inclusive development. The chapter shows that social policies, social protection in particular (social security and social assistance), wield much potential to improve lives and livelihoods in African countries, thus serving as the basis for inclusive development. It posits that Africa is an agrarian community, with many of its people using farming either as the main or additional source of livelihood and income. However, resource constraints and unfavourable weather patterns often frustrate agricultural activities in a continent that is characterised by poor people constantly facing food insecurity. Social policies that deliberately target the productive activities of these people could help boost production and provide the base for inclusive development. In Europe, social policies were used as responses to the crises of capitalist industrialisation. However, in Africa, social policies, including social protection, were adopted as part of the ambitious social development programmes despite these countries being underdeveloped.

This chapter demonstrated how, in agrarian Africa, social policies can be used to improve agricultural production; implement equitable resource distribution and redistribution; improve social protection, social cohesion, and cooperation; and facilitate nation-building. This approach, if well implemented, could pave the way for the realisation of inclusive development.

References

Adesina, J. O. (2007). *Social policy in sub-Saharan African context in search of inclusive development*. Palgrave Macmillan.

Adesina, J. O. (2015). Return to a wider vision of social policy: Rereading theory and history. *South African Review of Sociology, 46*(3), 99–119.

Adesina, J. M. (2021). Why development and transformative social policy matter: Lessons of COVID-19 in Africa. *CODESRIA Bulletin Online, No., 10*, 1–9.

Albadarin, Y., Saqr, M., Pope, N., et al. (2024). A systematic literature review of empirical research on ChatGPT in education. *Discover Education, 3*, 60. https://doi.org/10.1007/s44217-024-00138-2

Amin, S. (1990). *Delinking: Towards a polycentric world*. Zed Books.

Anand, S., & Sen, A. (2000). The income component of the Human Development Index. *Journal of Human Development, 1*(1), 83–106.

Awortwi, N. (2018). Social protection is a grass roots reality: Policy reflections on community based social protection actors and services in Africa. *Development Policy Review, 36*(S2), 897–913.

Ayittey, G. (2005). *Africa unchained: The blueprint for Africa's future*. Palgrave Macmillan.

Barrett, C., & Maxwell, D. (2005). *Food aid after fifty years: Recasting its role*. Routledge.

Beierl, S., & Dodlova, M. (2022). Public works programmes and cooperation for the common good: Evidence from Malawi. *European Journal of Development Research, 34*, 1264–1284.

Béland, D., Foli, R., & Kpessa-Whyte, M. (2018). Social policy as nation-building: Identity formation, policy feedback, and social citizenship in Ghana. *Canadian Journal of African Studies, 52*(1), 19–36. https://doi.org/10.1080/00083968.2018.1425147

Bernstein, H. (2010). *Class dynamics of agrarian change*. Fernwood Publishing and Kumarian Press.

Bertocchi, G., & Canova, F. (2002). Did colonization matter for growth? An empirical exploration into the historical causes of Africa's underdevelopment. *European Economic Review, 46*(10), 1851–1871.

Beyene, S. D. (2023). The impact of food insecurity on health outcomes: Empirical evidence from sub-Saharan African countries. *BMC Public Health, 23*, 338. https://doi.org/10.1186/s12889-023-15244-3

Brobbey, P. (2010). *Theories of social development*. University of Ghana: Department of Sociology.

Burns, J., Hull, G., Lefko-Everett, K., et al. (2018). *Defining social cohesion*. SALDRU working paper number 216. SALDRU, UCT.

Cheru, F. (2009). Development in Africa: The imperial project versus the national project and the need for policy space. *Review of African Political Economy, 36*(120), 275–278.

Chikwendu, C. P., Alazor, C., & Agbanyim, S. C. (2024). African women in post-conflict nation-building in Africa since independence. *ESTAGA, 1*(3), 40–51.

Conradie, I. (2018). Social policy in South Africa: The challenges of poverty, inequality and exclusion. *Social Work and Society, 16*(2), 1–9.

Damachi, U. G. (1976). *Leadership ideology in Africa*. Praeger Publishers.

Devereux, S. (2013). Trajectories of social protection in Africa. *Development Southern Africa, 30*(1), 13–23. https://doi.org/10.1080/0376835X.2013.755871

Dörffel, C., & Schuhmann, S. (2022). What is inclusive development? Introducing the multidimensional inclusiveness index. *Social Indicators Research*. https://doi.org/10.1007/s11205-021-02860-y

Drahokoupil, J. (2004). Re-inventing Karl Polanyi: On the contradictory interpretations of social protectionism. *Czech Sociological Review, 40*(6), 835–849.

Esping-Andersen, G. (1990). *The three worlds of welfare capitalism*. Polity Press.

European Commission. (2023). *Euro area unemployment at 6.5%*. Retrieved March 3, 2025, from https://ec.europa.eu/eurostat/documents/2995521/17990094/3-30112023-BP-EN.pdf/943a8c59-bbb4-8c44-b4b7-54ce08b45a49

FAO, AUC, ECA, & WFP. (2023). *Africa—Regional overview of food security and nutrition 2023: Statistics and trends*. FAO. https://doi.org/10.4060/cc8743en

Germain, J. M. (2023). Beyond GDP: A welfare-based estimate of growth for 14 European countries and the USA over past decades. *Economie et Statistique/Economics and Statistics, 539*, 3–25. https://doi.org/10.24187/ecostat.2023.539.2095

Giannetti, B. F., Agostinho, F., Almeida, C. M. V. B., & Huisingh, D. (2015). A review of limitations of GDP and alternative indices to monitor human wellbeing and to manage eco system functionality. *Journal of Cleaner Production, 87*, 11–25. https://doi.org/10.1016/j.jclepro.2014.10.051

GoA. (2018). *Angola social protection public expenditure review (PER): Main report*. Republic of Angola.

GoL. (2020). *Food security projections (April–Sept 2020)*. Lesotho Vulnerability Assessment Committee.

Gumede, V. (2015). Exploring the role of thought leadership, thought liberation and critical consciousness for Africa's development. *Africa Development, 40*(4).

Gumede, V. (2019). Rethinking and reclaiming development in Africa. In Ndlovu-Gatsheni & B. Mpofu (Eds.), *Rethinking and unthinking development: Perspectives on inequality and poverty in South Africa and Zimbabwe*. Berghahn Books.

Gyimah-Brempong, K., & Traynor, T. L. (1999). Political instability, investment and economic growth in sub-Saharan Africa. *Journal of African Economies, 8*(1), 52–86.

Ikechukwu, O. B. (2016). African theories of development and the reality of underdevelopment. *International Journal of Development and Economic Sustainability, 4*(4), 12–19.

ILO. (2023). *World employment and social outlook trends 2023*. International Labour Organisation.

Jansen, A., Wang, R., Behrens, P., & Hoekstra, R. (2024). Beyond GDP: A review and conceptual framework for measuring sustainable and inclusive wellbeing. *Lancet Planet Health, 8*, e695–e705.

Khapoya, V. B. (2013). *The African experience* (4th ed.). Taylor and Francis.

Kpessa-Whyte, M. (2021). Transformative public policy design? The lifeworld of early postcolonial social policy in Ghana. *Scientific African, 12*, e00828. https://doi.org/10.1016/j.sciaf.2021.e00828

Liu, K., Wang, R., Behrens, P., Schrijver, I., Jansen, A., Rum, I. A., & Hoekstra, R. (2024). A comprehensive beyond-GDP database to accelerate wellbeing, inclusion, and sustainability research. *Scientific Data, 11*, 1166. https://doi.org/10.1038/s41597-024-04006-4

Martins, P. M. (2011). Aid absorption and spending in Africa: A panel cointegration approach. *Journal of Development Studies, 47*(12), 1925–1953.

Mensah, J. (2019). Sustainable development: Meaning, history, principles, pillars, and implications for human action: Literature review. *Cogent Social Sciences, 5*(1), 1–21. https://doi.org/10.1080/23311886.2019.1653531

Mhlanga, D., & Ndhlovu, E. (2021). Socio-economic and political challenges in Zimbabwe and the development implications for Southern Africa. *Journal of African Foreign Affairs, 8*(2), 77–100.

Mkandawire, T. (2001). *Social policy and development programme*. Paper number 7, June. United Nations Research Institute for Social Development.

Mkandawire, T. (2011a). Running while others walk: Knowledge and the challenge of Africa's development. *Africa Development, 36*(2), 1–36.

Mkandawire, T. (2011b). Thinking about developmental states in Africa. *Cambridge Journal of Economics, 25*(3), 289–313.

Moyo, S. (2005). Land and natural resource redistribution in Zimbabwe: Access equity and conflict. *African and Asian Studies, 4*(1–2), 187–223.

Moyo, S., Jha, P., & Yeros, P. (2012). Editorial. The agrarian question: Past, present and future. *Agrarian South: Journal of Political Economy, 1*(1), 1–10.

Murisa, T. (2007). *Social organisation and agency in the newly resettled areas of Zimbabwe: The case of Zvimba District*. Monograph, AIAS Monograph Series, AIAS, Harare, Zimbabwe.

Nasri, K., Amara, M., & Helmi, I. (2022). *The landscape of social protection in Tunisia*. Economic Research Forum. Working Paper No. 1592.
Ndhlovu, E. (2022). Socio-economic characterisation of resettled smallholder farmers in rural Zimbabwe. *Journal of Asian and African Studies, 57*(8), 1495–1510.
Ndhlovu, E. (2024a). Building farming resilience to climate change through social assistance in South-Eastern Zimbabwe. *International Journal of Research in Business and Social Science, 13*(10), 105–117.
Ndhlovu, E. (2024b). Rethinking autonomous development and social change in Africa. *International Journal of Social Science Research and Review, 7*(8), 257–271.
Ndhlovu, E. (2024c). Social protection instruments and strategies in Southern Africa. *International Journal of Innovation in Management Economics and Social Sciences, 4*(2), 44–59.
Ndhlovu, E. (2025). *Agriculture, autonomous development, and prospects for industrialization in Africa*. Springer.
Ndhlovu, E., & Majova, B. (2023). Women empowerment in the South African agri-business: Opportunities and constraints in the Gauteng Province. In D. Mhlanga & E. Ndhlovu (Eds.), *Economic inclusion in post-independence Africa: An inclusive approach to economic inclusion* (pp. 273–293). Springer.
Ndhlovu, E., & Mhlanga, D. (2023). Decolonization of development in early post-independence Africa. In D. Mhlanga & E. Ndhlovu (Eds.), *Post-independence development in Africa decolonisation and transformation prospects* (pp. 63–81). Springer.
Ndlovu-Gatsheni, S. (2012). *Coloniality of power in development studies and the impact of global imperial designs on Africa*. Inaugural Professorial Lecture, University of South Africa, 16, October.
Nielson, K. (1996). Conception of justice. In M. Hawkesworth (Ed.), *Encyclopedia of government and politics* (Vol. 1). Routledge.
Nkrumah, K. (1964). *Consciencism: Philosophy and ideology for decolonization and development with particular reference to the African revolution*. Heinemann Educational Books Ltd.
Nussbaum, M. C. (2004). Beyond the social contract: Capabilities and global justice. An Olaf Palme lecture. Delivered in Oxford 19 June 2003. *Oxford Development Studies, 32*(1), 3–18.
Nyerere, J. (1969). *Freedom and unity*. Oxford University Press.
Ojo, O., & Oshikoya, T. (1995). Determinants of long-term growth: Some African results. *Journal of African Economies, 4*(2), 163–191.
Palmatier, R. W., Houston, M. B., & Hulland, J. (2018). Review articles: Purpose, process, and structure. *Journal of the Academy of Marketing Science, 46*, 1–5. https://doi.org/10.1007/s11747-017-0563-4
Polanyi, K. (1944/2001). *The Great Transformation: The Political and Economic Origins of Our Time*. Boston, MA: Beacon Press.
Prados de la Escosura, L. (2015). World human development: 1870–2007. *Review of Income and Wealth, Series, 61*(2), 220–247.
Ravetti, C., & Cambini, C. (2021). *Energy use beyond GDP: A dynamic panel analysis with different development indicators*. Copenhagen Business School [wp]. Working Paper/Department of Economics. Copenhagen Business School No. 10-2021CSEI Working Paper No. 03-2021. https://research-api.cbs.dk/ws/portalfiles/portal/66383694/chiara_ravetti_et_al_energy_use_beyond_gdp_workingpaper_10_2021.pdf
Reyes, G. E. (2001). Four main theories of development: Modernization, dependency, word-system, and globalization. *Nómadas. Revista Crítica de Ciencias Sociales y Jurídicas, 4*(2), 109–124.
Sakala, B., Noyoo, N., & Mabundza-Dlamini, L. (2022). An exploration of social assistance programmes in Eswatini, South Africa and Zambia. *Southern African Journal of Social Work and Social Development, 34*(1), 1–17.
Savvides, A. (1995). Economic growth in Africa. *World Development, 23*(3), 449–458.
Seepamore, B. (2018). Indigenous social security systems: A South African perspective. In N. Noyoo & E. Boon (Eds.), *Indigenous social security systems in Southern and West Africa* (pp. 69–88). African Sun Media.

Simonova, M. D. (2019). Beyond GDP. Measuring what counts for economic and social performance. *Вестник МГИМО-Университета, 12*(5), 172–180. https://doi.org/10.24833/2071-8160-2019-5-68-172-180

Snyder, H. (2019). Literature review as a research methodology: An overview and guidelines. *Journal of Business Research, 104*, 333–339. https://doi.org/10.1016/j.jbusres.2019.07.039

Stiglitz, J., Fitoussi, J., & Duran, M. (2018). *Beyond GDP: Measuring what counts for economic and social performance*. OECD Publishing. https://doi.org/10.1787/9789264307292en

Surender, R. (2023). Social policy in Africa: Risks, protection, and dynamics. *Journal of International and Comparative Social Policy, 39*(3), 239–246.

Talmage, C. A. (2020). Community development, quality of life, and community well-being: Three fields ripe with opportunities for future research and practice. *Community Development Practice, 24*(1), 4. https://egrove.olemiss.edu/cdpractice/vol24/iss1/4

Taylor, V. (2021). *A social protection floor—A South African approach for social justice, Social Security Review 2021: Evolution of social security in South Africa—An agenda for action*. Department of Social Development, Pretoria. Retrieved October 17, 2024, from https://www.researchgate.net/publication/355844214_Social_Security_Review_2021_Evoluti7on_of_Social_Security_in_South_Africa_An_Agenda_for_Action

Todaro, M. P., & Smith, S. C. (2006). *Economic development* (8th ed.). Addison-Wesley.

Uchendu, F. N. (2018). Hunger influenced life expectancy in war-torn sub-Saharan African countries. *Journal of Health Population and Nutrition, 37*(1), 1–4.

UNDP. (2020). *Human Development Report 2020*. United Nations Development Programme.

Van Ittersum, M. K., Van Bussel, L. G., Wolf, J., Grassini, P., Van Wart, J., Guilpart, N., Claessens, L., De Groot, H., Wiebe, K., Mason-D'Croz, D., & Yang, H. (2016). Can sub-Saharan Africa feed itself? *Proceedings of the Natural Academy of Sciences, 113*(52), 14964–14969.

White, P & Devereux, P. (2011). *Social protection in Botswana—A model for Africa? Frontiers of Social Protection Brief*, Number 9, March. Regional Hunger and Vulnerability Project. www.wahenga.net

Part II
Sectoral Innovations

Chapter 5
From Denial to Engagement: Tanzania's Pandemic Response and Lessons for Inclusive Health Policy Development

Ambrose T. Kessy

5.1 Introduction

5.1.1 Contextualizing the Pandemic in Africa

The COVID-19 pandemic, caused by the novel coronavirus SARS-CoV-2, emerged in late 2019 and has rapidly escalated into a global health crisis. By October 2023, the World Health Organization (WHO) reported over 770 million confirmed cases and more than 6.9 million deaths worldwide (World Health Organization, 2023c). Africa, though initially reporting fewer cases and deaths than other continents, faced unique challenges that significantly impacted its pandemic response and outcomes. Africa's relatively youthful population was initially thought to mitigate the severe impact of the virus (Nkengasong & Mankoula, 2020). However, the continent's limited healthcare infrastructure, high prevalence of comorbidities such as HIV/AIDS and tuberculosis, and socioeconomic factors such as poverty and densely populated informal settlements heighten vulnerability (Mendelson & Madhi, 2020). Testing capacity is constrained, leading to underreporting and difficulties in tracking the spread of the virus (Adams et al., 2021).

The pandemic exacerbated existing inequalities, with marginalized communities bearing the brunt of both the health crisis and socioeconomic fallout from containment measures (United Nations Economic Commission for Africa, 2021). The informal economy, which employs a significant portion of the African workforce, was particularly affected by lockdowns and movement restrictions, leading to increased poverty and food insecurity (International Labour Organization, 2021). Health policy responses across African countries vary, reflecting differences in governance, resource availability, and public health capacities. Some nations

A. T. Kessy (✉)
The Law School of Tanzania (LST), Dar es Salaam, Tanzania

E. Ndhlovu, C. Chipenda (eds.), *Redefining the Social Policy Discourse*,
https://doi.org/10.1007/978-3-032-24067-5_5

implemented early lockdowns, mass testing, and contact tracing, while others struggled to enforce preventive measures because of economic constraints and public skepticism (Dyer, 2020). The importance of effective, inclusive health policies has become evident as key to mitigating the impact of the pandemic, protecting vulnerable populations, and maintaining social and economic stability.

The continent also faced challenges in terms of vaccine access and distribution. Vaccine inequity has been pronounced, with African countries receiving a fraction of the doses needed for widespread immunization (Moeti & Fitzgerald, 2021). Initiatives such as COVAX aimed to address this disparity, but logistical hurdles and vaccine hesitancy remain significant obstacles (Burki, 2021). As new variants of concern emerged, such as Delta and Omicron variants, Africa's vulnerability was highlighted. These variants brought about surges in cases and placed additional strain on healthcare systems (WHO, 2023). This situation underscored the interconnectedness of global health and the necessity for comprehensive and inclusive policies that address both immediate health needs and long-term systemic issues. Based on the provided background information, a key research question to be answered by this study is: How did the transition in political leadership from President Magufuli to President Hassan impact Tanzania's COVID-19 response policies and their implications for inclusive health policy development? This research question addresses the core focus of the study, examining the shift in Tanzania's pandemic approach and its effects on health policy inclusivity. It allows for exploration of contrasting leadership styles, policy changes, and outcomes for vulnerable populations.

5.1.2 Tanzania's Unique Position

Tanzania presents a distinctive case in Africa's pandemic narrative because of its initial response and subsequent policy shifts. Under the leadership of the late President John Pombe Magufuli, Tanzania adopted a controversial stance toward COVID-19. President Magufuli publicly downplayed the severity of the virus, questioned the reliability of testing kits, and promoted prayer and herbal remedies over scientifically proven interventions (The Guardian, 2020). In June 2020, he declared Tanzania "COVID-19-free," attributing this to divine intervention, and ceased publishing official COVID-19 data (BBC News, 2021). This denialist approach led to minimal intervention measures. The government discouraged mask-wearing and social distancing, and life continued relatively unchanged, with large public gatherings and events proceeding as normal (Fraser & Msemo, 2021). Tanzania's position contrasted sharply with neighboring countries and global health recommendations, raising concerns about regional implications and the potential for uncontrolled virus spread (Kavanagh et al., 2021).

Following President Magufuli's sudden death in March 2021, Vice President Samia Suluhu Hassan assumed the presidency. President Hassan initiated a significant shift in Tanzania's pandemic response. Acknowledging the presence of COVID-19, she formed a task force to evaluate the situation, resumed publishing data, and took steps to procure vaccines (Reuters, 2021). Tanzania joined the COVAX facility, aiming to vaccinate at least 60% of its population (Africa Centres for Disease Control Prevention, 2021). President Hassan's approach marked a re--engagement with international health organizations and a move toward transparency and evidence-based policies. Public health campaigns were launched to educate citizens on preventive measures, and efforts were made to improve testing and healthcare infrastructure (The East African, 2021). This policy shift had significant implications for public health and highlighted the influence of political leadership on health policy decisions.

This chapter delves into Tanzania's tumultuous journey through the COVID-19 pandemic, spotlighting its dramatic shift from denial to proactive engagement and the profound implications of this transformation on inclusive health policy development. By dissecting the strategies implemented under different leaderships, we assess their effectiveness and their tangible impact on various population groups—most notably the vulnerable and marginalized. Unveiling both the triumphs and missteps of Tanzania's approach, we extract crucial lessons poised to inform and shape future health policy development across Africa. Our analysis contributes a critical perspective to the broader discourse on health governance on the continent. It underscores the indispensable roles of strong leadership, transparency, community engagement, and strict adherence to international health guidelines. We ardently advocate for policies that are not only inclusive and equitable but also resilient ensuring that every segment of the population has unimpeded access to essential health services and vital information during crises. This chapter serves as a clarion call for policymakers to craft health policies that can robustly confront present challenges and are adaptable to future health emergencies.

5.1.3 The Tanzanian Health System

Tanzania's health system is a complex amalgamation of public, private for-profit, private not-for-profit, and traditional sectors, all contributing to healthcare delivery across the nation. The Ministry of Health (MoH) oversees this multifaceted system, aiming to ensure equitable access to quality healthcare services for all citizens (Ministry of Health, 2022b). The public healthcare sector operates on a hierarchical structure comprising primary, secondary, and tertiary levels. At the primary level, dispensaries and health centers offer outpatient services, maternal and child healthcare, and basic laboratory diagnostics. The secondary level includes district and regional hospitals that provide specialized services and serve as referral points for primary facilities. Tertiary care is delivered through national referral hospitals and specialized institutes equipped to handle advanced medical cases and provide training (Ministry of Health, 2022a).

The private sector plays a significant role, especially in urban areas, and consists of private for-profit facilities and faith-based or nongovernmental organizations (NGOs). These entities often fill gaps in service provision, particularly in rural and underserved regions (PharmAccess, 2021). Healthcare financing in Tanzania is sourced from government allocations, donor contributions, out-of-pocket payments, and health insurance schemes such as the National Health Insurance Fund (NHIF) and Community Health Funds (CHFs). Despite these mechanisms, only about 32% of the population is covered by health insurance, leaving many vulnerable to financial hardship due to health expenses (National Bureau of Statistics, 2022). Despite these apparent improvements, critics argue that Tanzania's response remained inadequate and delayed. Some contend that the resumed data reporting was selective and potentially underreported, questioning the true extent of the pandemic's impact. Additionally, skeptics point out that the shift in approach came too late, after significant damage had already been done to public trust and health infrastructure during the previous administration's denial phase.

Challenges persist within the health system, including a shortage of healthcare professionals, with doctor-to-population ratios falling below the World Health Organization (WHO) recommendations (World Health, 2021b; World Health Organization, 2021a, 2021b, 2021c). Infrastructure deficiencies are notable in rural areas, where facilities may lack essential equipment and supplies. Additionally, the quality of care varies due to resource constraints and gaps in training, impacting patient outcomes and trust in the healthcare system (The World Bank, 2022).

Historically, Tanzania has confronted significant public health challenges that have shaped its health policies and responses. The HIV/AIDS epidemic in the late 1990s saw prevalence rates reach 7%, prompting the government to implement the National Multisectoral Strategic Framework on HIV/AIDS and scale up antiretroviral therapy programs. As a result, the prevalence rate decreased to 4.5% by 2023, with improved access to treatment and prevention services (UNAIDS, 2023). Malaria remains endemic, accounting for substantial morbidity and mortality, particularly among children under 5. Interventions such as the distribution of insecticide-treated nets and indoor residual spraying have led to a 50% decline in malaria incidence since 2000, although challenges like insecticide resistance persist (President's Malaria Initiative, 2023).

Maternal and child health has been another focus area due to high mortality rates associated with limited access to skilled birth attendants and emergency obstetric care. Initiatives like the National Road Map Strategic Plan (One Plan III) have contributed to reducing the maternal mortality ratio from 556 per 100,000 live births in 2010 to 321 in 2022 (Ministry of Health, 2022b). Tuberculosis (TB) control efforts have also been intensified, with integrated HIV/TB services and enhanced diagnostic capabilities leading to a decrease in TB incidence, though multidrug-resistant TB remains a concern (World Health Organization, 2022).

5.1.4 Political Landscape and Leadership

The political landscape in Tanzania has significantly influenced health policy decisions, particularly under the leadership of President John Magufuli (2015–2021) and President Samia Suluhu Hassan (2021–present). President Magufuli was known for his anti-corruption stance and infrastructural development efforts, earning the nickname "The Bulldozer" (The Economist, 2019). His governance style was characterized by centralized decision-making and a focus on economic reforms, including industrialization and reduced government spending. However, he faced criticism for suppressing dissent and restricting media freedoms (Human Rights Watch, 2022). In terms of health policy, Magufuli increased investments in healthcare infrastructure but adopted a skeptical approach to the COVID-19 pandemic, questioning its severity and promoting prayer over scientific interventions (BBC News, 2021).

Upon President Magufuli's death in March 2021, Vice President Samia Suluhu Hassan assumed the presidency, marking a shift toward diplomatic engagement and openness. She improved relations with neighboring countries and international organizations, lifted bans on media outlets, and signaled a willingness to engage in political dialogue (Reuters, 2021). In health policy, President Hassan acknowledged the impact of COVID-19, formed a task force to address the pandemic, resumed reporting of COVID-19 data, and initiated vaccination campaigns (The Citizen, 2021). Her leadership emphasized maternal health and the launch of programs to improve healthcare access and address noncommunicable diseases (Ministry of Health, 2023).

Political leadership in Tanzania has thus played a crucial role in shaping health policy direction and implementation. Under President Magufuli, the denial of COVID-19's severity led to minimal preventive measures, halted testing, and discouraged mask usage, potentially contributing to underreporting of cases and public confusion (The Lancet, 2021). In contrast, President Hassan's acceptance of scientific evidence and collaboration with the WHO improved public awareness and restored international confidence, leading to increased vaccination rates (World Health Organization, 2022). Leadership styles influence public trust, resource allocation, and international relations, all of which are critical components in effective health policy implementation.

5.1.5 Global Health Policy Frameworks

Global health policy frameworks and international discourse have also significantly impacted national health policies in Tanzania. During pandemics, international health bodies like the WHO provide guidelines to coordinate global responses. The International Health Regulations (IHR) of 2005 serve as a legally binding framework for preventing and responding to public health risks (World Health, 2005). The WHO's COVID-19 Strategic Preparedness and Response Plan offers guidance on surveillance, testing, clinical management, vaccination, and risk communication (World Health Organization, 2023a). The United Nations' comprehensive response emphasizes solidarity, support for vulnerable countries, and equitable access to treatments and vaccines (United Nations, 2021). Initiatives like the COVAX facility aim to ensure global equitable access to COVID-19 vaccines, particularly for low- and middle-income countries (Gavi, 2023).

International discourse shapes national policies through guidelines, funding, technical support, and promoting global solidarity. Tanzania's initial divergence from WHO guidelines under President Magufuli led to concerns about regional health security and strained relationships with international partners (Amnesty International, 2021). President Hassan's re-engagement with international recommendations improved cooperation, facilitated vaccine access, and enhanced the effectiveness of the national response (Reuters, 2022). Adherence to global health policies enables enhanced preparedness, resource mobilization, and collective action, which are essential for addressing health crises effectively.

5.2 Literature Review

5.2.1 Pandemic Responses in African Contexts

The onset of the COVID-19 pandemic presented unprecedented challenges for African countries, testing the resilience of their health systems and governance structures. Studies have documented a spectrum of responses across the continent, reflecting variations in resources, political will, and public health capacities (Nkengasong & Mankoula, 2020). Many African nations implemented early containment measures, such as border closures, lockdowns, and curfews, aiming to curb the spread of the virus despite potential economic repercussions (Hopman et al., 2020). For instance, Rwanda swiftly enforced strict lockdowns and mass testing, resulting in relatively low infection rates and fatalities (Maruyama et al., 2021). South Africa, facing the highest number of cases on the continent, adopted aggressive testing and contact tracing, although it grappled with economic fallout and social unrest due to prolonged restrictions (Abdool Karim & Abdool Karim, 2021). Conversely, some countries faced difficulties implementing public health measures due to limited resources, conflict, or political instability (Makoni, 2021). Common challenges identified across African countries include inadequate healthcare infrastructure, shortages of medical supplies and personal protective equipment (PPE), limited testing capacity, and the prevalence of misinformation hindering public compliance (Ataguba, 2020). The pandemic also exacerbated existing health issues, such as HIV/AIDS, tuberculosis, and malaria, as resources were diverted to address COVID-19, disrupting routine healthcare services (Chanda-Kapata et al., 2020).

Vaccine access has been a significant hurdle, with African countries receiving a disproportionately small share of global vaccine supplies due to hoarding by wealthier nations and challenges within the COVAX facility (So & Woo, 2020). This inequity has delayed immunization efforts, leaving populations vulnerable to new variants and prolonging the pandemic's impact on the continent (Abdool Karim & de Oliveira, 2021).

5.2.2 Political Leadership and Health Policy

Political leadership plays a pivotal role in shaping health policy responses during crises, influencing not only the formulation and implementation of health policies but also their integration within broader social policy frameworks. Leadership styles and governance strategies determine resource allocation, public communication, and the prioritization of equity in addressing health challenges (Greer et al., 2020). Effective leadership fosters public trust, ensures coordination across sectors, and aligns health policy objectives with social policy imperatives. Conversely, weak or denialist leadership, as seen in some African contexts, can exacerbate vulnerabilities and deepen inequities during health crises.

In Tanzania, the contrasting leadership approaches of President John Magufuli and President Samia Suluhu Hassan illustrate the transformative potential of leadership in health and social policy integration. President Magufuli's skepticism toward COVID-19 science and international guidelines limited the government's ability to develop cohesive responses, compounding disparities among vulnerable populations. This approach also highlighted a broader issue across the continent: the challenge of balancing national sovereignty with international health imperatives (Amnesty International, 2021). President Hassan's administration marked a significant departure, characterized by re-engagement with international organizations, transparency in data reporting, and an emphasis on equity. Her leadership facilitated the integration of health policies with broader social welfare measures, such as targeted communication campaigns and expanded access to healthcare in underserved areas (UNICEF, 2022). This shift demonstrates how leadership can act as a catalyst for health policy transformation, ensuring alignment with social policy objectives like poverty alleviation, education, and gender equity.

Comparatively, other African nations provide further insights into the role of leadership in integrating health and social policies. For example, Ethiopia's Prime Minister Abiy Ahmed emphasized community-based health initiatives as part of a broader social policy framework addressing poverty and education disparities during the pandemic (Van der Merwe, 2020). Rwanda's government, under President Paul Kagame, linked public health measures with economic stability initiatives, demonstrating the importance of cross-sectoral policy coordination (Maruyama et al., 2021). These examples highlight the importance of leadership in fostering policy synergies that address the structural determinants of health and ensure more inclusive outcomes.

The Tanzanian experience underscores that effective leadership during health crises requires a multidimensional approach that integrates health policies with social policy imperatives. Leadership must prioritize vulnerable populations, build trust through transparency, and engage communities in co-creating solutions. Furthermore, aligning national policies with regional and international frameworks enhances resource mobilization and knowledge sharing, which are critical for addressing systemic challenges in health and social welfare systems. By viewing health policy as an integral component of broader social policy, leaders can ensure more resilient, equitable, and inclusive outcomes during crises.

5.2.3 *Inclusivity in Health Policies*

Inclusive health policies are critical in ensuring equitable access to healthcare services while addressing the broader social determinants of health, such as housing, education, and employment, which influence health outcomes (Marmot et al., 2020). For example, Ghana's incorporation of health interventions into social safety nets during the pandemic highlights the importance of integrated approaches (Jerving, 2020). Such frameworks are particularly relevant in African contexts, where socioeconomic disparities are stark and often exacerbated during health crises (Ahmed et al., 2020). Effective inclusivity in health policies requires not only equitable healthcare provision but also alignment with broader social policies to address systemic inequities holistically. These policies aim to eliminate barriers that marginalized groups face in accessing care, thereby promoting fairness and social justice within the health system. Inclusivity involves the active engagement of communities in both the development and implementation of health policies, ensuring that the needs and voices of all population segments are considered and addressed. One of the fundamental dimensions of inclusive health policies is equity in access. This principle emphasizes that healthcare services should be available, accessible, and affordable to everyone, particularly vulnerable populations who are often disproportionately affected by health disparities (Braveman et al., 2017). By focusing on equity, health systems can address social determinants of health that contribute to unequal outcomes, such as poverty, education, and living conditions.

Another essential aspect is cultural competence, which involves recognizing and respecting the diverse cultural beliefs and practices related to health within different communities. Health policies that incorporate cultural competence can improve patient-provider interactions, enhance patient satisfaction, and lead to better health outcomes by ensuring that services are delivered in a culturally sensitive manner (Betancourt et al., 2003). This approach helps in building trust between healthcare providers and patients, which is crucial for effective healthcare delivery.

Community participation is also a vital component of inclusive health policies. Involving community members in decision-making processes enhances the relevance and acceptance of health interventions. When communities are actively engaged, policies are more likely to be tailored to the specific needs and preferences of the population, leading to increased utilization of services and empowerment of individuals (Rifkin, 2014). This participatory approach fosters a sense of ownership and responsibility among community members, which can contribute to the sustainability of health initiatives. Transparency and accountability are critical for building trust and facilitating compliance with health policies. Open communication about policy objectives, processes, and outcomes allows the public to stay informed and engaged. Transparency helps in demystifying health interventions, reducing misinformation, and encouraging community support (Braveman et al., 2017). Accountability mechanisms ensure that policymakers and healthcare providers are responsible for delivering on their commitments, which enhances the effectiveness of health policies.

The importance of inclusivity is underscored by evidence linking equitable health policies to better health outcomes and increased resilience against health crises (Ottersen et al., 2014). Inclusive policies help reduce disparities by ensuring that interventions reach those most in need, thereby enhancing social cohesion and stability within societies. For instance, during the COVID-19 pandemic, countries that prioritized inclusivity were more effective in mitigating impacts among vulnerable groups. By addressing the unique challenges faced by these populations, such as limited access to information, healthcare services, and economic support, these countries were able to reduce transmission rates and improve health outcomes (Ahmed et al., 2020). In conclusion, inclusive health policies are essential for achieving universal health coverage and improving overall population health. By focusing on equity, cultural competence, community participation, and transparency, health systems can address the diverse needs of their populations, reduce health disparities, and build resilience against future health crises.

Thus, despite growing literature on COVID-19 responses in Africa, gaps remain in understanding the nuanced effects of political leadership transitions on health policy inclusivity. Few studies have conducted in-depth analyses of how changes in leadership during a health crisis influence policy direction, implementation, and outcomes for marginalized populations (Kavanagh et al., 2021). Moreover, there is a need for case-specific studies that examine the interplay between leadership styles, policy shifts, and inclusivity within particular national contexts. Such research can inform broader policy frameworks by providing insights into best practices and lessons learned from real-world experiences (Gilson, 2022). Tanzania's unique journey from denial to engagement presents an opportunity to fill this gap, offering valuable lessons for other nations facing similar challenges.

5.3 Methodology

This study employed a mixed method. Quantitative design was used to collect data from a households' survey, while a qualitative case study approach was used to thoroughly examine Tanzania's pandemic response and its implications for inclusive health policy development. Qualitative research is particularly suited for exploring complex social phenomena where contextual understanding is crucial (Creswell & Poth, 2018). The case study design enabled the identification of the subtleties of policy development and implementation that are influenced by sociocultural factors and political leadership. This method also facilitated the identification of patterns and themes that quantitative methods may have overlooked, thereby providing a more comprehensive and in-depth comprehension of the subject matter.

A survey of 3098 respondents was conducted for the exploration of participants' perspectives, experiences, and interpretations, which are essential for understanding the intricacies of health policy dynamic. A purposive sampling strategy was employed to select 20 interview participants from each region who were most knowledgeable and experienced regarding Tanzania's pandemic response. The criteria for selection included relevance, ensuring participants had direct involvement or significant insight into the health policies and their implementation; diversity, representing different genders, ages, regions (urban and rural), and professional backgrounds to capture a wide range of perspectives; and availability, participants' willingness and ability to participate within the study's timeframe. To enhance diversity and reach, snowball sampling was also utilized, where initial participants referred others who met the inclusion criteria (Naderifar et al., 2017). This approach was particularly useful for accessing hard-to-reach or marginalized groups, ensuring that the sample reflected a broad spectrum of experiences and viewpoints.

To ensure a robust and triangulated analysis, the study utilized multiple data collection methods, including document analysis, media reports, households survey, and key informant interviews (KIIs). Document analysis involved a systematic examination of relevant texts to extract meaningful information and insights (Bowen, 2009). The study reviewed a variety of documents such as policy papers, government reports, official statements, health sector strategic plans, and international guidelines. Sources included publications from the Tanzanian Ministry of Health, the World Health Organization (WHO), the African Union, and other pertinent bodies. This method provided a foundational understanding of the official narratives, policy shifts, and institutional frameworks guiding Tanzania's pandemic response. Analyzing media reports offered insights into public discourse, societal reactions, and the dissemination of information regarding the pandemic (Krippendorff, 2013). The study examined local and international media coverage from reputable outlets like *The Citizen*, *The East African*, *BBC News*, and *Reuters*. This analysis helped to understand how policies were communicated to the public, the portrayal of leadership actions, and the role of media in shaping public perception and behavior during the pandemic.

Semi-structured interviews were conducted with a purposive sample of stakeholders to gather in-depth qualitative data (Kallio et al., 2016). Participants included healthcare professionals such as doctors, nurses, and public health officials involved in the frontline response to COVID-19; policymakers including government officials and advisors who played roles in formulating and implementing health policies; civil society representatives comprising members of nongovernmental organizations (NGOs), community leaders, and activists engaged in health advocacy and community outreach; and community members from various regions and demographics to capture diverse experiences and perceptions. The semi-structured format allowed for flexibility in questioning while ensuring that key topics were covered, facilitating rich and detailed responses (Patton, 2015).

The data collected were analyzed using thematic analysis, a method for identifying, analyzing, and reporting patterns within data (Braun & Clarke, 2006). The process involved several steps: familiarization with the data by reading transcripts and notes multiple times, coding by systematically highlighting significant features of the data, theme development by collating codes into potential themes that capture important aspects of the data in relation to the research questions, reviewing themes to ensure they accurately represent the data and are coherent, defining and naming themes clearly for clarity and analysis, and writing up by integrating the themes into a comprehensive narrative that addresses the research objectives. Qualitative data analysis software, such as NVivo 14, was utilized to organize and manage the data efficiently, facilitating rigorous and systematic analysis (Edhlund & McDougall, 2019).

5.4 Findings

5.4.1 Initial Denial and Minimal Intervention

5.4.1.1 Government's Stance Under President Magufuli

In the early stages of the COVID-19 pandemic, Tanzania adopted a unique and controversial approach under the leadership of President John Pombe Magufuli. Unlike many countries that implemented strict measures to contain the virus, President Magufuli downplayed the severity of COVID-19, casting doubt on the efficacy of testing kits and the legitimacy of reported cases (BBC News, 2020). In a public address in May 2020, he questioned the reliability of imported test kits by alleging that nonhuman samples had tested positive, suggesting flaws in testing procedures (The Guardian, 2020).

The government ceased publishing official COVID-19 statistics in April 2020, after reporting 509 cases and 21 deaths (World Health, 2021a). President Magufuli declared Tanzania "COVID-19-free" in June 2020, attributing this to national prayers and divine intervention (Al Jazeera, 2020). He encouraged the use of herbal remedies and steam therapy, dismissing scientific recommendations for preventive measures such as mask-wearing and social distancing (Reuters, 2022). The administration discouraged testing and reporting of cases, leading to a significant underestimation of the pandemic's impact within the country (Amnesty International, 2021). Interviews with key stakeholders in Mtwara and Mbeya had the following to share:

> President Magufuli's administration sparked widespread skepticism about COVID-19 vaccines, leading to a divided perception among the populace, with some believing them unnecessary or harmful. (KIIs in Mtwara)

> The late president's initial stance on rejecting vaccines and emphasizing natural remedies influenced public attitudes significantly. His reluctance to promote vaccination left a legacy of doubt and mistrust, even after the government shifted its approach. (KII in Mbeya)

5.4.1.2 Public Perception and Behavior

The government's dismissal of the pandemic influenced public perception and behavior significantly. With official narratives downplaying the risk, many Tanzanians did not adopt preventive practices recommended by global health authorities (Human Rights Watch, 2021). The lack of transparent communication fostered an environment where misinformation proliferated, including beliefs that COVID-19 was a foreign problem or that faith alone could provide protection (BBC News, 2020).

Table 5.1 illustrates the perceptions of individuals across eight regions regarding the effectiveness of COVID-19 vaccines in preventing severe illness or death. Overall, nearly half of the respondents express some level of agreement, with 32.2% agreeing and 15.4% strongly agreeing, totaling 47.6%. A significant portion remains neutral at 39.3%, indicating uncertainty or ambivalence toward the vaccines' efficacy. Those who disagree comprise a smaller segment, with 8.2% disagreeing and 4.9% strongly disagreeing, summing up to 13.1% who express skepticism or disbelief in the vaccines' effectiveness.

Analyzing regional differences reveals notable variations in perceptions. Tabora exhibits the highest combined agreement at 63.5% (39.4% agree and 24.1% strongly agree), suggesting strong confidence in vaccine efficacy. In contrast, Njombe shows the lowest combined agreement at 26.8% and the highest combined disagreement at 28.1% (14.8% disagree and 13.3% strongly disagree), indicating considerable doubt or mistrust. Mbeya has the highest neutrality rate at 54.3%, pointing to a large undecided population. These disparities underscore the need for region-specific public health strategies to address varying levels of vaccine acceptance and to enhance educational efforts where skepticism is prevalent.

Misinformation was further exacerbated by restrictions on media and suppression of voices attempting to raise awareness about the pandemic's realities (Reporters

Table 5.1 Do you believe that COVID-19 vaccines are effective in preventing severe illness or death?

	Do you believe that COVID-19 vaccines are effective in preventing severe illness or death?				
Region	Agree (%)	Disagree (%)	Neutral (%)	Strongly agree (%)	Strongly disagree (%)
Arusha	27.4	3.9	41.1	23.2	4.5
Mbeya	25.3	11.4	54.3	3.5	5.6
Morogoro	38.1	7.4	43.0	9.0	2.5
Mtwara	37.6	10.7	39.5	8.8	3.3
Njombe	20.7	14.8	45.0	6.1	13.3
Shinyanga	36.3	4.6	32.0	24.9	2.2
Singida	33.7	7.5	27.7	26.4	4.7
Tabora	39.4	4.0	29.4	24.1	3.2
Total	32.2	8.2	39.3	15.4	4.9

Source: Field survey ($N = 3098$)

Without Borders, 2021). Some healthcare workers reported being pressured not to wear personal protective equipment (PPE) to avoid causing public alarm (The Lancet, 2021). The combination of official denial and limited access to accurate information hindered public engagement in preventive behaviors, potentially facilitating the silent spread of the virus. These factors collectively created an environment of uncertainty and misinformation, significantly impacting the public's perception and response to the pandemic as recounted by the following key stakeholders:

> Public attitudes towards vaccination varied, with concerns about safety, efficacy, benefits, side effects, and campaign motives. Stigmatization and misinformation hindered uptake. (KII in Tabora)

> Cultural and traditional practices influenced public behavior, including skepticism about foreign-made vaccines and reliance on local remedies. Religious and social leaders played crucial roles in shaping public opinion, either promoting or discouraging vaccination. (KII in Shinyanga)

5.4.2 Leadership Transition and Policy Shift

5.4.2.1 Ascension of President Samia Suluhu Hassan

The sudden death of President Magufuli in March 2021 marked a significant turning point in Tanzania's pandemic response. Vice President Samia Suluhu Hassan assumed the presidency, becoming the country's first female head of state (CNN, 2021). President Hassan signaled a departure from her predecessor's approach by acknowledging the existence of COVID-19 and the need for a coordinated response (Reuters, 2021). In her initial public addresses, she emphasized the importance of science and data in addressing the pandemic. President Hassan formed a special task force to evaluate the COVID-19 situation and provide recommendations, demonstrating a commitment to transparency and evidence-based policymaking (The Citizen, 2021). This shift was well received by the international community and domestic stakeholders who had advocated for a more proactive response. President Hassan's approach marked a significant shift in Tanzania's COVID-19 strategy, moving from denial to active management of the pandemic. Her evidence-based policies and commitment to transparency helped rebuild trust with international partners and the Tanzanian public. This change in leadership and pandemic response strategy positioned Tanzania to better address the health crisis and its economic impacts. While President Hassan's approach to COVID-19 was initially praised, some critics argued that the sudden policy shift could create confusion and undermine public trust in government communications. Despite these mixed reactions, President Hassan's policy changes represented a crucial turning point in Tanzania's approach to the pandemic, setting the stage for more effective public health measures and international cooperation as supported by the interviews in Tabora and Morogoro regions:

> President Samia Suluhu's transition to COVID-19 policy marked a shift towards vaccines and proactive measures, causing confusion among the public due to conflicting messages. (KII in Tabora)

On the contrary, some other respondents argued that:

> The new administration's focus on vaccination faced challenges like communication gaps, logistical issues, and inequitable access, which slowed the progress of vaccination campaigns. (KII in Morogoro)

The comment above shows that there were concerns that the formation of a new task force might delay immediate action needed to address the ongoing health crisis, potentially wasting valuable time in the fight against the pandemic.

5.4.2.2 Adoption of Proactive Measures

Under President Hassan's leadership, Tanzania re-engaged with international health organizations, including the WHO. The government resumed publishing COVID-19 data in July 2021 after a 13-month hiatus, reporting 858 cases and 29 deaths at that time (World Health Organization, 2021a). Tanzania joined the COVAX facility to facilitate access to vaccines and began efforts to procure doses for the population (t. V. A. Gavi, 2021). The administration launched public health campaigns to promote awareness about COVID-19, encouraging citizens to adopt preventive measures such as mask-wearing, hand hygiene, and social distancing (The East African, 2021). Testing capacities were enhanced, and data transparency improved, allowing for better monitoring of the pandemic's trajectory within the country (UNICEF, 2021a). Despite these apparent improvements, critics argue that Tanzania's response remained inadequate and delayed. Some contend that the resumed data reporting was selective and potentially underreported, questioning the true extent of the pandemic's impact (Carlitz et al., 2021; World Health Organization, 2021a). Additionally, skeptics (Pipkin, 2022; Yamanis et al., 2023) point out that the shift in approach came too late, after significant damage had already been done to public trust and health infrastructure during the previous administration's denial phase.

5.4.3 Present Proactive Engagement

5.4.3.1 Health Infrastructure Improvements

Recognizing the need to strengthen the health system, the government initiated investments in healthcare facilities and equipment. Efforts were made to increase the availability of hospital beds, oxygen supplies, and intensive care units to manage severe COVID-19 cases (Ministry of Health, 2022b). Training programs were implemented to equip healthcare workers with skills to handle the pandemic effectively, including infection prevention and control measures (World Health

Organization, 2022). The government allocated additional funding to the health sector to bolster capacity, particularly in regions with limited resources. Partnerships with international donors and organizations facilitated the acquisition of medical supplies and technical assistance (The World Bank, 2022). These improvements aimed not only to address the immediate crisis but also to enhance the overall resilience of the health system for future challenges. Despite these efforts, critics argue that the improvements were insufficient and unevenly distributed, leaving many rural and marginalized communities underserved (Chakawata, 2022). Some contend that the focus on COVID-19 preparedness diverted resources from other essential health services, potentially leading to long-term negative impacts on overall public health (Carlitz et al., 2021; Pipkin, 2022). Additionally, concerns have been raised about the sustainability of these improvements, questioning whether the enhanced capacity will be maintained once the immediate crisis subsides.

5.4.3.2 Vaccination Campaigns

A cornerstone of Tanzania's proactive engagement has been the initiation of vaccination campaigns. In July 2021, the country received over one million doses of the Johnson & Johnson vaccine through the COVAX facility (Reuters, 2022). President Hassan publicly received the vaccine, setting a precedent and encouraging the population to follow suit (The Citizen, 2021). Strategies for vaccine acquisition included bilateral agreements and participation in international initiatives to secure sufficient doses for the population. The government prioritized frontline healthcare workers, the elderly, and individuals with comorbidities in the initial phases of vaccine rollout (Ministry of Health, 2022a). Public outreach efforts were intensified to combat vaccine hesitancy, involving community leaders, religious figures, and civil society organizations to disseminate accurate information (UNICEF, 2022).

Educational campaigns utilized various media platforms to reach diverse audiences, emphasizing the safety and efficacy of vaccines. Mobile clinics and outreach programs were deployed to improve access in remote and rural areas (The East African, 2021). By October 2023, Tanzania had made significant progress, with approximately 30% of the eligible population vaccinated, reflecting a substantial increase from the previous year (World Health Organization, 2023b). However, not all the measures taken by the government are well known by the local people. For example, Table 5.2 indicates that a substantial majority of individuals across the surveyed regions are unaware of government measures supporting COVID-19 vaccination. Overall, 77.6% of respondents reported not knowing of any such measures, while only 22.4% affirmed awareness. Regions like Mbeya and Mtwara exhibit particularly low levels of awareness, with 94.0% and 93.8% of respondents, respectively, stating they are uninformed about government initiatives. Even in areas with relatively higher awareness, such as Arusha (39.4% aware) and Singida (37.0% aware), less than half of the population is informed.

Table 5.2 Do you know of any government measure in support of COVID-19 vaccination (e.g., policy on vaccination for travelers, etc.)?

Region	Do you know of any government measure in support of COVID-19 vaccination? No (%)	Yes (%)	Total (%)
Arusha	60.6	39.4	100.0
Mbeya	94.0	6.0	100.0
Morogoro	85.5	14.5	100.0
Mtwara	93.8	6.2	100.0
Njombe	76.7	23.3	100.0
Shinyanga	71.5	28.5	100.0
Singida	63.0	37.0	100.0
Tabora	71.4	28.6	100.0
Total	**77.6**	**22.4**	**100.0**

Source: Field survey ($N = 3098$)

5.5 Discussion

5.5.1 Marginalization During Initial Response

The Tanzanian government's response to the COVID-19 pandemic presents a compelling case study of how political leadership can significantly influence health policy and the well-being of vulnerable populations. Initially, under President Magufuli's administration, the government's denial and minimal intervention had profound adverse effects on marginalized groups, including the elderly, individuals with comorbidities, and those in rural areas (Human Rights Watch, 2021). The lack of acknowledgment of the pandemic's severity led to the absence of preventive measures, leaving these populations exposed to heightened health risks. The elderly and those with underlying health conditions such as HIV/AIDS, tuberculosis, and non-communicable diseases were particularly vulnerable. Tanzania, with an estimated 1.6 million people living with HIV/AIDS (UNAIDS, 2022), faced disruptions in healthcare services, leading to interruptions in treatment and care for these patients (Amnesty International, 2021). The government's stance impeded efforts to adapt healthcare delivery to protect immunocompromised individuals, exacerbating existing health disparities.

Access disparities between urban and rural areas were further accentuated. Rural communities, already struggling with limited healthcare infrastructure, were left without adequate information or means to protect themselves (African Union, 2021). The suppression of information and restriction of media freedoms hindered the dissemination of critical health information (Reporters Without Borders, 2021), disproportionately affecting marginalized groups such as women, people with disabilities, and those living in poverty (UNICEF, 2021b). This period highlighted the detrimental impact of government denial and systemic vulnerabilities on the most at-risk populations.

5.5.2 *Changes in Leadership and Post-policy Shifts*

Following the leadership transition to President Samia Suluhu Hassan, there was a notable shift in policies affecting vulnerable populations. The government implemented measures to improve healthcare access, acknowledging the importance of protecting those at heightened risk (World Health Organization, 2022). Reintegration of community health workers and strengthening of primary healthcare services, especially in underserved areas, became a priority. The inclusion of community leaders in health initiatives enhanced outreach and education efforts. Engaging religious leaders, traditional healers, and local influencers helped disseminate accurate information about COVID-19 and promote preventive practices (UNICEF Tanzania, 2022). This approach leveraged existing social structures to build trust and encourage compliance with health guidelines among diverse communities.

Healthcare facilities received support to adapt services to the needs of vulnerable populations, including implementing infection prevention and control measures and prioritizing vaccinations for high-risk groups (Ministry of Health, 2022b). Mobile clinics and outreach programs expanded access to testing, treatment, and vaccination, particularly in remote areas. Additionally, collaboration with international organizations addressed socioeconomic challenges exacerbated by the pandemic. Strengthened social protection programs supported households affected by economic disruptions, focusing on women and children (World Bank, 2022). These initiatives aimed to mitigate the indirect impacts of the pandemic on food security, education, and overall well-being.

Personal narratives and case studies underscore the real-world impact of the government's policy shift on vulnerable populations. In the rural district of Kigoma, community health volunteers reported increased engagement and cooperation from residents following targeted communication efforts involving local leaders (The Citizen, 2022). In urban centers like Dar es Salaam, healthcare workers noted improvements in managing patients with chronic conditions, highlighting the importance of maintaining essential health services during a pandemic (Amref Health, 2022). Women entrepreneurs benefited from social protection measures, enabling them to sustain their businesses amid economic challenges and contribute to community resilience (World Bank, 2022).

5.5.3 *Misinformation and Hesitancy*

Ongoing misinformation and vaccine hesitancy in some communities indicate the need for continued efforts to address misconceptions (The East African, 2022). Sustained commitment is necessary to ensure that all segments of the population are adequately protected and supported. The importance of transparent communication emerged as a critical lesson. Initially, the lack of transparency under

President Magufuli's administration led to public skepticism and hindered the adoption of preventive behaviors (Human Rights Watch, 2021). Effective communication strategies, including regular updates through multiple channels and the use of local languages, enhance understanding and adherence among diverse populations (UNICEF, 2020). President Hassan's administration improved transparency by resuming the publication of COVID-19 data and conducting public awareness campaigns, which bolstered public trust and compliance with health guidelines (Reuters, 2021). Acknowledging uncertainties and being honest about challenges can counter misinformation and strengthen public health efforts (Garrett, 2020).

5.5.4 Community Involvement, Leadership, and Alignment with International Guidelines

Community involvement and leadership are crucial for the successful implementation of health policies, especially in culturally diverse societies. Local leaders and organizations act as bridges between government initiatives and the populace, ensuring that policies are tailored to community needs and are culturally sensitive (van Niekerk et al., 2017). In Tanzania, engaging community influencers was instrumental in disseminating accurate information about COVID-19 and encouraging vaccination uptake (UNICEF, 2022). Participatory approaches empower communities, promote ownership of health interventions, and enhance sustainability (Rifkin, 2014).

Alignment with international guidelines provides access to global expertise, resources, and support mechanisms. Tanzania's initial departure from international guidelines limited its access to international support and may have exacerbated the pandemic's impact (Amnesty International, 2021). Re-engagement under President Hassan facilitated vaccine acquisition through the COVAX facility and provided technical assistance (t. V. A. Gavi, 2021). Balancing national sovereignty with international cooperation enhances effectiveness and ensures comprehensive responses (Katz et al., 2019).

5.5.5 Building Resilient Health Systems and Implications for Other African Countries

Building resilient health systems capable of withstanding crises is essential. Investments in healthcare infrastructure, workforce capacity, and supply chains enhance preparedness and response capabilities (Kruk et al., 2015). Strengthening disease surveillance systems and early warning mechanisms can significantly improve a health system's ability to detect and respond to emerging threats.

Developing robust emergency response protocols and conducting regular simulations can help healthcare workers and institutions better prepare for potential crises. Additionally, fostering cross-sector collaboration and international partnerships can enhance knowledge sharing and resource mobilization during times of crisis. Post-policy shift investments in Tanzania focused on improving healthcare facilities and training healthcare workers (Ministry of Health, 2022a), aiming not only to address immediate needs but also to strengthen the overall health system. Integrating services, promoting universal health coverage, and ensuring equitable access to care are vital components of resilience (World Health Organization, 2020). Investing in health information systems and surveillance enhances the ability to detect and respond to emerging health threats promptly (World Bank, 2021). While emergency preparedness and international collaboration are important, they may not always be sufficient in addressing complex health crises. Overreliance on simulations and protocols could potentially lead to rigid responses that fail to adapt to unique, unforeseen circumstances. Furthermore, cross-sector partnerships and international cooperation, though beneficial, might introduce bureaucratic complexities and conflicting priorities that could hinder swift, decisive action during emergencies.

Tanzania's experience illustrates the significant impact of political leadership on health policy and the well-being of vulnerable populations during a health crisis. The initial denial of the pandemic had detrimental effects, while the subsequent policy shift under President Hassan led to improvements in healthcare access, communication, and community engagement. The critical lessons learned emphasize the importance of transparent communication, community involvement, alignment with international guidelines, and building resilient health systems to ensure inclusive health policies that protect all segments of the population.

The transition from denial to proactive engagement during the COVID-19 pandemic provides critical insights for other African nations grappling with similar public health crises. This section critically examines the transferability of Tanzania's strategies, potential barriers, and broader implications for health governance and social policy integration across the continent. For instance, Tanzania's approach under President Samia Suluhu Hassan demonstrated how evidence-based leadership can reverse policy denialism and rebuild public trust. The formation of a COVID-19 task force, resumption of data reporting, and alignment with international health bodies (e.g., WHO and COVAX) facilitated access to vaccines and technical support (World Health Organization, 2022). These actions highlight the importance of strong political will and international collaboration, lessons applicable to countries such as Zimbabwe and Malawi, which have faced challenges in obtaining vaccines due to strained diplomatic relations or governance issues (Nkengasong et al., 2020; Yaya et al., 2022). Furthermore, Tanzania's emphasis on community involvement, such as engaging religious leaders and local influencers, proved effective in promoting vaccine uptake and combating misinformation (UNICEF, 2022). This strategy could be replicated in culturally diverse nations like Nigeria and Kenya, where trust in traditional and community leaders often surpasses confidence in formal government institutions (Van der Merwe, 2020). However, its success hinges on tailoring communication to local contexts and ensuring sustained funding for grassroots initiatives.

While Tanzania's strategies offer valuable lessons, structural barriers may limit their applicability in other African countries. For instance, fragile health infrastructures in nations such as the Democratic Republic of Congo (DRC) and South Sudan present challenges for scaling up testing and vaccination efforts (Kruk et al., 2015). Unlike Tanzania, which leveraged existing partnerships with international donors, these countries face difficulties in attracting external resources due to prolonged conflict and political instability. Furthermore, vaccine inequity remains a pressing issue. Tanzania's initial delay in joining COVAX hindered vaccine acquisition, but subsequent efforts to align with international frameworks improved access (Gavi, 2021). This experience underscores the critical role of global health diplomacy. For other nations, such as Ethiopia and Chad, addressing vaccine inequity will require stronger regional cooperation and advocacy to counteract the effects of vaccine nationalism by wealthier countries (So & Woo, 2020).

Despite Tanzania's progress, significant equity gaps persisted, particularly in rural and underserved areas. Limited awareness of government vaccination programs, as shown in survey data, indicates that marginalized populations were not fully reached (Carlitz et al., 2021). Similar challenges are prevalent across Africa, where disparities in healthcare access disproportionately affect rural residents, women, and people living with disabilities (Braveman et al., 2017; Marmot et al., 2020). These inequities highlight the necessity of integrating health and social policies to address structural determinants of health. For example, South Africa's integration of cash transfers and food security programs during the pandemic provides a model for linking social protection with health outcomes (Seekings & Nattrass, 2020). By combining health governance with broader social policy measures, African nations can build resilience to future crises while promoting equity. Tanzania's experience also underscores the importance of regional collaboration in addressing public health crises. Platforms such as the Africa Centres for Disease Control and Prevention (Africa CDC) provide opportunities for knowledge sharing, coordinated responses, and resource pooling (Nkengasong et al., 2020). Tanzania's alignment with regional and global health initiatives after a period of isolation demonstrates the value of cooperative frameworks for addressing cross-border challenges, such as virus transmission and vaccine distribution.

5.6 Conclusion

Tanzania's pandemic response underscores the profound interdependence between health policy and broader social policy, particularly in contexts marked by socioeconomic disparities and political transitions. The initial denial of COVID-19's severity under President Magufuli left marginalized populations exposed to heightened risks, undermining trust in public institutions and deepening systemic inequities. In contrast, the policy shift under President Hassan demonstrated the transformative potential of evidence-based leadership, international collaboration, and inclusive governance. Her administration's emphasis on transparency, community involvement, and health equity not only improved pandemic outcomes but also laid the foundation for integrating health governance into a broader social policy framework.

These experiences reveal critical lessons for African nations aiming to build resilient and inclusive systems. Transparent communication and community participation are indispensable for fostering trust and ensuring the cultural and contextual relevance of policies. Aligning national priorities with international frameworks can facilitate resource mobilization while preserving local agency. Crucially, the Tanzanian case emphasizes the need for integrated social policies that address the structural determinants of health, including poverty, education, and access to services.

Looking forward, Tanzania must sustain and institutionalize these advances to safeguard progress and resilience against future crises. For the continent as a whole, Tanzania's journey offers valuable insights into balancing global health imperatives with Africa-centered social policy aspirations, contributing to a future where equitable and inclusive systems are the cornerstone of development.

References

Abdool Karim, S. S., & Abdool Karim, Q. (2021). COVID-19 vaccine affordability and accessibility. *The Lancet, 397*(10281), 1996–1998. https://doi.org/10.1016/S0140-6736(21)00719-3

Abdool Karim, S. S., & de Oliveira, T. (2021). New SARS-CoV-2 variants—Clinical, public health, and vaccine implications. *New England Journal of Medicine, 384*(19), 1866–1868. https://doi.org/10.1056/NEJMc2100362

Adams, J., MacKenzie, M. J., & Amegah, A. K. (2021). The conundrum of low COVID-19 mortality burden in sub-Saharan Africa: Myth or reality? *Global Health: Science and Practice, 9*(3), 433–443. https://doi.org/10.9745/GHSP-D-21-00172

Africa Centres for Disease Control Prevention, A. (2021). *Africa CDC COVID-19 vaccine dashboard*. https://africacdc.org/covid-19-vaccines/dashboard/

African Union, A. (2021). *Impact of COVID-19 on Africa's health system*. https://au.int/en/pressreleases/20210607/impact-covid-19-africas-health-system

Ahmed, F., Ahmed, N.-E., Pissarides, C., & Stiglitz, J. (2020). Why inequality could spread COVID-19. *The Lancet Public Health, 5*(5), e240. https://doi.org/10.1016/S2468-2667(20)30085-2

Al Jazeera. (2020). *Tanzania's Magufuli says prayer has healed the country of coronavirus*. https://www.aljazeera.com/news/2020/6/8/tanzanias-magufuli-says-prayer-has-healed-the-country-of-coronavirus

Amnesty International, A. (2021). *Tanzania: Government must end denial of COVID-19 pandemic*. https://www.amnesty.org/en/latest/news/2021/02/tanzania-government-must-end-denial-of-covid-19-pandemic/

Amref Health, A. (2022). *COVID-19 response in Tanzania*. https://amref.org/covid19/tanzania

Ataguba, J. E. (2020). COVID-19 pandemic, a war to be won: Understanding its economic implications for Africa. *Applied Health Economics and Health Policy, 18*(3), 325–328. https://doi.org/10.1007/s40258-020-00580-x

BBC News. (2020). *Coronavirus in Tanzania: The country that's rejecting the virus*. https://www.bbc.com/news/world-africa-52723594

BBC News. (2021). *John Magufuli: Tanzania's president dies aged 61 after Covid rumours*. https://www.bbc.com/news/world-africa-56437852

Betancourt, J. R., Green, A. R., Carrillo, J. E., & Owusu Ananeh-Firempong, I. (2003). Defining cultural competence: A practical framework for addressing racial/ethnic disparities in health and health care. *Public Health Reports*.

Bowen, G. A. (2009). Document analysis as a qualitative research method. *Qualitative Research Journal, 9*(2), 27–40. https://doi.org/10.3316/QRJ0902027

Braun, V., & Clarke, V. (2006). Using thematic analysis in psychology. *Qualitative Research in Psychology, 3*(2), 77–101.

Braveman, P., Arkin, E., Orleans, T., Proctor, D., & Plough, A. (2017). *What is health equity? And what difference does a definition make?* https://www.rwjf.org/en/library/research/2017/05/what-is-health-equity-.html

Burki, T. (2021). Challenges in the rollout of COVID-19 vaccines worldwide. *The Lancet Respiratory Medicine, 9*(4), e42–e43. https://doi.org/10.1016/S2213-2600(21)00129-6

Carlitz, R., Yamanis, T., & Mollel, H. (2021). Coping with denialism: How street-level bureaucrats adapted and responded to COVID-19 in Tanzania. *Journal of Health Politics, Policy and Law*.

Chakawata, W. (2022). Africa's response to COVID-19: A governmentality in disguise masterclass? *International Review of Sociology, 32*(1), 147–173. https://doi.org/10.1080/03906701.2022.2028403

Chanda-Kapata, P., Ntoumi, F., Kapata, N., et al. (2020). Tuberculosis, HIV/AIDS and malaria health services disruption in sub-Saharan Africa. *The Lancet Global Health, 8*(11), e1450–e1451. https://doi.org/10.1016/S2214-109X(20)30315-5

CNN. (2021). *Tanzania swears in Samia Suluhu Hassan as first female president after death of Magufuli*. https://www.cnn.com/2021/03/19/africa/tanzania-samia-suluhu-hassan-intl/index.html

Creswell, J. W., & Poth, C. N. (2018). *Qualitative inquiry and research design: Choosing among five approaches*. Sage Publications.

Dyer, O. (2020). COVID-19: Africa's case numbers are rising rapidly, but mortality remains low. *BMJ, 370*, m2839. https://doi.org/10.1136/bmj.m2839

Edhlund, B. M., & McDougall, A. G. (2019). *NVivo 12 essentials*. Lulu Press.

Fraser, A., & Msemo, J. (2021). COVID-19 in Tanzania: A coup or continuity? *Review of African Political Economy, 48*(169), 481–491. https://doi.org/10.1080/03056244.2021.1963505

Garrett, T. M. (2020). COVID-19, wall building, and the effects on Migrant Protection Protocols by the Trump administration: The spectacle of the worsening human rights disaster on the Mexico-U.S. border. *Administrative Theory & Praxis*, 1–9. https://doi.org/10.1080/10841806.2020.1750212

Gavi, t. V. A. (2021). *Tanzania joins COVAX facility to access COVID-19 vaccines*. https://www.gavi.org/news/media-room/tanzania-joins-covax-facility-access-covid-19-vaccines

Gavi, t. V. A. (2023). *COVAX: Working for global equitable access to COVID-19 vaccines*. https://www.gavi.org/covax-facility

Gilson, L. (2022). *Health policy and systems research: A methodology reader*. World Health Organization. https://www.who.int/alliance-hpsr/resources/publications/9789241503136/en/

Greer, S. L., King, E. J., Massard da Fonseca, E., & Peralta-Santos, A. (2020). The comparative politics of COVID-19: The need to understand government responses. *Global Public Health, 15*(9), 1413–1416. https://doi.org/10.1080/17441692.2020.1783340

Hopman, J., Allegranzi, B., & Mehtar, S. (2020). Managing COVID-19 in low-and middle-income countries. *JAMA, 323*(16), 1549–1550.

Human Rights Watch. (2021). *Tanzania: Covid-19 denial risks health crisis*. https://www.hrw.org/news/2021/02/03/tanzania-covid-19-denial-risks-health-crisis

Human Rights Watch. (2022). *World Report 2022: Tanzania*. https://www.hrw.org/world-report/2022/country-chapters/tanzania-and-zanzibar

International Labour Organization, I. (2021). *COVID-19 and the world of work: Impact and policy responses in Africa*. https://www.ilo.org/africa/information-resources/publications/WCMS_778491/lang%2D%2Den/index.htm

Jerving, S. (2020). Ghana's response to COVID-19 pandemic praised. Devex. https://www.devex.com/news/ghanas-response-to-covid-19-pandemic-praised-97052. Accessed on 13.03.2021

Kallio, H., Pietilä, A.-M., Johnson, M., & Kangasniemi, M. (2016). Systematic methodological review: Developing a framework for a qualitative semi-structured interview guide. *Journal of Advanced Nursing, 72*(12), 2954–2965. https://doi.org/10.1111/jan.13031

Katz, R., Graeden, E., Kerr, J., & Eaneff, S. (2019). Tracking the flow of funds in global health security. *Ecohealth, 16*(2), 298–305. https://doi.org/10.1007/s10393-019-01402-w

Kavanagh, M. M., Erondu, N. A., Tomori, O., et al. (2021). Access to lifesaving medical resources for African countries: COVID-19 testing and response, ethics, and politics. *The Lancet, 397*(10281), 1735–1738. https://doi.org/10.1016/S0140-6736(21)00727-2

Krippendorff, K. (2013). *Content analysis: An introduction to its methodology*. SAGE Publications.

Kruk, M. E., Myers, M., Varpilah, S. T., & Dahn, B. T. (2015). What is a resilient health system? Lessons from Ebola. *The Lancet, 385*(9980), 1910–1912.

Makoni, M. (2021). Tanzania refuses COVID-19 vaccines. *The Lancet, 397*(10274), 566. https://doi.org/10.1016/S0140-6736(21)00362-7

Marmot, M., Allen, J., Boyce, T., Goldblatt, P., & Morrison, J. (2020). Health equity in England: The Marmot review 10 years on. *The Health Foundation*.

Maruyama, H., Franks, J., Laki, D., Msumi, O., Makyao, N., Rwabiyago, O. E., Rabkin, M., Kagashe, M. J., & El-Sadr, W. M. (2021). Bringing HIV services to key populations and their communities in Tanzania: From pilot to scale. *Journal of the International AIDS Society, 24*, e25718. https://www.ncbi.nlm.nih.gov/pmc/articles/PMC8242965/pdf/JIA2-24-e25718.pdf

Mendelson, M., & Madhi, S. A. (2020). COVID-19 vaccines in Africa. *BMJ, 371*, m4707. https://doi.org/10.1136/bmj.m4707

Ministry of Health, MoH. (2022a). *Annual Health Sector Performance Report 2021/2022*. https://www.moh.go.tz

Ministry of Health, MoH. (2022b). *Health Sector Strategic Plan V (HSSP V) July 2021–June 2026*. https://www.moh.go.tz

Ministry of Health, MoH. (2023). *National Non-Communicable Disease Strategy 2023–2028*. https://www.moh.go.tz

Moeti, M., & Fitzgerald, J. (2021). Tackling COVID-19 vaccine inequity in Africa. *BMJ, 374*, n2101. https://doi.org/10.1136/bmj.n2101

Naderifar, M., Goli, H., & Ghaljaie, F. (2017). Snowball sampling: A purposeful method of sampling in qualitative research. *Strides in Development of Medical Education, 14*(3), e67670. https://doi.org/10.5812/sdme.67670

National Bureau of Statistics, T. (2022). *Tanzania in figures 2021*. https://www.nbs.go.tz

Nkengasong, J. N., & Mankoula, W. (2020). Looming threat of COVID-19 infection in Africa: Act collectively, and fast. *The Lancet, 395*(10227), 841–842. https://doi.org/10.1016/S0140-6736(20)30464-5

Nkengasong, J. N., Ndembi, N., Tshangela, A., & Raji, T. (2020). COVID-19 vaccines: How to ensure Africa has access. *Nature, 586*(7833), 197–199. https://doi.org/10.1038/d41586-020-02774-8

Ottersen, T., Norheim, O. F., & World Health Organization Consultative Group on, E., & Universal Health, C. (2014). Making fair choices on the path to universal health coverage. *Bulletin of the World Health Organization, 92*(6), 389. https://doi.org/10.2471/BLT.14.139139

Patton, M. Q. (2015). *Qualitative research & evaluation methods*. SAGE Publications.

PharmAccess, F. (2021). *Tanzania health sector*. https://www.pharmaccess.org

Pipkin, S. (2022). *An analysis of the Tanzanian public response to Tanzanian leadership during the COVID-19 pandemic*. The University of Mississippi. https://egrove.olemiss.edu/hon_thesis/2705

President's Malaria Initiative, P. (2023). *Tanzania Malaria Operational Plan FY 2023*. https://www.pmi.gov/where-we-work/tanzania/

Reporters Without Borders. (2021). *Tanzania*. https://rsf.org/en/tanzania

Reuters. (2021). *Tanzania's new president lifts media ban imposed by predecessor*. https://www.reuters.com/world/africa/tanzanias-new-president-lifts-media-ban-imposed-by-predecessor-2021-04-06/

Reuters. (2022). *Tanzania receives over 1 million COVID-19 vaccines from the US*. https://www.reuters.com/world/africa/tanzania-receives-over-1-million-covid-19-vaccines-us-2022-02-20/

Rifkin, S. B. (2014). Examining the links between community participation and health outcomes: A review of the literature. *Health Policy and Planning, 29*, 98–106. https://doi.org/10.1093/heapol/czu076

Seekings, J., & Nattrass, N. (2020). Policy responses to Covid-19 in South Africa. *World Development, 137*, 105179. https://doi.org/10.1016/j.worlddev.2020.105179

So, A. D., & Woo, J. (2020). Reserving coronavirus disease 2019 vaccines for global access: Cross sectional analysis. *BMJ, 371*.

The Citizen. (2021). *President Samia receives Covid-19 jab, urges Tanzanians to ignore hearsay*. https://www.thecitizen.co.tz/tanzania/news/national/president-samia-receives-covid-19-jab-urges-tanzanians-to-ignore-hearsay-3499662

The Citizen. (2022). *Community leaders boost COVID-19 fight in Kigoma*. https://www.thecitizen.co.tz/tanzania/news/national/community-leaders-boost-covid-19-fight-in-kigoma-3769212

The East African. (2021). *President Samia shifts Tanzania's Covid-19 response*. https://www.theeastafrican.co.ke/tea/news/east-africa/president-samia-shifts-tanzania-covid19-response-3364656

The East African. (2022). *Tanzania battles vaccine hesitancy amid COVID-19 surge*. https://www.theeastafrican.co.ke/tea/news/east-africa/tanzania-battles-vaccine-hesitancy-3732912

The Economist. (2019). John Magufuli's war on wasteful spending in Tanzania. https://www.economist.com/middle-east-and-africa/2019/02/16/john-magufulis-war-on-wasteful-spending-intanzania. Accessed on 22.02.2025

The Guardian. (2020). *Tanzania's president says country is virus free. Others warn of disaster*. https://www.theguardian.com/world/2020/jun/08/tanzania-president-coronavirus-free-warning

The Lancet. (2021). Tanzania's COVID-19 response: Did denialism lead to excess deaths? *The Lancet, 397*(10277), 770. https://doi.org/10.1016/S0140-6736(21)00561-X

The World Bank, WB. (2022). *Tanzania gender assessment* (AUS0002736).

UNAIDS. (2022). *Country factsheets: United Republic of Tanzania*. https://www.unaids.org/en/regionscountries/countries/unitedrepublicoftanzania

UNAIDS. (2023). *Country factsheets: United Republic of Tanzania*. https://www.unaids.org/en/regionscountries/countries/unitedrepublicoftanzania

UNICEF. (2020). *COVID-19 communication strategy*. https://www.unicef.org/tanzania/media/2466/file/COVID-19%20Communication%20Strategy.pdf

UNICEF. (2021a). *Addressing witchcraft in Tanzania: Case study of a promising approach*.

UNICEF. (2021b). *Tanzania COVID-19 Situation Report*. https://www.unicef.org/tanzania/reports/covid-19-situation-report

UNICEF. (2022). *Community engagement in COVID-19 vaccination*. https://www.unicef.org/tanzania/stories/community-engagement-covid-19-vaccination

UNICEF Tanzania. (2022). *Community engagement in COVID-19 vaccination*. https://www.unicef.org/tanzania/stories/community-engagement-covid-19-vaccination

United Nations Economic Commission for Africa, UNECA. (2021). *Building forward for an African green recovery*. https://www.uneca.org/publications/building-forward-for-an-african-green-recovery

United Nations, UN. (2021). *UN comprehensive response to COVID-19*. https://www.un.org/en/coronavirus/un-response

Van der Merwe, J. (2020). Ethiopia's COVID-19 response. *Institute for Security Studies*. https://issafrica.org/iss-today/ethiopias-covid-19-response

van Niekerk, L., Chater, R., Naydenova, E., Lim, J., Chamas, L., Manderson, L., Gilson, L., Peeling, R., Hartigan, P., Bonnici, F., World Health, O., Research, U. N. U. W. B. W. S. P. f., & Training in Tropical, D. (2017). *Social innovation in health: case studies and lessons learned from low- and middle-income countries*. World Health Organization. https://apps.who.int/iris/handle/10665/259187

World Bank, W. (2021). *Building resilient health systems: Policies for inclusive health*. https://documents.worldbank.org/en/publication/documents-reports/documentdetail/955221619896131521/building-resilient-health-systems-policies-for-inclusive-health

World Bank, WB. (2022). *Tanzania economic update: Empowering women is smart economics*. https://www.worldbank.org/en/country/tanzania/publication/tanzania-economic-update

World Health, O. (2021a). *Impact on the ground: WHO's action in countries, territories and areas*. World Health Organization. https://apps.who.int/iris/handle/10665/344464

World Health, O. (2021b). *WHO guideline on health workforce development, attraction, recruitment and retention in rural and remote areas: Web annexes*. World Health Organization. https://apps.who.int/iris/handle/10665/341112

World Health Organization, WHO. (2005). *The World health report: 2005: Make every mother and child count*. World Health Organization.

World Health Organization, WHO. (2020). *Risk Communication and Community Engagement (RCCE) action plan guidance COVID-19 preparedness & response*. https://www.who.int/publications/i/item/risk-communication-and-community-engagement-(rcce)-action-plan-guidance

World Health Organization, WHO. (2021a). *Tanzania stops publishing COVID-19 data*. https://www.afro.who.int/news/tanzania-stops-publishing-covid-19-data

World Health Organization, WHO. (2021b). *United Republic of Tanzania resumes reporting COVID-19 cases*. https://www.who.int/news/item/28-07-2021-united-republic-of-tanzania-resumes-reporting-covid-19-cases

World Health Organization, WHO. (2021c). *United Republic of Tanzania: Health workforce*. https://www.who.int/data/gho/data/countries/country-details/GHO/tanzania-united-republic-of

World Health Organization, WHO. (2022). *WHO supports Tanzania's COVID-19 response efforts*. https://www.afro.who.int/countries/united-republic-of-tanzania/news/who-supports-tanzanias-covid-19-response-efforts

World Health Organization, WHO. (2023). *WHO health emergency dashboard: WHO (COVID-19) homepage*. Retrieved November 23, 2023, from https://covid19.who.int/region/afro/country/tz

World Health Organization, WHO. (2023a). *COVID-19 strategic preparedness and response plan (SPRP 2023)*. https://www.who.int/publications/i/item/WHO-WHE-SPP-2023.1

World Health Organization, WHO. (2023b). *COVID-19 vaccination in the United Republic of Tanzania*. https://covid19.who.int/region/afro/country/tz

World Health Organization, WHO. (2023c). *WHO Coronavirus (COVID-19) dashboard*. https://covid19.who.int/

Yamanis, T., Carlitz, R., Gonyea, O., Skaff, S., Kisanga, N., & Mollel, H. (2023). Confronting 'chaos': A qualitative study assessing public health officials' perceptions of the factors affecting Tanzania's COVID-19 vaccine rollout. *BMJ Open, 13*(1), e065081.

Yaya, S., Otu, A., & Labonte, R. (2022). COVID-19 and social determinants of health in Africa. *World Development, 140*, 105274. https://doi.org/10.1016/j.worlddev.2020.105274

Chapter 6
Political Economy of COVID-19 Social Protection in Southern Africa

Emmanuel Ndhlovu and Daniel F. Meyer

6.1 Introduction

The advent of the coronavirus disease pandemic (COVID-19) in late December 2019 once again reinforced the state's leading role in ensuring the welfare of the people. The lethal nature of COVID-19 called for immediate measures to save lives and suppress the virus's transmission. The measures include travel and gathering restrictions, including in markets and workplaces, prohibition of activities other than those identified as essential, isolation, and quarantine of the infected or suspected. Countries with operational social protection (policies and programmes meant to reduce the effect of poverty on vulnerable people) quickly replenished them to respond to the challenge. However, in sub-Saharan Africa, where over 70% of the population relies on farming and informal sector activities for livelihoods and income, the COVID-19 response strategy turned into enormous social, economic, and political issues (Jatula & Conshello, 2021; Mutekwe & Vanyoro, 2021). Most countries in the region have yet to establish state-run or coordinated social protection systems. One of the subregions where social protection is used as a reactionary measure is Southern Africa. Except for South Africa which has the entire social protection system run and overseen by the government through the Department of Social Development and administered by the South African Social Security Agency (SASSA) (Nzabamwita & Ndhlovu, 2024), the social protection of most countries in the subregion is tilted towards immediate interventions, particularly following

E. Ndhlovu (✉)
School of Public Management, Governance and Public Policy, College of Business and Economics, University of Johannesburg, Auckland Park, South Africa

D. F. Meyer
School of Public Management, Governance and Public Policy, College of Business and Economics, University of Johannesburg, Auckland Park, South Africa
e-mail: dfmeyer@uj.ac.za

E. Ndhlovu, C. Chipenda (eds.), *Redefining the Social Policy Discourse*,
https://doi.org/10.1007/978-3-032-24067-5_6

natural disasters such as floods, wildfires, and drought, as well as man-made factors such as war and conflict. The donor community provides much of the social assistance, not the state (Ndhlovu, 2024a). With the arrival of COVID-19 around March 2020, social assistance, particularly cash transfers and food parcel distribution, became widespread. However, studies exploring the nature of the political economy that emerged with this social assistance remain limited. This has political and practical implications. A clear understanding of the dynamics of how the social protection system interacts with the political level in the country is needed. It is also vital to stress the complexity behind decisions, how social protection programmes are designed, and the latent associations with democracy and accountability. Political parties and ideology are imperative aspects to consider.

Existing studies on social protection during COVID-19 either celebrate how certain governments or their partners were able to roll out interventions almost immediately (Ajam, 2020; Bhorat et al., 2021; Brauner et al., 2021) or criticise them for complacency and inaction (Adesina, 2020; Chineka & Kurevakwesu, 2021; Nhapi, 2021). For instance, South Africa is praised by some scholars for having taken advantage of its well-established social protection system to expand social assistance (Haurovi, 2021; Mhlongo et al., 2022). However, it is also criticised for its small size of transfers (Adesina, 2021), incompetence, rampant corruption, and poor targeting of beneficiaries (Casale & Shepherd, 2020). In Zimbabwe, existing studies mainly document how the pandemic became an excuse by the ruling party to thwart the opposition and expand its dwindling political base through partisan social assistance targeting (Mutekwe & Vanyoro, 2021). In Mozambique, Nuvunga et al. (2024) report that COVID-19 further weakened the government's social protection challenges, with the country's social security provided by the National Institute of Social Action (INAS) through the following programmes: Basic Social Subsidy Programme (PSSB) and Productive Social Action Programme (PASP), cash transfers; Direct Social Support Programme (PASD), food vouchers, in-kind transfers, and service payments; and Social Action Social Services Programme (PSSAS), assistance in social units and social services. Social protection interventions during COVID-19 were constrained by inconsistent funding from the government and poor economic management and performance (Arthur et al., 2023). Similar findings have been reported in Malawi (Hemsteede, 2024; Lambon-Quayefio et al., 2024) and Zambia (Banda & Thelma, 2024).

Focusing on the non-contributory social assistance aspect of social protection, this chapter (i) identifies COVID-19-specific social assistance programmes in Southern Africa, (ii) examines how social assistance programmes have been implemented, and (iii) also explores how power dynamics, interest groups, and institutional frameworks shape social protection programme choices and implementation. The chapter points to major social protection loopholes that may affect the region's political stability and undermine prospects for attaining Sustainable Development Goals (SDGs) within Southern Africa, particularly those related to poverty, hunger, and equality. The chapter argues that in a subregion that is predominantly agrarian and yet characterised by high poverty levels, strong social protection would help the populace deal with hunger and poverty through farming. In addition, deliberately targeting women in programmes could also increase food consumption and health access chances within households. The chapter, therefore, wields much potential to highlight policy gaps and weaknesses that governments, policymakers, politicians, and non-governmental actors can address in the distribution of social protection, particularly in crisis contexts.

The chapter proceeds as follows: After the introduction, the following section presents the conceptual and literature review. This is followed by a discussion of the political economy framework as the theoretical framework for the study. After that, a country-by-country COVID-19 review of social assistance programmes, their implementation, and how power dynamics, interest groups, and institutional frameworks shaped social protection programme choices and implementation is made. A brief discussion of the findings is made, leading to the conclusions.

6.2 Social Protection: A Conceptual Framework and Literature Review

There is no single definition of social protection, particularly in Africa. Existing definitions and descriptions of the concept by African institutions, governments, and other institutions working on Africa draw from local and international experiences and views. However, despite lacking an African definition, the common trend in available definitions is the objective of responding to issues of risk, vulnerability, and extreme poverty. Social protection policies are generally viewed as policies designed to prevent and reduce targeted people's poverty, vulnerability, and social exclusion. Onyina and Baye (2024) offer a broader standpoint and conceptualise social protection policies as all types of public and private initiatives regarding the transfer of income to the poor shield vulnerable people from being poor and thus improve the rights and social strata of the marginalised.

Typically, several studies define social protection as a set of policies and programmes meant to reduce the effect of poverty on vulnerable people (Abramo et al., 2019; Arthur et al., 2023; Haurovi, 2021). The African Union (AU), through a document prepared by Taylor (2008), defined social protection as 'a package of policies and programmes with the aim of reducing poverty and vulnerability of large segments of the population … this is undertaken through a "mix" of policies/programmes that promote efficient labour markets, reduce people's exposure to risks, and contribute to enhancing their capacity to protect and cover themselves against lack of or loss of adequate income, and basic social services' (Taylor, 2008). In another AU study led by Taylor, social protection is defined as 'a range of public (government funded) measures that gives support to all citizens and helps individuals, households, and communities to better manage risks and participate actively in all spheres of life' (AU, 2010). This definition focuses on formal policies and programmes. This study adopts the definitions by the AU.

The author is not oblivious to the weaknesses of these definitions. The AU definitions have long been criticised for not considering informal social protection systems and limiting themselves to public funding that assists people (Holmes & Lwanga-Ntale, 2012). Informal social protection systems encompass community-based or traditional approaches to social protection, often based on kinship, friendship, or community links, taking place within households, groups, or other social networks, and they fill the gaps left by the absence of or inadequacies in formal social protection interventions (Holmes & Lwanga-Ntale, 2012). This is achieved through distributing risk within a community, family, or other social network. According to Holmes and Lwanga-Ntale (2012), this inadequacy in the AU's definition limits the extent to which informal social protection can be conceptualised within the broader body of the approach and may stifle potential innovation. Despite these weaknesses, the author believes the definitions are sufficient to conceptualise social protection in an African context.

There are four types of social protection: social assistance, social insurance, labour market programmes, and community-based social protection. Social assistance programmes are non-contributory and involve cash or in-kind (e.g. food transfers and clothing) provided to vulnerable individuals or households (Cirillo et al., 2021). In Africa, social assistance predominantly comprises cash transfers, social pensions, public works programmes, and in-kind transfers (Ncube & Murray, 2024).

Social insurance schemes are contributory programmes in which beneficiaries make regular financial contributions to join a scheme that will assist them in the event of a shock (Chawurura, 2020). The most popular of these programmes is the health insurance. Other types of social insurance include contributory pensions, unemployment insurance, funeral assistance, and disaster insurance. In Southern Africa, social insurance is primarily linked to the formalised labour market and seldom reaches the poor and informal workers—most of the population (Chikoko et al., 2022). It is also argued that social insurance programmes are too expensive for the poor and should be supplemented with social assistance (Ndhlovu, 2024b).

Labour market programmes comprise interventions that protect poor people who can work (Leburu & Skhosana, 2024). These interventions can be both active and passive: active programmes include training and skills development and employment counselling, while passive interventions include unemployment insurance, income support, and changes to labour legislation, such as establishing a minimum wage or safe working conditions (Zizzamia, 2020). These interventions can run alongside many social assistance and cash transfer programmes and be integrated into longer-term development strategies.

The concept of 'social protection' is very familiar in Southern Africa, where it primarily consists of non-contributory support being provided to the most vulnerable people in society to shield them from experiencing extreme poverty (Ndhlovu, 2024b). However, the success of social protection programmes depends on governments' support and active role in funding programmes and creating an enabling environment for other relevant actors to roll out such programmes.

6.3 Political Economy: A Theoretical Framework

The concept of political economy generally refers to the collective and interrelating effects of economic and political practices. The concept captures the link between politics and economic activities or practices. In the mainstream social discourses, Karl Marx was among the first philosophers to use the concept of political economy to interpret social, economic, and political developments. Marx (1973) used the political economy to understand the patterns of thinking that inform conduct and how people approach issues. As an ideology, political economy defines how a society deals with its fundamental political and economic challenges of what to produce, how to produce it, who to produce it, and how to achieve economic growth to decrease joblessness and poverty (Essia, 2012). In using the political economy approach, Marx sought to explain the phenomenon and flag how social relations need to be considered during progress.

Political economy subscribes to both communism and socialism. Socialism is based on either public or direct administration of the production resources and the distribution of resources by the workers themselves. Communism is when socialism transforms to the level where the 'privacy' of capital and assets finally vanishes and ceases to be imperative for the economy to function correctly (Bose, 2011). Marx (1973) argued that in deploying the political economy approach in the interpretation of the phenomenon, it is vital, as a starting point, to consider both the political and economic conditions of the country under study—its location, its population, its various classes, production patterns, export and import, consumption, and commodity prices. This could generate an understanding of the 'rich totality of many determinations and relations' as well as the 'concrete' understanding, which appears through 'iterations between conceptual abstractions and detailed empirical observation: [T]he concrete is concrete because it is the concentration of many determinations, hence unity of the diverse' (Marx, 1973, p. 41). This outlook on the political economy makes understanding phenomena such as social protection programmes in Southern Africa vital.

Notwithstanding the criticism against Marxism, the political economy approach demonstrates the important role of governments in shaping social, economic, and political developments in countries. The political economy theory is recommended for its ability to provide a refined, wide-ranging, and in-depth analysis that advances conventional 'economic concepts and theories to a point where the practical reality behind them (i.e., historical and social contents) become real and subject to critical attack' (Essia & Okoi, 2013, p. 63). The political economy theory can also be used to understand the 'many determinations and relations' (Marx, 1973, p. 41) inherent in social protection.

Several African scholars have drawn attention to the political economy of resource distribution through citizen-politician linkages in the acquisition and delivery of public goods by nation-states and access to the same goods by citizens (Bratton, 2007). These linkages (which may finally become informal channels for accountability and legitimation) are viewed as systems of reciprocity between politicians and ordinary citizens (Chabal, 1999). The linkages are not, of necessity, built upon the honesty and decency of the occupation of political offices, and neither are they based on legitimate channels of how state resources are acquired, but rather they are based on the political economy of how politicians fulfil their obligations to meet the expectations of their political capital (Ndhlovu, 2021).

Chabal and Daloz (1999) posit that most channels powerful oligarchs deploy to distribute state resources are biased. These powerful oligarchs' distribution of scarce resources is mainly enforced through formal state structures, which they abuse to reward their loyal supporters in top-down arrangements. These arrangements may, however, also be bottom-up as the same politicians intercede for the provision of and access to state resources between the state and its citizens (Chamunogwa, 2012). Shivji (2017) argues there are no straight cages between the two as they may be interlinked. It could be interesting to deploy the political economy theory in the analysis and interpretation of how social assistance was rolled out at the height of one of the deadly global pandemics (COVID-19) in Southern Africa. The political economy theory could be handy in understanding the various basic scopes inherent in social assistance programmes, ranging from the political membership of recipients; the sources of redistribution authority; the nature of institutions undertaking social assistance; channels and reasons for social assistance programmes; and competitions, exchanges, and negotiations within the acquisition and redistribution system. Therefore, despite the numerous criticisms, Marx's political economy theory provides a more helpful framework for understanding the provision of social assistance during COVID-19 in Southern Africa.

6.4 Methods and Materials

This chapter is based on a country-by-country review of social assistance interventions during COVID-19 in Southern Africa. Five countries, namely, South Africa, Zimbabwe, Mozambique, Malawi, and Zambia, were purposefully selected as case studies based on their politics, economic development, and the reputation of their social protection approach to COVID-19. The review draws on literature in both grey and academic formats. Reports by the governments of the selected countries, reports and reviews by non-governmental organisations, research and policy institutions such as the United Nations and its subsidiaries, the World Bank and the IMF, and the opinions by private citizens and academic publications were reviewed. The search keywords that were used were 'COVID-19', 'coronavirus', 'COVID-19 pandemic', 'COVID-19 response' and 'COVID-19 social assistance'. The search words were searched against each of the selected countries to ensure that each country was covered adequately in the analysis.

Search showed that the COVID-19 pandemic had received significant policy and academic attention. Many documents were identified for analysis all in the countries. Surprisingly, there were many academic records than all other types of texts. This was vital since it provided the authors to utilise peer-reviewed information.

Content analysis—a study of what is contained in the text—was used to understand the types of COVID-19 social assistance provided by countries, how they were implemented, and the nature of the political economy that emerged with such programmes. Although characterised by replicability and generalisability limitations, this approach afforded the author to provide an overview of the political economy nature of COVID-19 social assistance in the subregion.

6.5 Political Economy of COVID-19 Food Distribution and Cash Transfers

This review shows that cash transfers and food parcels were the primary social assistance interventions implemented in the selected countries in Southern Africa. This observation satisfies the first research objectives, which focused on identifying COVID-19-specific social assistance programmes in Southern Africa. The following country-specific sub-sections show how social assistance programmes have been implemented and how power dynamics, interest groups, and institutional frameworks shape social protection programme choices and implementation. Therefore, the sub-sections satisfy the second and third objectives of the review.

6.5.1 South Africa

In South Africa, the government was at the centre of COVID-19 social protection interventions, with minimal tolerance for the participation of private actors. The government wholly accepted the responsibility of safeguarding the lives and livelihoods of the people. The COVID-19 national lockdown and other associated prohibitions had disrupted most people's livelihoods and sources of income, ignited hikes in commodity prices due to limited access to them, and thus left people more vulnerable to the impact of the pandemic (Gustafsson & Nuga-Deliwe, 2020; Noyoo, 2021). To shield the poor from plunging into abject poverty, the government used available resources and infrastructure to expand its existing social protection programmes, particularly social assistance, and immediately boosted cash transfer and food packages to ensure that vulnerable groups, especially the poor and the unemployed, had sufficient resources to deal with the social and economic costs of the pandemic (Ndhlovu & Gumede, 2023). In South Africa, the state and its political structures play the primary role in providing social assistance. South Africa is one of the leading democracies in Africa and the second biggest economy, following Nigeria. These attributes worked well in mobilising the support of strategies by all relevant actors, especially opposition political parties and the private sector, and the resources needed to respond to the pandemic. This confirms the political economy nature of COVID-19 social assistance in the country.

South Africa rolled out a social relief and economic support package of R500 billion, amounting to around 10% of its GDP. From this amount, it pulled a special COVID-19 Social Relief of Distress Grant. Utilising available state social infrastructure, such as the social grant pay-points of the SASSA, vulnerable people continued to receive their grants, which were now accompanied by new relief measures. For instance, the existing Child Support Grant was raised from R440 to R740 per child in May 2020, while every caregiver received R500 a month from June to October 2020. Beneficiaries of other grants received an R250 top-up per month (Ramaphosa, 2020). About 17 million people vetted and qualified by the previous year in 2019 benefited from this social protection scheme (Rogan & Skinner, 2020). The South African government also implemented an R350 Special COVID-19 Social Relief of Distress Grant for non-grant recipients. These grants sustained most of the poor (Gustafsson & Nuga-Deliwe, 2020; Nyashanu et al., 2020). The Department of Social Development escalated access to food relief from an initial 1.1 million to 3.3 million vulnerable households impacted by the lockdown (Noyoo, 2021). The department also partnered with the Solidarity Fund, non-governmental organisations, and community-based organisations to distribute 250,000 food parcels nationwide (Ndhlovu & Gumede, 2023). At first, the government prohibited private actors from participating in food distribution, but it withdrew after public criticism (Hart et al., 2022).

Some private citizens and faith-based organisations opened soup kitchens, distributed food parcels, or volunteered their time to provide services to the vulnerable (Noyoo, 2021). Some private citizens specifically targeted foreign nationals, particularly those undocumented and who were being excluded from government programmes (Nzabamwita & Dinbabo, 2022). This ensured the survival of various categories of people who could not benefit from government-funded programmes.

However, the use of government structures and facilities to distribute resources was soon accompanied by a type of political economy that compromised the effectiveness of social assistance. Rampant corruption, patronage tendencies, and mismanagement at the local level negatively affected the food distribution, leading to the abandonment of the programme (Mudau, 2022). With the political economy of affection exercised by politicians and local leadership structures, food parcels became inadequate to curb hunger while the special grant was below the food poverty line (Adesina, 2021). The COVID-19 Social Relief of Distress Grant was intended for unemployed people in poor households. About ten million people were eligible for the ZAR350 grant per person per month. Around 60% of rejected applicants were eligible for grants, prompting researchers to conclude that the grant was poorly targeted and 'beleaguered by both errors of inclusion and errors of exclusion' (Moses & Wollard, 2023, p. 160). Thus, the grant and the food distribution became inadequate to support measures to ensure an adequate standard of living (Institute of Economic Justice (IEJ), 2021).

The significant challenges experienced within the South African social assistance system were related to the political economy that emerged with the targeting of recipients, with local politicians often redirecting food parcels to either their supporters or to their businesses for resale (Mudau, 2022). The use of political structures in distributing social assistance, particularly food parcels, resulted in unequal or lack of distribution. Reliance on government infrastructure for cash transfers also resulted in extensive delays in the payment of the special grant, with political interference often implicated in the delays (Nzabamwita & Dinbabo, 2022). Two months after the special grant was introduced, only ten people had been paid. At the same time, many eligible recipients struggled to access it until it was terminated (Institute of Economic Justice (IEJ), 2021). Targeting was also biased towards citizens, leaving out non-citizens, mostly legal poor migrants, with little support despite being equally affected by the pandemic (Nzabamwita & Ndhlovu, 2024). This confirms Chabal and Daloz (1999), who posited that most of the channels deployed by powerful oligarchs in distributing state resources are biased and mainly enforced through formal state structures, which they abuse to be able to reward their loyal supporters in top-down arrangements. Thus, since foreigners are not part of the potential political capital, they do not attract the compassion of politicians who lead and control social protection programmes.

6.5.2 Zimbabwe

COVID-19 arrived when Zimbabwe did not have a well-functioning social protection system. The country's response to the pandemic unfolded in intense social, economic, and political crises, with an economy that has been the worst in the subregion. Thus, the primary instruments employed by Zimbabwe were selective temporary food relief and cash transfers (Ncube & Murray, 2024). It also rolled out healthcare measures and provided an economic stimulus package, pension benefits, and cushioning allowances (Tom & Chipenda, 2020; Zamchiya et al., 2020). Food relief provision and cash transfers were based on the country's existing social food programmes, and recipients were, in theory, determined through means-testing and targeting (Chipenda & Tom, 2021). The government also played a central role in the distribution of assistance, with civil society organisations (local and international) supporting the government's response. With non-governmental organisations and the private sector suspected of funding opposition political parties, the ZANU-PF-led government closely monitored and censored relief distribution. All relief needed to be submitted to the state, which would then be distributed on behalf of well-wishers. This enabled the embattled regime to place itself at the centre of COVID-19 interventions and thus demonstrate its care for the people. Thus, for the ruling ZANU-PF, the COVID-19 pandemic was an opportunity to prove political relevance and gain electoral support by providing emergency social protection through formal government structures.

Zimbabwe's response to the COVID-19 social policy was primarily based on statutory instruments (SIs), economic recovery, and stimulus packages. This political economy approach generated enormous socio-economy implications. The general response was a Presidential Declaration of COVID-19 as a national disaster on 27 March 2020 (Mhlanga & Ndhlovu, 2020). Following the declaration, government ministries adopted initiatives that included budget reprioritisation, reconfiguring existing social protection initiatives, and developing new assistance programmes. Instead of declaring a state of emergency, which the parliament would have approved within 14 days, the president declared the pandemic a mere national disaster, thus retaining presidential powers to control COVID-19 interventions and avoid bureaucratic bottlenecks that would call for more transparency and accountability. This created opportunities for politically infused responses, thus subjecting COVID-19 responses to the political economy of affection, particularly in distributing food and cash transfers.

Notwithstanding coverage and insufficiency concerns, direct grain provision was rolled out such that, by April 2020, eight rural districts, most of which ZANU-PF (Zimbabwe African National Union-Patriotic Front) strongholds, had received 7114 metric tonnes of grain to reduce the food insecurity worsened by the pandemic. The initiative was later extended to urban areas, although cash payments substituted the grain allocation scheme. In the cities of Harare and Bulawayo, 21,431 food-deficient households, most of which were ZANU-PF-aligned, received a total of ZWL 3,870,089 in the first 2 weeks of the COVID-19 lockdown (Chipenda & Tom, 2021; Ndhlovu & Mhlanga, 2023).

In addition, ZWL 4,756,409 was paid to 26,140 beneficiary households in eight districts where, apparently, ZANU-PF had the most electoral votes in the previous election (Bindura, Kariba, Gwanda, Beitbridge, Kwekwe, Plumtree, Shurugwi, and Mvuma) (Ministry of Finance and Economic Development (MoFED), 2020). The programme had political connotations and was branded as the government's 'harmonised cash transfer programme', a terminology emanating from the previous 'harmonised presidential elections' in which Emmerson Mnangagwa was controversially declared the winner. Cronyism and nepotism manifested when food parcels were dispersed based on relationships rather than need. This was reported both in urban and rural areas. Both politicians from ZANU-PF and opposition political party officials engaged in these tendencies (Ndhlovu & Mhlanga, 2023). The distribution of social assistance along party lines undermined citizenry trust in the COVID-19 containment and reconstruction efforts by the government and, thus, ignited violent political protests across the country. ZANU-PF, therefore, used its position in government to instrumentalise COVID-19 social assistance (Mutekwe & Vanyoro, 2021), with ZANU-PF officials using social assistance to accumulate political capital through distributing food packages, while, in most cases, opposing political parties were denied the opportunity to do the same for their supporters.

Regarding cash transfers, Zimbabwe's COVID-19 cash transfer programme allocated ZWL 2.4 billion in the stimulus package. By 13 October 2020, 202,077 people had benefited. The government ran and funded the cash transfer programme through ZANU-PF, which was then used for populist motives (Chipenda & Tom, 2021). The targeting was not apparent. ZANU-PF's patronage politics often exercised. Initially, the government indicated that the beneficiaries would be identified from its social protection beneficiary databases and that the usual means-testing would identify new beneficiaries. The Minister of Finance and Economic Development also reported that some beneficiaries would be identified through a 'sophisticated algorithm' (Mudzingwa, 2020). However, in fact, lists of beneficiaries were received from the Minister of Women Affairs, local authorities, and informal traders. Lack of clarity on beneficiary selection created opportunities for patronage politics (Ndhlovu & Mhlanga, 2023). Thus, ZANU-PF capitalised on both the shattered economy and the pandemic to reach out to the electorate and prove that it still cares.

6.5.3 *Mozambique*

Because of the COVID-19 interventions adopted in Mozambique, from April to June 2020, 62,700 workers had their livelihoods and income negatively affected by the suspension of contracts, while another 77,489 workers were affected by the termination of contracts, and 43,579 by the closure of companies (National Statistics Institute, 2020). Many households across the country have also had their livelihoods disrupted (Nuvunga et al., 2024). The pandemic arrived when Mozambique was already facing a state-citizen crisis in which part of the population lacked trust in the central government for numerous reasons, including the 'hidden debt' scandal, regional inequalities, food insecurity and chronic malnutrition, a sharp depreciation of the national currency against the US dollar, and political-military instability in the central region of the country and the insurgency in the province of Cabo Delgado (Presidência da República, 2020). However, despite these situations, the government placed itself at the centre of social protection, confirming the political and economic nature of COVID-19 interventions. On 30 March 2020, a state of emergency was declared (Presidential Decree No. 11/2020) to allow the government to implement measures to combat the spread of COVID-19. The Minister for Gender, Children, and Social Action advised the National Institute of Social Action (INAS) and the Provincial Social Affairs Services to execute a sectorial plan at the national level to mitigate the negative socio-economic impacts of the pandemic on the most vulnerable population groups because COVID-19 interventions (United Nations Children's Fund (UNICEF) & International Labour Organisation (ILO), 2020). Basic Social Security was then provided by INAS through the Basic Social Subsidy Programme and Productive Social Action Programme, cash transfers; Direct Social Support Programme, food vouchers, in-kind transfers, and service payments; and Social Action Social Services Programme, assistance in social units and social services (República de Moçambique—Sistema de Protecção Social, 2019).

Mozambique allocated about US$ 240 million to transfers to families to guarantee social protection to the most vulnerable households during the pandemic (Nuvunga et al., 2024). Cash transfers were the highest priority in allocating social protection, absorbing about 34% of all COVID-19 funds. Due to resource constraints, Mozambique allowed external actors to fund non-contributory social protection (MGCAS, 2020a). The budget deficit forced the government to request US$ 700 million from development cooperation partners to finance the COVID-19 social protection plan. However, due to the lack of confidence of development partners in the government, as a result of the 'hidden debt' scandal and the lack of improvement in good governance, particularly in dealing with corruption and promoting transparency and accountability, the government failed to obtain financial support (Salvucci & Tarp, 2024). The government, however, mobilised substantial resources through loans and grants (Mulher, 2020). For instance, through its US$ 309 million credit approval document to contribute to the COVID-19 Response Plan (PRC19) budget, the International Monetary Fund (IMF) (2020) expressed a desire to see its financing (as a conditional loan) to not only help the country in the fight against COVID-19 but also to have a catalytic effect on Mozambique's other development partners. Thus, the IMF merged transparency and accountability mechanisms into the agreement, including the publication of large public procurement contracts and the carrying out and publication of audits.

Mozambique adopted the Basic Social Security programmes managed by INAS to provide cash transfers. The Ministry of Gender, Children, and Social Action detailed this measure as non-contributory social protection executed through external funds and that, further to extending cash transfers to new beneficiaries (for 6 months), 3-month cash transfers would also be provided to PSSB and PASP beneficiaries (MGCAS, 2020b). The criteria for receiving cash transfers were that households had to be in poverty and vulnerable and live in urban, peri-urban, and border areas. Priority was given to households headed by older people, people with chronic and degenerative diseases, people with disabilities, children, pregnant women without a source of income, and women with six or more dependents (Oliveira & Cravo, 2021; Salvucci & Tarp, 2024). Furthermore, households that hosted internally displaced persons were also given priority (Mulher, 2020). In addition, internally displaced people/refugees, disabled people, children in challenging situations, older people in absolute poverty, disabled people in absolute poverty, people with chronic and degenerative diseases, people benefiting from Basic Social Security programmes, women/women heads of households, and people living on the streets were also targeted as beneficiaries (MGCAS, 2020a). The Ministry of Gender, Children, and Social Action also spearheaded the distribution of food baskets corresponding to 3 months of food to existing beneficiaries through the Direct Social Support Programme (PASD), initially for 18,438 beneficiaries (Nuvunga et al., 2024).

With the dominance of the government in the distribution of cash transfers and food, however, numerous challenges soon emerged, highlighting how political economy can easily result in run-away projects. The report identified two major inconsistencies, 'Status of Commitments under COVID-19 – Maputo' (December 2020). First, it was noted that although the periodic broadcasting of information was acknowledged as a good practice, the updates provided by the government only focused on the funds received to implement the response plan to COVID-19 and did not report on every fund applied in response to COVID-19 in the country. The Front for Liberation of Mozambique (FRELIMO) Government only emphasised aspects that gave it political leverage. It failed to report on interventions by development cooperation partners, whose actions had the same goals. Second, as COVID-19 worsened, the government shifted focus from the poor and vulnerable towards sustaining businesses (Betho et al., 2022; Salvucci & Tarp, 2024). Initially, 34% (USD 240 million) of the total financing volume was allocated to social protection. However, only 7% (USD 38.9 million) was actually channelled to social protection. The shift in budget prioritisation weakened the original focus on targeting vulnerable groups, possibly with funds being redirected to businesses owned by politicians, their families, or supporters. Nuvunga et al. (2024) argue that the change in funds allocation signals a change in the government's understanding of the allocation priorities for COVID-19, which had sacrificed the funding provided to protect vulnerable households from financing the state budget above what was initially planned. Explanations for this shift were not provided.

It has also been argued that since social protection programmes targeted poor and vulnerable populations (Ministry of Economy and Finance (MEF), 2020), the most critical allocations were supposed to be directed to the provinces with a higher percentage of poor and vulnerable households (Barletta et al., 2021). However, social protection allocations to the most disadvantaged provinces were not equal, thus exposing bias towards provinces and areas that are FRELIMO's electoral strongholds. Allocations to provinces with the country's highest poverty rates remained insufficient to meet the needs of their vulnerable populations (Nuvunga et al., 2024). For instance, although the national average allocation per poor person was about MZN 464 per year, the allocation for the poor was only MZN 327 in Niassa and MZN 375 in Nampula, while in Maputo it was MZN 959 per year (UNICEF & ILO, 2020).

Despite the lack of sufficient resources to make meaningful social protection interventions, Mozambique could still exhibit political economy actions by directly monitoring the activities of the private sector and non-governmental organisations. The government also sources funds from international institutions and private partners to respond to the pandemic. The political economy of affection emerged as provinces and districts dominated by poor and vulnerable people did not receive commensurate social protection attention, while perceived FRELIMO strongholds such as Maputo received even higher payouts. This situation speaks to the instrumentalisation of COVID-19 social assistance to make political statements and to sustain political capital through preferential resource distribution.

6.5.4 Malawi

On 4 April 2020, the Malawian government implemented a partial lockdown, including the closure of borders. The High Court prohibited the implementation of a complete lockdown due to a lack of social protection measures to help the poor and vulnerable (Magalasi, 2021). In response to the livelihoods of the people, particularly the self-employed and peasant populations, that had been disrupted by the pandemic and associated lockdown, the government implemented an Emergency Urban Cash Intervention (CUCI), amounting to about $50 million (0.6% of GDP) to support small businesses in major urban areas (GOM, 2021). The intervention targeted peri-urban areas covering about 172,337 households, each receiving a monthly sum of K35,000.00. This was the only new government policy response to COVID-19. The CUCI programme, which was an unconditional transfer that provided a critical safety net for the most marginalised urban Malawians, first focused on urban areas. It aimed to decrease the adverse impact of the pandemic on livelihoods, human capital accumulation, and essential consumption (Zidana & Luka, 2023). Targeting about 35% of the country's urban population—about 200,000 people—the programme's admissibility criteria focused on households that were likely to lose livelihoods due to the pandemic (Sahle & Kayaitsa, 2023). Each vulnerable household received a monthly transfer of MK 35,000 for 3 months in 2020.

There are, however, mixed views on the impact of CUCI in Malawi. The United States Agency for International Development (USAID) reported the programme as highly effective in providing immediate relief to vulnerable households in informal settlements during the pandemic. It is argued that the cash transfers were a lifeline for individuals whose income and livelihood sources were disrupted, enabling them to meet their basic needs (USAID, 2020). Additionally, the cash transfers radically impacted the gender equity front (Charlot & Mwamlima, 2024). On the other hand, academic assessments showed that emergency cash transfers had a minimal impact on offsetting the increases in poverty and inequality because they were inadequate (Magalasi, 2021). Across Africa, the outcomes were reported for Nigeria, South Africa, and Zimbabwe (Adesina, 2021; Chipenda & Tom, 2021).

However, like in other African countries where COVID-19 interventions were implemented, a political economy soon emerged regarding the targeting of recipients. The targeting of urban areas (mainly Zomba, Lilongwe, Mzuzu, Lilongwe, and Blantyre) betrays the ruling party's bias towards its supporters who are concentrated in urban areas while neglecting rural dwellers who are experiencing more extreme poverty (Chawinga et al., 2023). The Malawi government, which also has the backing of traditional leadership, allowed community leaders, mainly customary and block chiefs, to be involved in the selection process of beneficiaries. Customary chiefs are influential, despite the Chiefs Act 1967, which prohibited them from exercising their authority in urban areas. These community leaders decided who should benefit from the programme, characteristically exercising the political economy of affection by registering members of a particular political party, their family members, and those who bribed them and excluding the most deserving households (Mangal et al., 2021). The programme must utilise a community-level decision-making approach rather than individual leaders to select recipients. Governance institutions such as the ward civil protection committee should instead be used in these social protection programmes to offset the disruptive role of preferential political economy in the distribution of resources.

There was also the political economy of gender targeting in CUCI. This aligns with the 2030 Agenda for Sustainable Development, which promotes social protection systems and measures for all. The argument was that by targeting female family members, the programme would allow women to start small-scale businesses or make financial decisions based on the funds they received (Mangal et al., 2021). This tended to empower and liberate many vulnerable community members.

However, money disbursements through mobile technology soon created challenges since many women used their husbands' phones as they lacked their own. Men were reported as abusing these funds (USAID, 2020). In addition, some men who were extremely poor and deserving were also excluded from the programme, thus making them more vulnerable to COVID-19 socio-economic impacts.

The Malawian COVID-19 social assistance also faced numerous execution political economy-related challenges, including administrative and financial mismanagement by politicians and politically connected administrators. These challenges resulted in delayed payments, funds benefitting unintended individuals or groups, and sluggish programme roll-out.

6.5.5 Zambia

In Zambia, COVID-19 cases were first recorded in March 2020. Like other countries across the world, the government implemented response mechanisms to deal with various issues. In addition to the prohibition of gatherings and travels and national lockdowns, on 28 July 2020, the Government of Zambia launched the COVID-19 Emergency Cash Transfer (ECT) programme through the Ministry of Community Development and Social Services and the Ministry of Labour and Social Security (Ministry of Community Development and Social Services (MCDSS), 2020). The United Nations agencies and non-governmental organisations such as Plan International, ChildFund, and the Zambia Red Cross Society support the initiative. The programme was implemented through three shock-responsive social protection strategies: the vertical expansion strategy, the horizontal expansion strategy, and the alignment strategy.

Under the vertical expansion strategy, the ECT was given to all households that were already on the Social Cash Transfer (SCT) programme in the implementation districts (Pruce, 2021). This was because the traditional SCT was inadequate to support poor households during the emergency (World Bank, 2020). The SCT covered households with members who were severely disabled, chronically ill and on palliative care, headed by a single, headed by a child, or with an elderly member (Zambia Institute for Policy Analysis and Research (ZIPAR), 2020).

The horizontal expansion strategy considered additional beneficiaries who would be covered temporarily depending on the availability of resources (Viberti et al., 2021). Additional beneficiaries included those who had become vulnerable due to the COVID-19 pandemic (MCDSS, 2022a, 2022b). This included orphans and vulnerable children, informal workers, and other high-risk groups related to disability, chronic illness, and gender (ZIPAR, 2020).

Under the alignment strategy, recipients were linked to other relevant services during the emergency, such as information on health and COVID-19, support for personal hygiene, information on nutrition, support to protection from violence and abuse, and support to disability-specific needs during the emergency (MCDSS, 2022a, 2022b; Viberti et al., 2021).

Targeting was done by the National Cash and Livelihood Technical Working Group (NCWG) in coordination with MCDSS and district authorities, who assigned priority districts to different humanitarian agencies that were responding to the crisis (MCDSS, 2022a, 2022b). These agencies included UNICEF and the World Food Programme (WFP). Thus, although various agencies rolled out interventions, the government continued to play the overseer role. This confirms the political economy of COVID-19 response interventions in Zambia. Following assessments by agencies, food was identified as the priority need for the affected population. It was also observed that COVID-19 arrived when the government paused social protection support for communities. This caused many households to be unable to meet their most pressing needs when the pandemic struck (Kumar & Sakthivel, 2020). The discontinuity of social protection cash transfers had mainly been due to a lack of government funds. To cater to the most vulnerable households, agencies quickly aligned their programme with the government's pre-existing Social Cash Transfer programme, which had been discontinued. The National Society (NS) used the government's existing recipient lists as a starting point, identifying the most vulnerable households within communities, and then completed further verification and validation of these lists. In alignment with the government's Social Cash Transfer programme, the National Society also opted for cash as the response modality and used mobile money as the payment mechanism. This shows how agencies were quickly roped into reinforcing the political economy of COVID-19 interventions.

The government dominated the response process by getting involved in the screening processes. The screening was conducted in partnership with the MCDSS to ensure the selected recipients appeared on the government's Social Cash Transfer register. The programme required recipients to appear on both the Zambia Red Cross Society (ZRCS) recipient list and the MCDSS' Social Cash Transfer list before qualifying as beneficiaries. Representatives from the MCDSS participated in the registration, validation, and cash distribution process (ZRCS, 2021). The ZRCS also reinforced the political economy of the COVID-19 response in Zambia when it coordinated with the World Food Programme to ensure alignment between the programmes and recipient lists at each stage. During the registration and screening, ZRCS used the KOBO Mobile Data Collection tool; the National Society already had experience using this tool for data management. 1903 households met the primary and economic qualifiers set and were on MCDSS' Social Cash Transfer list, 1216 in Chililabombwe and 687 in Kalulushi; these households were registered and validated by ZRCS. A monthly transfer value of ZMW 410 (US$ 18) was set. The transfer was provided in two tranches. The NCWG agreed on the transfer amount and that it was in line with the MCDSS' Social Cash Transfer programme. The cash was distributed via MTN Zambia, a mobile phone company. Some recipients received their cash transfer from a designated ZRCS cash distribution centre.

6.6 Discussion

The review indicates that in all the selected five countries, the state played the central role in the implementation of social assistance programmes, thus confirming the views of Marx (1973), who argued that since it is the state that holds the central power of social, political, and economic relations, the state remains at the centre of socio-economic and political activities unfolding in any country. In all the five countries discussed above, the state remained influential in determining the provision of social assistance. Even in Malawi, Zimbabwe, and Mozambique countries, where the state did not have adequate funds to run assistance programmes, the state remained influential in determining how non-governmental actors and other agencies assisted the people. Therefore, the state played an active and primarily commendable role in responding to the impact of COVID-19 in Southern Africa. The effectiveness of the response was, however, closely tied to the condition of the economy and politics of each country. As argued by Onyina and Baye (2024), despite inadequate resources, countries managed to prevent and reduce poverty, vulnerability, and social exclusion among their people. However, this view is disputed by Magalasi (2021), who found that in Malawi social assistance had no meaningful effect on improving the situation of the people.

The review, however, also showed that in the face of inadequate resources, the state or government officials often tilt towards the preferential distribution of resources, thus exhibiting the political economy of affection in resource distribution. The pandemic allowed politicians to act in ways that maximised their self-interests by distributing scarce resources such as food and cash.

There were differences in how countries responded to COVID-19 through social assistance. For instance, although it had its challenges, South Africa managed to roll out support more smoothly and efficiently than other countries in the region. A well-developed democracy ensured the social protection system was well monitored, with officials and institutions held accountable for their actions. The country also had a functional social protection system, which had just expanded to respond to COVID-19. Thus, South Africa used its strong economy and political stability to respond effectively.

In contrast, politically unstable countries such as Zimbabwe and Mozambique, where accountability systems did not exist, allowed politicians to use the pandemic to thwart criticism and to use the available resources to boost their political capital by practising affection in the distribution of resources (Ndhlovu & Mhlanga, 2023). Non-governmental organisations were also monitored and forced to surrender aid to the government, which would then be distributed on their behalf (Viberti et al., 2021). This was particularly the case in Zimbabwe, where social assistance was rolled out along political party lines by the ruling party, with the ZANU-PF and various opposition political parties targeting their members in resource distribution.

Overall, the nature of COVID-19 and the type of interventions that were needed required the authoritative intervention of the state to implement lockdowns, enforce rules and regulations, provide resources, ensure that resources reach the neediest, and enforce accountability. Individual or private stakeholders could not carry out these activities without being at loggerheads with national constitutions. Thus, using its powers, the state could override specific rules and regulations and redirect resources to save lives and livelihoods. Therefore, Marx (1973) and other political economists (Chamunogwa, 2012; Essia & Okoi, 2013; Pruce, 2021) argue that the state will always remain central to social and economic relations.

6.7 Conclusion

This chapter explored the types of COVID-19 social assistance provided to countries, how they were implemented, and the nature of the political economy that emerged with such programmes. Underpinned by a review of documents in grey and academic formats, the review showed that the state was the major player in COVID-19 social protection interventions in Southern Africa. The review showed that cash transfers and food distributions were the major types of social assistance in the subregion. The state played a key role in implementing national lockdowns, enforcing rules and regulations, and sourcing and rolling out resources to save the lives and livelihoods of the people. This confirms the argument of Karl Marx and other political economists on the role of the state in social, economic, and political relations. The state also provided platforms for non-governmental organisations and private citizens to assist the poor and vulnerable. Overall, the state was the overseer of all COVID-19 social assistance programmes.

However, as would be anticipated, particularly in countries with economic and political crises such as Mozambique and Zimbabwe, COVID-19 interventions were soon used to boost political capital, punish opponents, and reward local supporters. In Zimbabwe, the ruling ZANU-PF and the then main opposition party, the Movement for Democratic Change, exercised the political economy of affection in the distribution of resources by preferring their supporters. In South Africa, some political officials prioritised citizens regardless of the devastating and sudden nature of the pandemic. At the same time, the most vulnerable and poor migrants, including those who were in the country legally, were sidelined.

This chapter confirms the importance of the state in responding to disruptive and devastating crises such as COVID-19. It, however, recommends that in times of crisis, the political economy of resource distribution be exercised for the collective good and that political leaders should not use social assistance during crises for point-scoring and amassing political capital. Future studies should explore how social assistance during crises can be managed in manners that do not allow politicians to use state resources for personal gain.

References

Abramo, L., Cecchini, S., & Morales, B. (2019). *Social programmes, poverty eradication and labour inclusion. Lessons from Latin America and the Caribbean. Lessons from Latin America and the Caribbean*. ECLAC Books, No. 155 (LC/PUB.2019/5-P), Santiago, Economic Commission for Latin America and the Caribbean (CEPAL).

Adesina, J. O. (2020). Policy merchandising and social assistance in Africa: Don't call dog monkey for me. *Development and Change*, 562–582.

Adesina, J. M. (2021). Why development and transformative social policy matter: Lessons of COVID-19 in Africa. *CODESRIA Bulletin Online*, No. 10, 1–9.

Ajam, T. (2020). More eyes on COVID-19: Perspectives from economics: The economic costs of the pandemic—And its response. *South African Journal of Science, 116*(7/8), Art. #8490, 2 pages. https://doi.org/10.17159/sajs.2020/8490

Arthur, W., Seekings, J., & Siachiwena, H. (2023). *Social protection in urban Zambia: The view from below*. Institute for Democracy, Citizenship and Public Policy in Africa (IDCPPA): Working Paper No. 33.

AU. (2010). *Social protection in Africa: An overview of the challenges*. AU Commission.

Banda, C., & Thelma, C. C. (2024). The effectiveness of social cash transfer on poverty reduction among the beneficiaries: A case of Lumezi District in Eastern Province, Zambia. *Global Scientific, Journals, 12*(3), 22–43.

Barletta, G., Castigo, F., Egger, E. M., Keller, M., Salvucci, V., & Tarp, F. (2021). *The impact of COVID-19 on consumption poverty in Mozambique*. 2021/94. WIDER Working Paper. https://doi.org/10.35188/UNU-WIDER/2021/034-4

Betho, R., Chelengo, M., Jones, S., Keller, M., Mussagy, I. H., van Seventer, D., & Tarp, F. (2022). The macroeconomic impact of COVID-19 in Mozambique: A social accounting matrix approach. *Journal of International Development, 34*, 823–860.

Bhorat, H., Oosthuizen, M., & Stanwix, B. (2021). Social assistance amidst the COVID-19 epidemic in South Africa: A policy assessment. *South African Journal of Economics, 89*, 63–81.

Bose, P. (2011). *Marx's capital and its relevance today*. PRAGOTI. Retrieved October 11, 2018, from http://www.pragoti.in/node/4271

Bratton, M. (2007). Formal vs. Informal institutions in Africa. *Journal of Democracy*, 18(3), 96-110.

Brauner, J. M., Mindermann, S., Sharma, M., Johnston, D., Salvatier, J., Gavenciak, T., Stephenson, A. B., Leech, G., Altman, G., & Mikulik, V. (2021). Inferring the effectiveness of government interventions against COVID-19. *Science, 371*, eabd9338.

Casale, D., & Shepherd, D. (2020). *The 'gendered effects of the ongoing lockdown and school closures in South Africa: Evidence from NIDS-CRAM waves 1 and 2*. Working Paper 5. NIDS-CRAM Wave 2 Reports.

Chabal, P. (2009). Africa: The politics of suffering and smiling. University of KwaZulu-Natal Press.

Chabal, P., & Daloz, J. (1999). *Africa works: Disorder as political instrument*. Indiana University Press.

Chamunogwa, A. R. (2012). *Political economy analysis of fast track land reform programme in Zimbabwe* (Unpublished master's dissertation, University of Sussex).

Charlot, C., & Mwamlima, B. (2024). *Evaluation of the African Risk Capacity payment process for the 2021–22 agricultural season in Malawi*. Africa Risk Capacity.

Chawinga, W., Singini, W., Phuka J, et al. (2023). Combating coronavirus disease (COVID-19) in rural areas of Malawi: Factors affecting the fight. *African Journal of Primary Health Care & Family Medicine, 15*(1), a3464. https://doi.org/10.4102/phcfm.v15i1.3464

Chawurura, T. (2020). *Review of targeting mechanisms of social assistance programmes in Zimbabwe and towards social cohesion in communities*. Development Solution.

Chikoko, W., Mwapaura, K., Zvokuomba, K., Kabonga, I., & Chinyenze, P. (2022). Contemporary social protection programmes among the vulnerable elderly in Zimbabwe: Review of the capability approach. *Journal of Public Policy in Africa, 9*(2), 23–36.

Chineka, T. S., & Kurevakwesu, W. (2021). Challenges for child welfare and development during the COVID-19 pandemic in Zimbabwe. *African Journal of Social Work, 11*, 209–215.

Chipenda, C., & Tom, T. (2021). *Zimbabwe's social policy response to Covid-19: Temporary food relief and cash transfers* (CRC 1342 Covid-19 Social Policy Response Series, 23). Universität Bremen, SFB 1342 Globale Entwicklungsdynamiken von Sozialpolitik/CRC 1342 Global Dynamics of Social Policy. https://nbn-resolving.org/urn:nbn:de:0168-ssoar-72580-2

Cirillo, C., Palermo, T., & Viola, F. (2021). *Non-contributory social protection and adolescents in lower- and middle-income countries: A review of government programming and impacts.* Innocenti Working Paper 2021-07, UNICEF Office of Research—Innocenti, Florence.

Essia, U. (2012). Entrepreneurial culturing of formal education programmes in Nigeria. *Journal of Sustainable Society, 1*(2), 52-62.

Essia, U., & Okoi, E. I. (2013). Marxist political economy and the crisis of relevance in contemporary world. *International Journal of Sustainable Development and World Policy, 2*(5), 59–76.

Government of Malawi & World Bank. (2021). Malawi COVID-19 impact monitoring report high frequency phone survey for period May to September 2020. In NSO & M. Gustafsson (Eds.), *How does South Africa's Covid-19 response compare globally? A preliminary analysis using the new OxCGRT dataset. Stellenbosch Economic Working Paper WP07/2020.* University of Stellenbosch.

Gustafsson, M., & Nuga-Deliwe, C. (2020). How is the COVID-19 pandemic affecting educational quality in South Africa? Evidence to date and future risks, Working Papers 23/2020, Stellenbosch University, *Department of Economics*. https://ideas.repec.org/p/sza/wpaper/wpapers358.html.

Hart, T. G., Davids, Y. D., Rule, S., Tirivanhu, P., & Mtyingizane, S. (2022). The COVID-19 pandemic reveals an unprecedented rise in hunger: The South African Government was ill-prepared to meet the challenge. *Scientific African, 16*, e01169. https://doi.org/10.1016/j.sciaf.2022.e01169

Haurovi, M. (2021). Social protection and a developmental state in South Africa. *Journal of Public Administration and Development Alternatives (JPADA), 6*(1), 13–27.

Hemsteede, R. (2024). Power relations in Malawi's social cash transfer programme: The flip side of domination. *European Journal of Development Research, 36*, 194–215. https://doi.org/10.1057/s41287-023-00598-6

Holmes, R., & Lwanga-Ntale, C. (2012). Social protection in Africa: A review of social protection issues in research. *A PASGR Scoping Study.*

Institute of Economic Justice. (2021). *Social protection during COVID-19: The rights to social security and an adequate standard of living*. Institute of Economic Justice.

International Monetary Fund. (2020). *External Sector Report: Global Imbalances and the COVID-19 Crisis. Washington, DC.* https://www.imf.org/-/media/files/publications/esr/2020/english/text.pdf.

Jatula, V., & Conshello, S. (2021). Democratic deficits and underdevelopment in Mozambique. *Inkanyiso: Journal of Humanities and Social Sciences, 13*(2), 242–255.

Kumar, A. A., & Sakthivel, R. (2020). The impact of social cash transfer on rural livelihood in Zambia. *Shanlax International Journal of Management, 8*(1), 1–7.

Lambon-Quayefio, M., Peterman, A., Handa, S., Molotsky, A., Otchere, F., Mvula, F., et al. (2024). Unconditional cash transfers and safe transitions to adulthood in Malawi. *World Development, 175*, 106483. https://doi.org/10.1016/j.worlddev.2023.106483

Leburu, L., & Skhosana, R. (2024). Wrestling with the present to re-imagine the future of social entrepreneurship in social work: A South African context. *African Journal of Inter/Multidisciplinary Studies, 6*(1), 1–13. https://doi.org/10.51415/ajims.v6i1.1292

Magalasi, C. (2021). *The short-term distributional impact of COVID-19 in Malawi*. EROMOND Working Paper Series EM 07/21.

Mangal, T., Whittaker, C., Nkhoma, D., et al. (2021). Potential impact of intervention strategies on COVID-19 transmission in Malawi: A mathematical modelling study. *BMJ Open, 11*, e045196. https://doi.org/10.1136/bmjopen-2020-045196

Marx, K. (1973). *Grundrisse*. Penguin.
MCDSS. (2022a). *Manual of operations for social cash transfer schemes*. Department of Social Welfare Headquarters.
MCDSS. (2022b). *Social cash transfer factsheet*. https://www.unicef.org/zambia/media/2571/file/Zambia-SCT-factsheet-2022.pdf086
MGCAS. (2020a). *INCLUDE equity COVID-19 case study questionnaire*. (CDD, Interviewer). https://www.surveymonkey.com/r/Preview/?sm=u3L3sj0ZNWxVZFBw99EBpQ5cH_2B_2FXC0UE0eAj8OlKA5TT3YpxGXl_2FyOZZnlVScZic
MGCAS. (2020b). *Plano de Resposta à COVID-19 em Moçambique—Protecção Social* [Response plan to COVID-19 in Mozambique—Social protection]. Ministério do Género, Criança e Acção Social.
Mhlanga, D., & Ndhlovu, E. (2020). Socio-economic implications of the COVID-19 pandemic for smallholder farmer livelihoods in Zimbabwe. *The Saharan Journal, 1*(2), 174–199.
Mhlongo, N. Z. D., Gumbo, T., & Musonda, I. (2022). Inefficiencies in the delivery of low-income housing in South Africa: Is governance the missing link? A review of literature. *IOP Conference Series: Earth and Environmental Science, 1101*(5), 052004. https://doi.org/10.1088/1755-1315/1101/5/052004
Ministry of Community Development and Social Services. (2020). National Policy on Persons with Disabilities [Internet]. Lusaka. Available from: https://research.zapd.org.zm/assets/80da057a-fda3-4508-934f-abbb886daac7
Ministry of Economy and Finance. (2020). *Ponto de situação dos compromissos no âmbito da COVID-19* [Status of Covid-19 commitments]. Ministério da Economia e Finanças. https://www.mef.gov.mz/index.php/covid-19/927%2D%2D229/file
Moses, M., & Wollard, I. (2023). The role of temporary social grants in mitigating the poverty impact of COVID-19 in South Africa. In P. Fourie & G. Lamb (Eds.), *The South African response to COVID-19: The early years* (pp. 156–177). Routledge.
Mudau, P. (2022). The implications of food-parcel corruption for the right to food during the Covid-19 pandemic in South Africa. *ESR Review, 23*(2), 4–9.
Mudzingwa, F. (2020). Mthuli Ncube and His "Sophisticated Algorithms" for corona relief funds—Another privacy disaster looming? *TechZim*, 26 April 2020. Retrieved April 17, 2021, from https://www.techzim.co.zw/2020/04/mthuli-ncube-his-sophisticated-algorithms-for-corona-relief-funds-another-privacy-disaster-looming/
Mulher, F. (2020). *As dimensões de género da COVID-19 em Moçambique* [The gender dimensions of COVID-19 in Mozambique]. Fórum Mulher. http://forumulher.org.mz/project/impacto-do-covid-19-na-vida-das-mulheres/
Mutekwe, P., & Vanyoro, K. P. (2021). Politicising 'COVID-19': An analysis of selected ZANU-PF officials' 2020-2021 media statements on the pandemic in Zimbabwe. *Acta Academica, 53*(2), 12–37.
National Statistics Institute. (2020). *Resultados do inquérito sobre o impacto da COVID-19 nas empresas* [Survey results on the impact of COVID-19 on businesses]. Instituto Nacional de Estatística. http://www.ine.gov.mz/estatisticas/estatisticas-sectoriais/resultados-doinquerito-sobre-impacto-da-covid-19-nas-empresas.pdf
Ncube, T., & Murray, U. (2024). Reflections on government-led social assistance programmes under Zimbabwe's National Social Protection Policy Framework: A social contract lens. *International Social Security Review, 77*(3), 59–97.
Ndhlovu, E. (2021). *Land, agrarian change discourse and practice in Zimbabwe: Examining the contribution of Sam Moyo* (PhD thesis, University of South Africa, Pretoria).
Ndhlovu, E. (2024a). Africa food system recovery challenges from COVID-19 amidst other key global challenges. *IAHRW International Journal of Social Sciences Review, 12*(3), 298–306.
Ndhlovu, E. (2024b). Social protection instruments and strategies in Southern Africa. *International Journal of Innovation in Management Economics and Social Sciences, 4*(2), 44–59.

Ndhlovu, E., & Gumede, V. (2023). Reflections on South Africa social policy response to the COVID-19 pandemic. In V. Gumede & E. Ndhlovu (Eds.), *Development in Africa: Remembering Thandika Mkandawire* (pp. 128–155). Africa Century Editions Press.

Ndhlovu, E., & Mhlanga, D. (2023). The politicization of the COVID-19 pandemic in Zimbabwe and implications on the attainment of the SDGs. In D. Mhlanga & E. Ndhlovu (Eds.), *Post-independence development in Africa decolonisation and transformation prospects* (pp. 365–385). Springer.

Nhapi, T. G. (2021). Examining the robustness of Zimbabwean social protection mechanisms in the context of Corona Virus Disease 2019 (COVID-19). *African Journal of Social Work, 11*(4), 240–249.

Noyoo, N. (2021). *South Africa's social policy response to Covid-19: Relief measures in an unequal society*. (CRC 1342 Covid-19 Social Policy Response Series, 21). Universität Bremen, SFB 1342 Globale Entwicklungsdynamiken von Sozialpolitik/CRC 1342 Global Dynamics of Social Policy. https://nbn-resolving.org/urn:nbn:de:0168-ssoar-72578-7

Nuvunga, A., Collet, A., & Sinoia, D. (2024). Mozambique. In A. Altaf, D. Tsikata, G. D. Torvikey, & M. Dekker (Eds.), *Equity in COVID-19* (EADI Global Development Series). Palgrave Macmillan. https://doi.org/10.1007/978-3-031-58588-3_10

Nyashanu, M., Simbanegavi, P., & Gibson, L. (2020). Exploring the impact of COVID-19 pandemic lockdown on informal settlements in Tshwane Gauteng Province, South Africa. *Global Public Health, 15*(10), 1443–1453.

Nzabamwita, J., & Dinbabo, M. (2022). International migration and social protection in South Africa. *Cogent Social Sciences, 8*(1). https://doi.org/10.1080/23311886.2022.2144134

Nzabamwita, J., & Ndhlovu, E. (2024). Social grants impact on household livelihoods: Empirical evidence from international migrants in Cape Town, South Africa. *E-Journal of Humanities, Arts and Social Sciences, 5*(14), 2521–2537.

Oliveira, M. C., & Cravo, T. A. (2021). Social protection and cash transfers in Mozambique: Between international consensus and local agency. *Cadernos de Estudos Africanos, 41*. https://doi.org/10.4000/cea.6474

Onyina, K., & Baye, R. S. (2024). Impact of social protection policies on inclusive growth in sub-Saharan Africa: Evidence from bias-corrected dynamic panel. *Cogent Economics & Finance, 12*(1), 1–26. https://doi.org/10.1080/23322039.2024.2409421

Presidência da República. (2020). *Comunicado de imprensa do conselho* [Council press release]. Gabinete de Imprensa. https://www.presidencia.gov.mz/por/Actualidade/PR-dirige-Segunda-Reuniao-Ordinaria-do-CNDS

Pruce, K. (2021). *Zambia's social policy response to Covid-19: Protecting the poorest or political survival*? (CRC 1342 Covid-19 Social Policy Response Series, 9). Universität Bremen, SFB 1342 Globale Entwicklungsdynamiken von Sozialpolitik/CRC 1342 Global Dynamics of Social Policy. https://nbn-resolving.org/urn:nbn:de:0168-ssoar-71878-7

Ramaphosa, C. (2020). *President Cyril Ramaphosa: Additional Coronavirus COVID-19 economic and social relief measures*. https://www.gov.za/speeches/president-cyril-ramaphosa-additional-coronavirus-covid-19economic-and-social-relief

República de Moçambique—Sistema de Protecção Social. (2019). *1° Boletim Estatístico sobre Protecção Social* [1st Statistical Bulletin on Social Protection]. Government of Mozambique. http://www.mitess.gov.mz/node/631

Rogan, M., & Skinner, C. (2020). *The COVID-19 crisis and the South African informal economy: 'Locked out' of livelihoods and employment*. Working Paper 10. NIDS-CRAM Wave 1 Reports.

Sahle, E. N., & Kayaitsa, M. S. (2023). COVID-19 in Malawi: Civil society mobilization for socio-economic rights and constitutionality. *Africana Public Interest Journal, 1*, 58–70.

Salvucci, V., & Tarp, F. (2024). Assessing the impact of Covid-19 in Mozambique in 2020. *The European Journal of Development Research*, 803–840. https://doi.org/10.1057/s41287-023-00613-w

Shivji, G. I. (2017). The concept of the working people. *Agrarian South, 6*(1), 1–13.

Taylor, V. (2008). Social protection in Africa: An overview of the challenges. African Union. https://globalepri.org/wpcontent/uploads/2016/07/Taylor2008AUSocialProtectionOverview.pdf.

Tom, T., & Chipenda, C. (2020). Covid 19, lockdown and the family in Zimbabwe. *Journal of Comparative Family Studies, 51*(3–4), 288–300.

UNICEF & ILO. (2020). *Informe orçamental 2020 do sector da acção social* [2020 budget report for the social action sector]. United Nations Children's Fund. https://www.unicef.org/mozambique/media/2936/file/Budget%20Brief%202020:%20Social%20Action.pdf

USAID. (2020). *Whitepaper series—Operational innovations in cash grants to households in Malawi*. USAID. https://www.usaid.gov/sites/default/files/2022-05/Give_Directly_Malawi_Program_Report_2020.pdf

Viberti, F., Daidone, S., Pace, N., & Sitko, N. (2021) *Cash transfers and women's economic inclusion: Experimental evidence from Zambia*. Food & Agriculture Organisation. Retrieved November 12, 2023, from https://www.fao.org/3/cb7510en/cb7510en.pdf

World Bank. (2020). *Monitoring COVID-19 impacts on households in Zambia*. World Bank Group.

Zamchiya, P., Mavhinga, D., Gwinji, T., Chamunogwa, A., & Madhuku, M. (2020). *Zimbabwe's COVID19 lockdown: Ensuring the right to food for the poor*. Institute for Poverty, Land and Agrarian Studies.

Zidana, H., & Luka, Z. (2023). *Redefining urban social protection programmes in Malawi: Lessons from the Covid-19 emergency urban cash intervention (CUCI)*. African Cities Research Consortium. https://opendocs.ids.ac.uk/articles/online_resource/Redefining_Urban_Social_Protection_Programmes_in_Malawi_Lessons_from_the_Covid-19_Emergency_Urban_Cash_Intervention_CUCI_/26514658?file=49609287

ZIPAR. (2020). *Analytical Brief of the 2021 Social Sector Budget—Coping with the dual effects of a recession and a pandemic: Safeguarding lives and livelihoods*. Zambia Institute for Policy Analysis and Research.

Zizzamia, R. (2020). Is employment a panacea for poverty? A mixed-methods investigation of employment decisions in South Africa. *World Development, 130*, 1–15.

ZRCS. (2021). *Emergency cash transfer response during COVID-19 experience of Zambia Red Cross Society*. British Red Cross.

Chapter 7
In Pursuit of a Future Healthy Nation: Post-COVID-19 Kenyan Healthcare Social Policy Review and Re-alignment

Jeremiah M. Muneeni

7.1 Introduction

The COVID-19 pandemic served as a critical lens through which African governments were compelled to examine the adequacy of their healthcare social policies in addressing large-scale health emergencies. The existing policy frameworks within African health systems proved insufficient for managing highly contagious diseases like COVID-19. Well-designed healthcare social policies constitute the primary mechanism for reducing infection rates, transmission, and mortality during pandemics. However, most African governments were unprepared for an outbreak of this magnitude, having experienced relatively few similar health emergencies in recent decades.

This inadequacy necessitated urgent policy re-thinking and re-alignment within the health sector to ensure future outbreaks could be managed effectively before causing widespread devastation. This study investigates the healthcare policies that existed before COVID-19 and evaluates their effectiveness in minimizing disease spread and mortality rates. In addition, the study utilizes pre-COVID-19 healthcare policies as a baseline to demonstrate the policy transformations that occurred during and after the pandemic to enhance preparedness for future health emergencies.

The study employs social constructionism theory to understand the dynamic interactions between various stakeholders that influenced pandemic spread and informed the development of sustainable healthcare policies. Data sources include relevant literature, government gazette notices documenting policy reviews, and new policy enactments informed by COVID-19 experiences. Given the continent's

J. M. Muneeni (✉)
Department of Literature, Linguistics and Foreign Languages, School of Law, Arts and Social Sciences, Kenyatta University, Nairobi, Kenya
e-mail: muneeni.jeremiah@ku.ac.ke

E. Ndhlovu, C. Chipenda (eds.), *Redefining the Social Policy Discourse*,
https://doi.org/10.1007/978-3-032-24067-5_7

diversity, this study focuses on the Republic of Kenya while drawing broader implications for African healthcare policy development where applicable.

7.2 Contextualizing the Research into Transformative Social Policy

The nature of COVID-19 and the comprehensive governmental response it required necessitated a mixed-method research approach to obtain authentic data. The study combined qualitative data from secondary sources with questionnaires distributed among healthcare stakeholders in Kenya to capture policy evolution at different temporal points. Primary data was obtained through questionnaires administered to healthcare leaders, specifically medical superintendents from two major referral hospitals. The questionnaires focused on healthcare social policy development before, during, and after the COVID-19 outbreak. Purposive sampling targeted key stakeholders at Kenyatta University Teaching, Referral and Research Hospital (KUTRRH) and Moi Teaching and Referral Hospital (MTRH). These medical practitioners served as primary consultants in policy development during and after the COVID-19 pandemic. The selection of these referral hospitals was based on their central role in pandemic response decision-making and patient management.

To obtain comprehensive data, the study examined Kenyan healthcare social policies across three periods: pre-COVID-19 policies, emergency policies developed during the outbreak, and post-COVID-19 policies designed as precautionary measures for future pandemics, including universal healthcare coverage initiatives. This data was supplemented with historical literature on disease outbreaks globally and in Kenya specifically, focusing on government-developed policies to counter pandemics historically, with special attention to COVID-19 and post-pandemic policy development. Data analysis employed interpretivism for questionnaire responses, corroborated with secondary materials from books, journals, and online sources, especially the Kenya Gazette, referral hospital libraries, and parliamentary acts. These institutions directly engage with government health policies.

Understanding the devastating impact of COVID-19 requires examining the pandemic's genesis and the global response, especially in Kenya. In December 2019, a new coronavirus (SARS-CoV-2) emerged in Wuhan, China, causing acute respiratory syndrome (COVID-19) in humans (Zhou et al., 2020:1058). Within 3 months, the virus infected over 118,000 people and caused 4291 deaths across 114 countries. This devastation prompted the World Health Organization (WHO) to declare it a global pandemic on March 11, 2020, following reports of alarming infection and death rates worldwide (WHO, 2020). This declaration created opportunities for countries to enhance existing healthcare social policies and develop new ones aimed at controlling transmission of this unprecedented pandemic. The pandemic declaration shifted focus to Africa, a continent often characterized by economic challenges that complicate adherence to strict public health measures. Indeed, during the early stages of COVID-19 in 2020, Lone and Ahmad predicted that Africa, as the poorest continent with populations most vulnerable to infectious diseases and inadequate medical infrastructure, would be significantly affected by the ongoing outbreak (Lone & Ahmad, 2020:1300). Their prediction was based on observations that the pandemic had severely impacted the world's most developed countries and posed increasing threats to low- and middle-income nations.

Surprisingly, Africa navigated through poor health infrastructure and other challenges to record relatively low mortality rates compared to other continents (Chipenda, 2023). The existing and revised healthcare social policies, while imperfect, proved instrumental in helping the continent manage the pandemic threat. As of April 2023, global confirmed infections totalled 732,790,388, with 6,897,012 reported deaths. Africa recorded 5,382,790 confirmed infections and 140,681 confirmed fatalities by April 14, 2023. While acknowledging potential reporting errors due to home deaths and some countries like Tanzania suspending case reporting in 2020, the actual figures were significantly lower than predictions. Against a global population share of 16.72% in April, Africa's infection share was 0.7% and fatality share was 2.04%.

Given these relatively low infection and fatality rates, contrary to WHO projections, this study investigates possible explanations for this phenomenon and examines how these factors influenced government strengthening of healthcare social policies for future preparedness. Since Africa is large and diverse, the research focuses on Kenya while acknowledging that the continent shares similar resource endowments and cultural characteristics, especially among third-world countries with interdependent cultures. Consequently, the research findings can be extrapolated to project broader African healthcare challenges and solutions.

The study's central focus is understanding why the predicted high mortality rate of COVID-19 in Africa never materialized. What changed the trajectory? Given global predictions of higher death rates than observed, what did Africa do differently? Did existing healthcare social policies contribute to reduced spread and mortality rates? Did continuous social policy revision influence this phenomenon? Has the continent continued revising healthcare social policies post-COVID-19 to ensure future pandemics pose reduced threats to citizens?

7.3 Conceptual and Analytical Framework

Over the past 4 years since the COVID-19 outbreak, academia has intensively examined its effects on various human life and society sectors within social science domains (Roychowdhury et al., 2022). Governments worldwide have tested their pre-existing social policies' viability in combating the pandemic. During and after the outbreak, governments have revised their healthcare social policies to align with best practices for reducing pandemic impacts. This study seeks to understand how Kenyan healthcare responded with new social policies to reduce pandemic devastation and how policy review continues despite declining infections to ensure effective positioning for future pandemic mortality and spread reduction. The study conducts a micro-level analysis of interactions between COVID-19 pandemic stakeholders and examines how healthcare social policies guided navigation through the pandemic. Micro-level analysis is a sub-strand within social constructionism, specifically within the sociology of illness approach.

Social constructionism helps examine how illness is socially constructed, leading to understanding of how social forces shape our perspectives on and responses to health, illness, and healing (Roychowdhury et al., 2022). It explores how class, race, gender, language, technology, culture, political economy, and institutional and professional structures influence the knowledge base that shapes our assumptions about disease prevalence, incidence, treatment, and meaning. Although social construction is recognized for understanding society-disease relationships, it is rarely defined systematically and encompasses many often-divergent meanings.

The first, and most prominent, version draws from Spector and Kitsuse's (1975:75) work, asserting that social problems are 'activities of individuals or groups making assertions of grievances and claims concerning some putative condition'. This perspective focuses on social definition rather than actual condition existence. Early social construction scholars sought to break from positivist worldviews that considered social facts and problems as given, measurable entities. This perspective invited debate about phenomenal 'reality' degrees by questioning whether social problems are real or created through purposive action by social labellers and problem identifiers. To address this concern, the viewpoint further divided social constructionism into contextual and strict constructionism. Best (1989) distinguishes between these approaches, arguing that 'contextual constructionism' acknowledges actual conditions without primary emphasis, while strict constructionism restricts itself to claims made by social actors rather than putative conditions themselves. For strict constructionism, even statistics providing evidence for condition prevalence are suspect, since such statistics are also constructions deserving no more acceptance than other claims.

Regarding this distinction, Woolgar and Pawluch (1985) argue that social constructionists maintain 'realist' positions by selecting certain issues for attention. These issues have consistent reality but become important only when scholars choose to analyse them. Consequently, strict constructionism challenges sociological realism by emphasizing that social realities are accomplished through interactional and interpretive practices (Miller, 1993). Medical sociology in the social constructionism tradition therefore focuses on discourse, which is logical since symbolic interactionism emphasizes human agency in meaning exchanges. This chapter focuses on meaning and interpretation creation since people's interactions centre on social role performance and relationships with professional and institutional structures where interaction occurs. These interactions guide policymakers in developing policies aimed at reducing and, where possible, eradicating pandemics.

The second constructivism version draws more from European postmodern theory, particularly Foucault's (1965) *Madness and Civilization*. The postmodern approach shares with American constructionism a positivism critique but is unaccommodating to constructionist emphasis on given sets of social actors, groups, and institutions. Postmodernists argue that focusing on these institutional and actor sets represents false reliance on existing structures. The postmodernist alternative deconstructs language and symbols to show knowledge creation and explore ever-changing, indeterminate situation 'realities'. The third approach to social constructionism follows Latour (1987), who argues that scientific fact production results from mutually conceived scientist actions in laboratory work, combined with scientists' efforts to promote their work publicly and officially. This 'science in action' perspective is useful for medical sociology in examining professionals' disease discovery work and medical technology development, though less valuable for considering lay discovery elements and patient-medical world interactions.

Social construction of phenomena involves multiple social forces combining to create and modify phenomena. This involves examining how phenomena were identified and addressed, including social stratification system impacts and the roles of professionals, institutions, governments, media, pharmaceutical companies, and people with illnesses and disabilities and their families. This study uses a complementary approach, selectively borrowing from three social constructionism levels for data analysis. Spector and Kitsuse's (1975) viewpoint that early social construction scholars sought to break from positivist worldviews considering social facts and problems as given, measurable entities will help demonstrate how Africa approached COVID-19 as a social problem requiring social solutions through sound, appropriate social policy development. Similarly, the study borrows from Foucault and his follower Armstrong, who examines knowledge production resulting from changing social perceptions of space and time, including proper medical care locations (home and institutions). This argument will be applied to examine new pandemic-combating knowledge that enabled development of new healthcare social policies meant to augment existing ones in reducing pandemic mortality rates and spread.

Latour's social construction of illness, narrowing to three levels—micro-level (involving self-awareness, individual action, and interpersonal communication), meso-level (hospitals and education), and macro-level (nation's health status, healthcare system structure and political economy, and national health system)—is this study's main theory. Although all three levels are equally important and complementary for comprehensive understanding, scope limitations prevent individual treatment of each level. The study borrows from different micro-, meso-, and macro-level components in analysing interrelationships between medical intervention, development, and healthcare social policy implementation aimed at reducing pandemic impacts. The study relies heavily on the macro level through its three categories: nation's health status, healthcare system structure and political economy, and national health system. Education as a meso-level component will be incorporated while individual action, a micro-level component, complements them. The study borrows sparingly from Spector, Kitsuse, and Foucault, but the main focus lies in Latour's social construction of illness with macro-level emphasis.

7.4 Pandemic Response and Policy Change

Pandemics and epidemic outbreaks have consistently influenced governments to adjust policies to accommodate resulting changes. Pre-COVID-19 policies were not necessarily tailored to counter new pandemic phenomena. Consequently, the Kenyan government, like others worldwide, possessed existing healthcare social policies that required adaptation for the unprecedented challenge. Healthcare social policies can be categorized into three phases aligned with progressive changes introduced to counter the pandemic: pre-COVID-19 healthcare policies existing before the outbreak; short-term healthcare social policies developed during COVID-19 that were continuously readjusted as new challenges emerged; and post-COVID-19 healthcare social policies aimed at ensuring future pandemics do not catch Kenya unprepared while attempting to achieve universal healthcare coverage for all residents.

The two medical superintendents from Kenya's major referral hospitals, directly involved in reviewing existing healthcare policies, provided important perspectives regarding their advice to government and key roles in policy formulation during and after the pandemic. Social policy refers to government actions addressing social needs including employment, education, healthcare, housing, and sustenance. Academically, social policy studies how governments distribute resources to address societal needs. This study narrowed its focus to healthcare social policy to examine its establishment before, during, and after COVID-19 outbreak. Policy encompasses government actions but extends to systematized interventions and actions by various agents operating in the public realm beyond governments or state agents. In defining, formulating, and implementing policies, actors include the state, social movements, various social forces, and 'civil society'—that constellation of public realm actors with identifiable stakes, interests, and positions in media, social movements, and similar institutions (Aina, 1997:26–27).

In light of this the study systematically examined policies existing before the outbreak, measures implemented during peak devastation, and policies established to prevent or reduce future pandemic impacts. Kenya achieved independence from Britain in 1963. The first president immediately declared war on three issues troubling Kenyans: poverty, ignorance, and diseases. Despite this declaration, sound healthcare social policies were not prioritized adequately. Consequently, healthcare has remained a privilege for those affording private hospitals or overseas treatment. For the majority seeking medical attention from public hospitals, experiences have been suboptimal (Oyaya & Rifkin, 2003). Key persistent problems include diseases such as malaria, tuberculosis, measles, and HIV/AIDS. Additional challenges include health sector corruption, inadequate resource allocation, and ineffective policies making quality healthcare access difficult. In addition, frequent health worker strikes, unattained practitioner-to-population ratios, unequal health facility distribution, and expensive prescriptions beyond common citizens' reach have been problematic.

These challenges have hindered Kenya's attainment of Sustainable Development Goal 3, which seeks to ensure healthy living and well-being for all individuals. While recognizing the government's effort to establish the National Hospital Insurance Fund (NHIF) in 2004 as medical cover for all Kenyans aligned with WHO's universal health coverage goals, it is important to note that accessibility remains limited despite income-based payment structures (Oraro-Lawrence & Wyss, 2020). Most Kenyans in the informal sector cannot access or afford health insurance, leaving them dependent on relatives and friends during hospitalization. Since its inception, NHIF has struggled to serve the growing population and expanding healthcare needs. Statistically, NHIF has failed to bridge the healthcare gap. According to Ministry of Health data (2019), only 11% of Kenyans are covered by this insurance programme, leaving 89% without government-subsidized health coverage. With over 70% of Kenya's workforce in the informal sector, the majority are either ineligible or cannot afford government-set premiums for health insurance provisions.

Chuma and Maina observe that since most Kenyans access healthcare through out-of-pocket payments, many avoid hospitals unless in advanced disease stages (Chuma & Maina, 2012). Given the characteristics of COVID-19 and the need for urgent action, such delays could result in catastrophic disease spread and infection. Dependence on relatives and friends has partially addressed health sector gaps but sometimes proves ineffective since mobilizing help for less popular members is not always successful. This unstable health sector, lacking sound healthcare social policies, faced COVID-19 in early 2020. The sector had to respond in ways that would help the vulnerable population survive pandemic effects. Consequently, Kenya's health sector devised both short- and long-term healthcare social policies to cushion Kenyans from pandemic devastation.

On January 23, 2020, the WHO's International Health Regulations Emergency Committee advised all nations to prepare for the new coronavirus in their countries (then called 2019-nCoV, now known as SARS-CoV-2). The WHO message stated: 'All countries should be prepared for containment, including active surveillance, early detection, isolation and case management, contact tracing and prevention of onward spread of 2019-nCoV infection and to share full data with WHO'. This was followed by WHO's declaration of COVID-19 as a public health emergency of international concern on January 30, 2020. The WHO declaration meant time and resources needed diversion toward pandemic prevention and treatment. This called upon governments, including Kenya, to develop short-term policies for pandemic spread containment.

The Kenyan government adopted several pandemic response strategies: border closures and international travel bans with cargo exceptions; school/learning institution closures; social gathering and meeting bans; dawn-to-dusk curfews; religious place, bar, and restaurant closures; and 1.5-m social distancing in gathering spaces. These were sociocultural approaches to combating COVID-19 that, while not unique to Kenya, were adapted to each country's context. Despite these measures, Kenya was considered among countries adopting moderate rather than stringent measures to balance intervention benefits and costs. For instance, while other countries imposed strict lockdowns, Kenya opted for dawn-to-dusk curfews and only restricted movement in counties considered epidemic hotspots. Pharmaceutical interventions were also applied, including testing, contact tracing, isolation and treatment, quarantine, universal mandatory mask-wearing in public spaces, and hand and cough hygiene. Regarding individual economic status, the government intervened directly by supporting vulnerable citizens through measures including direct cash transfers, tax relief, increased economic investment, and revised health insurance schemes and hospital ownership to reduce pandemic effects (Ouma et al., 2020:1). For a low-middle-income country like Kenya, these were not sustainable healthcare social policy measures that could remain in place for future medical emergencies.

When responding to questionnaires, the Medical Superintendent at Kenyatta University Teaching and Referral Hospital (KUTRH) described the hurriedly developed policies as 'knee-jerk reactions' that are not economically maintainable for a government heavily reliant on borrowing and grants for healthcare funding. He called upon the government to find lasting solutions, such as offering insurance to all Kenyans based on earning percentages. Similar sentiments were echoed by Moi Teaching and Referral Hospital (MTRH) staff, who felt COVID-19 had opened avenues for Kenya to rethink medical insurance policy. He noted that, together with his KUTRH counterpart, they had provided comprehensive reports and recommendations to parliament for enacting laws to improve the contributory medical scheme (NHIF), making it compulsory and including everyone, including the informal sector. He felt such strategy would cushion Kenyans against existing illnesses like malaria, tuberculosis, and HIV/AIDS, as well as future pandemics, having learned that if infections had reached earlier predictions, mortality rates could have been very high due to high treatment costs.

7.5 Post-COVID-19 Transformation Opportunities

The COVID-19 outbreak served as a gauge for countries to reflect on their healthcare social policies and review them for effectiveness in fighting future pandemics and epidemics. Kenya participated in this re-alignment, recognizing that existing healthcare social policies were inadequate for fighting disease spread and resulting deaths. Urgently developed pandemic-fighting policies were equally unsustainable due to challenges characterizing low-middle-income economies. Therefore, the government needed to develop long-term healthcare social policies capable of addressing both existing national health threats and future pandemics.

Governments worldwide responded variously to the aftermath of COVID-19. For the Kenyan government, the pandemic triggered rethinking and re-aligning universal insurance coverage to include all Kenyan citizens in both contribution and coverage. This coverage draws from Kenya's Constitution Bill of Rights, which assures every person the highest health standard, including healthcare service rights. The government, through the Ministry of Health, embarked on creating a fund to ensure lower healthcare costs since many Kenyans would have healthcare access. Although this action was overdue, it was highly triggered by the COVID-19 outbreak, which exposed the government's social health insurance as inadequate for covering expenses from testing, contact tracing, isolation, and treatment. This inability resulted from the government's average 9% health budgetary allocation over the past decade, despite signing the Abuja Declaration requiring 15% health budget allocation (MoH, 2020).

The government, through parliamentary action, created the Social Health Insurance Fund Act No. 16 of 2023 (SHIFA), assented to on October 19, 2023, and effective November 22, 2023. The Act establishes the Social Health Authority and a framework for managing social health insurance to give effect to Article 43(1)(a) of the Constitution. The Social Health Insurance Regulations, 2024, were established to operationalize the Social Health Insurance Fund (SHIF), Primary Healthcare Fund (PHF), and Emergency, Chronic, and Critical Illness Fund (ECCIF). To ensure fund effectiveness, registration has been made mandatory for every Kenya resident as provided by SHIFA, with contribution modes depending on individual resident earnings. All Kenya residents, regardless of citizenship, are entitled to access health services under PHIF and must register as SHIF members. SHIF funding comes from contributions from every Kenyan household, non-Kenyan residents staying 12 months, national and county governments, National Assembly appropriations for indigent and vulnerable persons, and gifts, grants, innovative financing mechanisms, or donations. Contributions are structured as follows: salaried members contribute monthly as statutory deductions at 2.75% of gross salary deducted by employers; non-salaried members contribute at least KES 300 (approximately $2.50). The authority, collaborating with the ministry responsible for cooperatives and micro, small, and medium enterprises development and other financial institutions, shall provide premium financing for non-salaried members to pay annual SHIF contributions. SHIF replaces NHIF and assumes insurance fund control.

Regarding future pandemic combat, the government's approach to emergencies and chronic diseases through ECCIF establishment is noteworthy. ECCIF shall cover chronic illness management costs after SHIF depletion and emergency treatment costs. ECCIF funding comes from National Assembly appropriations, grants, gifts, donations, endowments, and other lawful sources. SHIF members and beneficiaries are entitled to ECCIF benefits upon depleting SHIF benefits. Regulations propose that beneficiaries suffering from chronic illness shall access treatment per the benefit package. While lauding the Kenyan government for this noble initiative ensuring resident insurance coverage, the question remains whether initial problems characterizing previous medical coverage were addressed to prevent recurring challenges with the new insurance.

7.6 Lessons for African Healthcare Policy

The COVID-19 pandemic taught Kenya, and indeed the world, that assumptions about existing healthcare's ability to combat pandemic emergencies may be false. The biggest lesson learned, especially by the Kenyan government, is the need for sound healthcare policies to address current and future health sector emergencies.

Kenya's government therefore learned several crucial lessons: First, there is need to commit to supplementing pandemic treatment for all patients by establishing emergency funds. Such measures ensure that vulnerable society members such as slum dwellers and low-income earners are guaranteed treatment. Pandemic treatment costs are extremely expensive, and economic disruption from lockdown measures has interfered with many people's incomes. Government treatment supplementation would mean quicker pandemic solutions as infected individuals would receive professional care rather than family member care that may further spread disease.

Second, the government should commit to streamlining public health sector service delivery. Experience from previous endemics and pandemics shows that disease management responsibility has been largely left to international organizations and NGOs with limited government input. These organizations have limitations, and full dependence on them undermines health sector service delivery. The government needs to take bold steps and increase resources toward establishing sufficient, well-equipped health facilities and allocating adequate resources for sector research.

Third, corruption denies citizens services and can itself be considered a national crisis. Numerous Health Ministry and National Hospital Insurance Fund corruption reports leave no doubt why the country's health sector struggles. However, the importance of a healthy nation in creating wealth cannot be overemphasized. Health sector corruption must be addressed swiftly and decisively. Corruption perpetrators should not escape punishment as this sets bad precedents for public resource protection. If not addressed, even SHIF's wonderful concept will encounter its predecessor's problems.

Fourth, there is need to make universal healthcare truly universal. Previous challenges undermining national health coverage access include a lack of transparency, limited facility access accepting national insurance, and unavailable key medications in accepting facilities, forcing insurance holders to use private pharmacies at higher costs. Universal healthcare means government should facilitate easy access to reasonably good-quality healthcare without excessive obstacles. Therefore, SHIF must be managed to ensure seamless healthcare access.

Finally, there is need to emphasize increased recruitment and welfare of health personnel to enable health sector efficiency. This involves addressing periodic strikes and brain drain that have become normalized in the health sector by ensuring health workers are well-paid and provided proper protective gear for safe working environments. This should be accompanied by increased health practitioner distribution across the country to expand access. In addition, the government needs to engage related practitioners such as social workers and community development officers to supplement medical workers in remote areas.

7.7 Conclusion: Transforming Crisis into Opportunity

The COVID-19 pandemic exposed the structural inadequacies inherent in African healthcare systems and catalysed unprecedented policy transformations that will define the continent's health security landscape for decades to come. The Kenyan government, representing a microcosm of broader African healthcare challenges, demonstrated that existing healthcare social policies prior to COVID-19 were critically insufficient to address large-scale health emergencies of pandemic proportions. The convergence of endemic challenges including systemic corruption, persistent brain drain, inadequate medical infrastructure, and chronic shortages of qualified personnel created a perfect storm that threatened to overwhelm healthcare systems across the continent.

This crisis, however, also revealed the remarkable adaptability and resilience of African governments when confronted with existential threats. The Kenyan government, alongside other African nations, demonstrated exceptional responsiveness by rapidly developing context-specific healthcare social policies that significantly mitigated transmission rates and mortality figures, outcomes that defied global predictions of catastrophic losses. This policy agility represents a paradigm shift from reactive to proactive healthcare governance, significantly altering the relationship between state capacity and public health emergency response. The pandemic's most significant legacy lies in its role as a catalyst for comprehensive healthcare system transformation. Kenya's transition from the NHIF to the SHIF represents more than administrative restructuring; it constitutes a reimagining of healthcare as a universal right rather than a privilege. This shift toward truly universal healthcare coverage, encompassing all residents regardless of socioeconomic status, signals a broader continental movement toward health equity and social justice.

From a theoretical perspective, the application of social constructionism theory has revealed how the pandemic served as a 'critical juncture' that disrupted established policy equilibria and created opportunities for transformative change. The micro-level individual adaptations, meso-level institutional responses, and macro-level policy reforms observed in Kenya illustrate the dynamic nexus between structure and agency in health policy transformation. This multilevel analysis demonstrates that effective pandemic response requires not merely technical interventions but a restructuring of the social contracts between states and citizens.

Looking toward the future, several critical research directions emerge from this analysis. First, longitudinal studies are essential to evaluate the long-term sustainability and effectiveness of newly implemented healthcare social policies, especially SHIF's capacity to deliver on its universal coverage promises. Second, comparative research across African contexts would illuminate how different political economies, cultural contexts, and institutional capacities shape pandemic preparedness and response strategies. Third, investigation into the political economy of health policy reform, especially the role of international organizations, donor influences, and domestic political dynamics, would provide important perspectives into the sustainability of post-pandemic transformations.

The emergence of the Emergency, Chronic, and Critical Illness Fund (ECCIF) represents an important innovative approach to pandemic preparedness that warrants further scholarly attention. This fund's design acknowledges that future health emergencies will require dedicated financial mechanisms that can be rapidly activated without disrupting routine healthcare services. However, critical questions remain about the fund's governance structures, accountability mechanisms, and capacity to respond to simultaneous multiple crises. This study also reveals the urgent need for addressing the political dimensions of healthcare transformation. While technical solutions like SHIF represent important advances, the persistence of corruption, inadequate resource allocation, and weak governance structures threatens to undermine even the most well-intentioned reforms. Future research must therefore examine how accountability mechanisms, transparency initiatives, and citizen participation can be strengthened to ensure that healthcare social policies translate into tangible improvements in health outcomes.

The continental implications of Kenya's experience extend beyond national borders. As African nations increasingly recognize that health security is interconnected, there is growing potential for regional cooperation in pandemic preparedness, knowledge sharing, and resource mobilization. The African Union's continental health initiatives, combined with national-level reforms like those observed in Kenya, suggest an emerging architecture for collective health security that could transform the continent's capacity to respond to future pandemics. Significant challenges however remain. The sustainability of expanded healthcare coverage depends critically on economic growth, fiscal discipline, and political stability, factors that remain uncertain in many African contexts. Moreover, the global health architecture continues to be dominated by external actors, raising questions about the long-term autonomy and sustainability of Africa-led health initiatives. Perhaps, most importantly, this study demonstrates that effective pandemic preparedness requires more than technical fixes; it demands a transformation of the social contract between states and citizens. The success of Kenyan healthcare reforms will ultimately depend not merely on financial mechanisms or administrative structures but on whether these changes can foster greater social cohesion, trust in government institutions, and collective commitment to public health as a shared responsibility.

In conclusion, while the COVID-19 pandemic exposed vulnerabilities in African healthcare systems, it simultaneously created unprecedented opportunities for transformative change. Kenya's experience suggests that crisis can indeed catalyse innovation, but realizing the full potential of these reforms will require sustained political commitment, adequate resource mobilization, and continuous adaptation to emerging challenges. The next decade will be crucial in determining whether the policy innovations born from this pandemic can establish Africa as a leader in global health security or whether the continent will remain vulnerable to future health emergencies. The stakes could not be higher; the health and wellbeing of over 1.3 billion Africans depends on getting this transformation right.

References

Aina, L. (1997). The state of civil society: Politics, government, and social organization in African cities. In *The urban challenge in Africa: Growth and management of its large cities* (pp. 411–446). United Nations University Press.

Best, J. (1989). Afterword. In *Images of issues: Typifying contemporary social problems* (pp. 243–252). Aldine de Gruyter.

Chipenda, C. (2023). Rethinking the role of social policy in responding to the Covid-19 pandemic in South Africa: Critical reflections using the intellectual contribution of Thandika Mkandawire. In V. Gumede & E. Ndhlovu (Eds.), *Development in Africa: Remembering Thandika Mkandawire* (pp. 197–231). Ace Press.

Chuma, J., & Maina, T. (2012). Catastrophic health care spending and impoverishment in Kenya. *BMC Health Services Research, 12*, 413. https://doi.org/10.1186/1472-6963-12-413

Foucault, M. (1965). Madness and civilization: A history of insanity in the Age of Reason (R. Howard, Trans.). Random House.

Latour, B. (1987). Science in action: How to follow scientists and engineers through society. Harvard University Press.

Lone, S. A., & Ahamad, A. (2020). Coronavirus disease 2019 (COVID-19): A literature review. *Journal of Infection and Public Health, 13*(5), 667–673.

Miller, G. (1993). Social constructionism and its critics: Assessing recent challenges. In *Reconsidering debates in social problems theories* (1st ed., pp. 535–547). Routledge.

Ministry of Health, The Republic of Kenya. (2020). *National and country health budget analysis*. Retrieved February 2, 2024, from http://www.healthpolicyplus.com/ns/pubs/1130611563_NationalandCountyBudgetAnalysis.pdf

Oraro-Lawrence, T., & Wyss, K. (2020). Policy levers and priority-setting in universal health coverage: A qualitative analysis of health-care financing agenda setting in Kenya. *BMC Health Services Research, 20*, 182. https://doi.org/10.1186/s12913-020-5041-x

Ouma, P. N., Masai, A. N., & Nyadera, I. N. (2020). Health coverage and what Kenya can learn from the COVID-19 pandemic. *Journal of Global Health, 10*(2), 020362. https://doi.org/10.7189/jogh.10.020362

Oyaya, C. O., & Rifkin, S. B. (2003). Health sector reforms in Kenya: An examination of district level planning. *Health Policy, 64*, 113–127. https://doi.org/10.1016/S0168-8510(02)00164-1

Roychowdhury, K., Bhanja, R., & Biswas, S. (2022). Mapping the research landscape of Covid-19 from social sciences perspective: a bibliometric analysis. *Scientometrics, 127*(8), 4547–4568. https://doi.org/10.1007/s11192-022-04447

Spector, M., & Kitsuse, J. I. (1975). Constructing social problems. New York: Aldine de Gruyter.

Woolgar, S., & Pawluch, D. (1985). Ontological gerrymandering: The anatomy of social problems explanations. *Social Problems*, 214–227.

World Health Organization. (2020). *Statement on the meeting of the International Health Regulations (2005) Emergency Committee regarding the outbreak of novel coronavirus 2019 (n-CoV) on January 23, 2020*. https://www.who.int/news-room/detail/23-01-2020-statement-on-the-meeting-of-the-international-health-regulations(2005)-emergency-committee-regarding-the-outbreak-of-novel-coronavirus-(2019-ncov)

Zhou, F., et al. (2020). Clinical course and risk factors for mortality of adult inpatients with COVID-19 in Wuhan, China: A retrospective cohort study. *Lancet*, 1054–1062.

Chapter 8
Ethical and Technological Innovations for Social Protection in Africa

Theshaya Naidoo

8.1 Introduction

The confluence of persistent socio-economic challenges across Africa has created an urgent imperative for reimagining social protection systems, particularly within the context of redefining social policy discourse in contemporary Africa. As Chipenda (Introduction in this volume) argues, contemporary Africa is at a critical juncture especially in relation to its social policy discourse, as the continent confronts an unprecedented convergence of global disruptions and local imperatives that demand transformative responses. This convergence necessitates moving beyond traditional social protection approaches that have been constrained by structural dependencies and donor-driven agendas, toward more integrated technological and ethical innovations that can address systemic challenges. Persistent challenges like weak infrastructure, poor governance, and recurring natural disasters undermine long-term development. Cyclone Idai hit Mozambique in 2019, triggering widespread flooding, cholera outbreaks, and significant economic setbacks simultaneously. In the Sahel, erratic rains and Boko Haram conflict disrupt livelihoods simultaneously; in southern Africa, ongoing power blackouts and COVID lockdowns leave families without basic needs. These crises exposed how fragile systems failed to respond quickly, leaving millions without essential support. Even before the COVID-19 pandemic, the combined impact of extreme weather and health emergencies had already driven nearly half a billion people in Africa into extreme poverty. These realities align with observations that the COVID-19 pandemic exposed the limitations of narrowly conceived social policies and highlighted the dangers of divorcing economic policy from social considerations (see Muneeni and Tekwa in this volume). There is an urgent need to rethink how support is designed and implemented, so that it can respond better to complex and overlapping crises.

T. Naidoo (✉)
School of Law, University of KwaZulu-Natal, Pietermaritzburg, South Africa

E. Ndhlovu, C. Chipenda (eds.), *Redefining the Social Policy Discourse*,
https://doi.org/10.1007/978-3-032-24067-5_8

Traditional approaches to social protection have proven inadequate in addressing the scale and complexity of contemporary challenges. This is in the context of the broader patterns of African social policy development, which are heavily influenced by donor agendas and external policy recommendations. Governments have traditionally relied on tools like cash transfers, public works, and food aid to respond to emergencies. However, in many African countries, these systems remain fragmented and reactive rather than integrated and proactive. This fragmentation exemplifies the residual social policies that inadequately address systemic challenges and often prioritise donor preferences over locally defined priorities. For instance, Nigeria's cash transfer pilots were limited to specific states and struggled to expand quickly during the COVID-19 crisis because the beneficiary registries were outdated. In Ethiopia, the well-established Productive Safety Net Programme (PSNP) still depends on lists updated only once a year, making it difficult to reach new households suddenly pushed into poverty by events like cyclones. In both cases, governments did not fully use digital tools such as mobile-based targeting or dynamic, real-time registries, which could have helped identify and assist affected populations more quickly and accurately.

This chapter examines how emerging digital technologies can transform social protection systems while addressing critical ethical considerations that have been largely overlooked in existing literature, contributing to the broader discourse on redefining social policy in contemporary Africa. The analysis reveals significant gaps in current knowledge, especially regarding the intersection of technological innovation and human rights protection in African contexts. While substantial research exists on individual technologies such as mobile money or blockchain applications, there is limited scholarly attention to their integrated application within comprehensive social protection frameworks that can address the structural constraints. Furthermore, ethical implications of large-scale technological deployment in settings with weak regulatory frameworks remain underexplored, creating a critical knowledge gap that this chapter seeks to address in alignment with the volume's focus on transformative social policy approaches.

The research questions guiding this exploration centre on three interconnected dimensions of technological integration in social protection systems, addressing the need for more adaptive and locally owned policy frameworks. First, how can emerging technologies address traditional social protection systems' inefficiencies and coverage gaps while ensuring sustainable implementation that reduces dependency on external donor-driven approaches? Second, what ethical considerations must be prioritised to ensure that technological adoption does not exacerbate existing inequities or create new forms of exclusion, especially given the continent's experience with structural dependencies that have historically undermined comprehensive social policy development? Third, how can African governments and institutions build the capacity to effectively implement and govern digital social protection systems in ways that protect beneficiary rights and promote inclusive development while enhancing local policy autonomy?

8.2 Conceptual Framework and Contextual Analysis

8.2.1 The Imperative for Technological Innovation in Social Protection

The socio-economic landscape of Africa has undergone significant transformations over the past two decades. The parameters of social policy have been redefined by rapid urbanisation, demographic shifts characterised by a youthful population, and a growing reliance on digital technology. Despite this progression, inequalities and vulnerabilities persist, as evident by statistics. The International Labour Organization (ILO) reports that only 17.4% of the African population is effectively covered by at least one form of social protection (Shawa, et al., 2024), compared to a global average of 46.9%. This disproportionate disparity demonstrates the necessity of transforming social protection systems to prioritise inclusivity, equity, and responsiveness.

Integrating technological innovations into social protection systems offers a transformative opportunity to address some of these gaps, yet the potential remains largely untapped due to the structural constraints that have historically limited African states' policy autonomy. Digital technologies, such as artificial intelligence (AI), blockchain, and mobile money, have the potential to introduce new systems and improve the operation of existing social protection systems. However, their implementation must be understood within the broader context of persistent neoliberalism and neocolonialism that continue to shape both social policy discourse and implementation across the continent. This is in a context where mobile money platforms like M-Pesa (Ndung'u, 2018) have revolutionised financial inclusion by facilitating accessibility to social transfers and savings tools for individuals in remote areas. Similarly, blockchain technology promises enhanced transparency and reduced corruption (Kshetri, 2021) in distributing social benefits. However, while these technology tools can address inefficiencies, their broad implementation is hindered by limited digital infrastructure, low literacy levels, and concerns about data privacy and cybersecurity, alongside the deeper structural dependencies that have historically undermined comprehensive social policy development.

This chapter adopts a multidisciplinary analytical framework that synthesises insights from social policy, technology studies, and data ethics, contributing to the broader theoretical discourse on transformative social policy approaches in Africa. Adopting a desktop research methodology, this chapter synthesises insights from a multidisciplinary lens. It comprehensively analyses how technological and ethical innovations can reshape Africa's social protection systems by reviewing existing literature, case studies, and empirical evidence while addressing the need for more integrated approaches that can simultaneously accelerate SDG achievement and lay foundations for post-2030 development priorities. This analysis is contextualised within Africa's unique socio-economic context. It is ideally positioned to propose actionable solutions that harmonise technological innovation with ethical imperatives, fostering resilient and inclusive social protection frameworks that can enhance local policy autonomy and reduce dependency on external donor-driven approaches. This approach enables a holistic understanding of the potential benefits and risks of integrating digital tools into these systems while contributing to the redefinition of social policy discourse in contemporary Africa.

8.2.2 *Africa's Complex Development Landscape*

8.2.2.1 Socio-economic and Institutional Challenges

From a socio-economic and institutional perspective, social protection programmes implemented within Africa face significant structural challenges that extend beyond mere resource constraints. These challenges are rooted in persistent inequality, broad poverty, and systemic exclusion (Hoddinott et al., 2012). Chronic inequality across Africa means that social protection programmes often fail to reach those most in need, with evidence from large-scale data showing that benefits tend to flow toward urban areas and male-headed households. At the same time, the rural poor especially women and people with disabilities are frequently left out. Many lack fixed addresses, formal identification, or access to the registration processes needed to enrol in these programmes. Even when governments try to expand coverage, local elites or village leaders can divert support toward connected families, undermining fairness. These problems are not just operational; they are systemic. Weak governance, limited transparency, and underdeveloped data systems continue to reinforce exclusion and limit the effectiveness of social programmes.

Fragmentation remains a significant obstacle to adequate social protection, with many countries operating dozens of small-scale relief and welfare programmes rather than building unified, coordinated safety nets. Often linked to specific donors or ministries, these programmes create uneven and inconsistent support. Poor households in one area may receive multiple aid forms, while others nearby get nothing. This creates serious inefficiencies, with resources wasted on administrative duplication and confusion among beneficiaries undermining trust in the system. These kinds of bureaucratic silos prevent social programmes from reaching their full potential and leave many of the most vulnerable people behind. Coordination challenges exacerbate these problems, creating patchwork coverage where eligibility depends more on programme access than actual need. Different types of support such as cash grants, food aid, and school feeding are managed separately, often by various ministries or donor agencies. This fragmentation highlights a deeper problem: social protection systems cannot function as a reliable safety net without a unified framework or shared database. Instead of filling gaps, they risk creating new ones.

The overall impact of these initiatives struggles to be broadly implemented due to a lack of integration, with many countries operating programmes in silos and creating gaps in coverage and inefficiencies in resource distribution. Consequently, without cohesive frameworks that reflect social protection and broader development goals, the systems fail to provide a comprehensive safety net. It becomes necessary to address these institutional weaknesses with the primary aim of fostering resilient and inclusive social protection programmes that prioritise adaptation to the dynamics of Africa's evolving socio-economic context.

8.2.2.2 Impact of Global Crises on Social Protection

Africa's already fragile social protection systems were severely tested by the onset of the COVID-19 pandemic, revealing critical gaps in crisis response capabilities that align with the broader challenges alluded to in the preceding sections. Lockdown measures caused economic contractions, leading to inevitable household income losses and increased unemployment across the continent (Martin et al., 2020). Due to the absence of formal social safety nets, informal workers, who represent a significant majority of Africa's labour force and rely on daily earnings (Schwettmann, 2020), were disproportionately impacted. The economic contractions increased poverty, forcing millions of Africans into extreme deprivation, and it demonstrated the inadequacy of residual social policies in addressing systemic challenges. With minimal or absent financial reserves, households struggled to manage daily expenses with the abrupt loss of income, directly resulting in higher rates of food insecurity, homelessness, and reduced access to essential services.

Governments introduced emergency measures, such as temporary cash transfers and food assistance, but these efforts highlighted the limitations of existing systems and the need for more adaptive governance approaches. For example, Nigeria and Kenya expanded their cash transfer programmes to reach more beneficiaries during the height of the pandemic (Awojobi et al., 2023). However, despite these efforts, the continent faced scale, speed, and inclusivity challenges. Emergency assistance failed to reach those most in need in many cases due to inadequate infrastructure and limited digital capacity. The crisis revealed a key weakness in existing responses: although cash transfers and food aid were used more widely than before, they often arrived too late or only reached people who were already registered in government systems. As a result, many newly vulnerable households were left out just when they needed help the most, highlighting that speed and inclusion remain significant gaps in current safety-net programmes, especially during fast-moving emergencies.

8.2.2.3 Limitations of Traditional Social Protection Systems

Despite positive intentions, social protection programmes have been criticised for creating dependency within impoverished regions rather than fostering sustainable development. International aid programmes provide insufficient investment in the local infrastructure and skills necessary for sustainable development, thus perpetuating a dependency cycle (Easterly & Pfutze, 2008). In Africa specifically, the absence of transparency and accountability structures within traditional social protection systems, often consumed by corruption and misappropriation of funds, impedes accurate allocation and distribution. Some charities only receive approximately 60% of donations (Goodbox, 2019), highlighting the need for new solutions to distribute global aid. Unpredictable political landscapes, political instability, and bureaucratic hurdles often perpetuate these barriers. The lack of oversight can lead to the diversion of resources away from their intended purposes, reducing the overall impact of these programmes and creating a disconnect between their intended effect. Funding for social protection programmes that originate from outside African borders has often been characterised by a top-down approach that fails to account for local contexts and needs. International philanthropy is frequently rife with ineffective allocation and use of resources, since these aid organisations are subject to diverse regulatory environments, cultural contexts, and logistical and geographical barriers to ensure that aid reaches its intended beneficiaries while mitigating overhead costs (Meer, 2023). These barriers are perpetuated by unpredictable political landscapes, political instability, and bureaucratic hurdles, which can all impede the delivery of funding (The Global Philanthropy Environment Index 2018, n.d.).

However, the potential adverse consequences of integrating cryptocurrency within social protection programmes must be acknowledged alongside their benefits. While these technologies are generally favoured for their emphasis on the security of financial transactions, the anonymity associated with these transactions can facilitate illegal activities such as money laundering and tax evasion (Kethineni & Cao, 2020). This may potentially adversely impact the reputation of social protection programmes facilitated using digital currencies. Technological barriers to widespread adoption are apparent since there is likely minimal technical expertise and infrastructure to integrate cryptocurrency into charitable organisations operating within these regions (Sadhya & Sadhya, 2018). Since traditional models of philanthropy rely heavily on established financial and logistical networks, remote and underserved communities are often excluded or receive minimal aid distribution. Limited financial infrastructure translates to the inaccessibility of banking services to potential beneficiaries, frequently a prerequisite for granting aid through conventional channels.

8.3 Digital Technologies in Social Protection

8.3.1 Technological Innovations: Addressing Systemic Barriers

International philanthropy has historically faced significant barriers that limit its effectiveness in addressing global resource disparities. These challenges are especially acute in regions with minimal financial infrastructure where informal financial institutions operate beyond conventional banking systems. These interconnected challenges create a compelling case for exploring alternative financial technologies that can bypass traditional banking limitations while maintaining security and transparency. Traditional financial systems impose multiple interconnected barriers that systematically undermine philanthropic effectiveness. Traditional financial systems impose excessive transaction fees, regulatory constraints, and processing delays that substantially impede philanthropic efforts. In addition, currency fluctuations and multiple intermediaries create opportunities for fund misappropriation and reduce aid value during cross-continental transfers (Maclean et al., 2021; Albrecht et al., 2019). These systemic challenges are especially acute in developing regions such as Africa, where the absence of robust financial infrastructure compounds the difficulties of effective aid distribution.

Cryptocurrency emerges as a transformative solution that directly addresses these barrier categories. It offers a decentralised alternative that operates on blockchain technology, a distributed ledger system that records transactions across multiple computers without central authority control (Hussain et al., 2022; Sarmah, 2018). Blockchain's distributed ledger system directly addresses the intermediary problem by eliminating the need for correspondent banks and payment processors. The inherent transparency and security of blockchain technology stems from its immutable structure. In this system, each block contains a chronological list of transactions that cannot be altered once added to the chain, creating an effective mechanism for fraud prevention and trust building among users and donors (Yadav, 2022). This immutability directly addresses weaknesses in traditional aid distribution systems, especially the potential for transaction manipulation or falsification that has historically plagued social protection efforts. By eliminating intermediaries and reducing costs, these digital assets create new pathways for engaging previously excluded communities. This expansion enables peer-to-peer transactions that extend beyond mere transaction efficiency to encompass targeting of historically underserved populations. These digital assets enable engagement with communities through social protection programmes that traditional aid models have systematically overlooked. This approach supports sustainable livelihood development through blockchain-facilitated microfinance initiatives for small-scale entrepreneurs and farmers (Ortega-Rodríguez et al., 2020).

The advanced cryptographic security utilising public and private key systems ensures that funds are accessible only to intended recipients while maintaining publicly verifiable transaction records. This creates an immutable audit trail that addresses transparency and accountability challenges in traditional social protection efforts (Weichbroth et al., 2023; Kizza, 2024). While cryptocurrency offers significant advantages, successful implementation requires addressing key challenges. These include price volatility based on network factors and economic conditions, despite most cryptocurrencies possessing finite supply characteristics that theoretically protect against long-term inflation. This necessitates comprehensive financial literacy programmes to ensure safe and effective utilisation alongside the learning curve associated with cryptocurrency adoption (Baur et al., 2018).

8.3.2 Financial Inclusion and Innovation

While cryptocurrency offers technological solutions to systemic barriers in philanthropy, its broader impact lies in advancing financial inclusion. Financial inclusion serves as an essential prerequisite for effective social protection programmes. Financial inclusion encompasses ensuring access to essential financial services for marginalised communities, prioritising affordable, accessible, and relevant financial products for demographics traditionally excluded from conventional banking systems (Tarigan & Mawardi, 2023). Effective financial inclusion requires not only service access but also financial literacy development to enable communities to save securely, access credit, and invest strategically, thereby improving economic stability and quality of life. Traditional barriers to financial inclusion can be systematically categorised into interconnected areas. Barriers include the absence of banking infrastructure in remote areas, socio-economic factors including limited education and low income, lack of formal identification, and cultural and gender biases that prevent marginalised groups, particularly women, from accessing financial services (de Luna Martinez, 2007; Grohmann & Menkhoff, 2021).

Cryptocurrency and blockchain technology systematically address these barriers. They eliminate the need for traditional banking infrastructure. This is an advantage in regions with limited financial services, while the expansion of cryptocurrency through mobile technology enhances accessibility, as mobile devices are more widely available in developing regions than traditional banking services (Chu, 2018). The absence of intermediaries reduces transaction costs and delays while improving accessibility and affordability. In addition, blockchain's immutable ledger fosters trust, especially valuable in developing nations where corruption and fraud propensity is higher. Unlike traditional banks, which require extensive formal documentation for account opening, select cryptocurrencies can be accessed with minimal identification requirements. This accessibility enables participation of marginalised individuals in the financial system and addresses systematic barriers faced by women and other marginalised demographics (Demirgüç-Kunt & Klapper, 2013; Lewis, 2018; Julius Otusanya, 2011).

Successful cryptocurrency integration, however, within social protection programmes requires addressing regulatory and environmental considerations. From regulatory perspectives, the legal status of cryptocurrencies varies significantly across developing nations, ranging from strict prohibition to comprehensive regulation. This variation necessitates supportive legal frameworks prior to cryptocurrency integration within social protection programmes. The environmental impact of cryptocurrency mining, especially proof-of-work consensus mechanisms, raises sustainability concerns through significant computational power requirements for transaction validation and network security. However, more sustainable alternatives such as proof-of-stake (PoS), delegated proof-of-stake (DPoS), and proof-of-authority (PoA) consensus mechanisms eliminate mining requirements, consuming over 99% less energy than proof-of-work systems (Wang et al., 2022; Shifferaw & Lemma, 2021).

Disruptive innovation theory provides essential theoretical grounding for understanding cryptocurrency's transformative potential. This theory describes how new technologies significantly alter existing markets through accessibility, targeting of underserved markets, incremental improvement, and new value network creation. Cryptocurrency exemplifies this transformation through its evolutionary capacity from Bitcoin's initial purpose as a decentralised digital currency alternative to Ethereum's introduction of smart contracts. Smart contracts are self-executing contracts with agreement terms directly written into code, enabling automated and transparent fund distribution in philanthropic operations. These innovations address specific pain points in traditional philanthropy, including high transaction costs and transparency deficits, while enabling peer-to-peer transactions that remove intermediaries and allow direct aid distribution. In addition, decentralised finance (DeFi) platforms further demonstrate scalability by offering lending, borrowing, and trading services without traditional intermediaries (Chu, 2014; Tern, 2021; Christensen et al., 2018).

8.3.3 Practical Applications and Real-World Implementation

Despite implementation challenges, innovative initiatives demonstrate cryptocurrency's practical benefits in addressing financial inclusion and enhancing charitable effectiveness across diverse contexts. BitGive exemplified organisational innovation by providing aid through bitcoin distribution for public health and environmental sustainability projects, with its GiveTrack platform enabling real-time transparent donation tracking that directly addresses fund management concerns in social protection programmes (Shin et al., 2020). The Pineapple Fund demonstrated cryptocurrency's philanthropic potential when a bitcoin investor donated $55 million worth of bitcoin to various charitable organisations, supporting causes from healthcare to environmental conservation; meanwhile, in Uganda, cryptocurrency platforms have facilitated cash transfers in rural communities, with organisations such as the Binance Charity Foundation providing blockchain-based cash assistance to vulnerable groups while fostering transparency and minimising transaction costs (Zhauniarovich et al., 2019).

Kenya has witnessed the introduction of community inclusion currencies (CICs) based on blockchain technology, creating localised digital currencies that encourage goods and services trading without external currency reliance, thereby enhancing financial inclusion in impoverished regions during economic uncertainty. Nonprofit organisations utilising platforms like Stellar deliver cash assistance to low-income African families through mobile money transfers, leveraging cryptocurrency's speed, transparency, and low transaction fees for cross-border transfers while ensuring prompt benefit delivery and reduced administrative overhead (Zhan, 2019). Blockchain-based platforms addressing refugee needs provide secure digital identities, addressing a core barrier to financial inclusion by enabling refugees to receive aid, open bank accounts, and access credit necessary for rebuilding lives and fostering economic independence without requiring extensive formal documentation that traditional banking infrastructure demands (Cheesman, 2022).

Ethereum's smart contract functionality enables conditional aid delivery through self-executing contracts with agreement terms directly integrated into code, promoting transparency and accountability in aid distribution by ensuring funds are released only when prescribed conditions are fulfilled (Buterin, 2016). This capability enables philanthropic organisations to prescribe fund dispersal conditions, ensuring intended use while enhancing accountability and building trust among donors and beneficiaries. Ripple's low transaction fees and fast processing times make it valuable for disaster aid scenarios requiring efficient cross-border fund transfers, while Stellar facilitates financial inclusion by providing low-cost financial services to impoverished regions, supporting cross-border transactions and micropayments for effective aid distribution (Harðarson, 2018; Khan et al., 2019).

8.3.4 *Artificial Intelligence and Mobile Money: Complementary Technologies*

Artificial intelligence (AI) represents a complementary technological advancement that significantly improves social protection system efficiency. It utilises advanced algorithms and machine learning models to analyse large datasets, identify patterns, make decisions, and predict future trends. This offers opportunities to optimise beneficiary targeting, resource allocation, and predictive analytics for future risk preparation. Traditional targeting methods rely on manual surveys and outdated databases, resulting in inaccuracies, delays, and exclusion errors. On the other hand, AI significantly improves this process through diverse data source analysis. This includes demographic, economic, and geographic information, with machine learning algorithms identifying poverty level trends, household income patterns, and vulnerability indicators by combining multiple datasets to ensure assistance reaches the most vulnerable individuals (Aery & Ram, 2017). In food insecurity contexts, AI-driven analysis of satellite imagery, weather data, and crop yield statistics can identify at-risk regions, enabling governments and humanitarian organisations to target cash transfers or food aid proactively before a crisis escalates; meanwhile, in urban areas, AI can analyse social and economic data from government records, utility bills, and mobile networks to identify low-income households requiring social assistance. AI-driven tools analyse real-time data to determine optimal resource allocation and distribution methods, with predictive analytics identifying areas at higher risk of economic shocks or natural disasters. This enables proactive, rather than reactive, resource allocation, while machine learning algorithms detect fraud and corruption within social protection systems by analysing financial transactions and beneficiary records to identify unusual patterns or inconsistencies (Boppiniti, 2021; Odilla, 2023).

Mobile money technology complements AI implementation by transforming financial service accessibility for millions of unbanked Africans, encompassing financial transactions and services conducted through mobile phones, including money transfers, bill payments, and savings or credit access (Hughes & Lonie, 2007). Unlike traditional banking systems, mobile money operates without physical infrastructure requirements, enhancing accessibility in remote and underserved areas. In Africa, where the continent experiences some of the world's highest financial exclusion rates, many individuals lack formal banking access due to limited bank branches, high transaction costs, and low literacy levels (Demirgüç-Kunt et al., 2015). Within social protection programmes, mobile money provides reliable and efficient cash transfer delivery methods, with governments and organisations transferring funds directly to mobile accounts. This eliminates intermediaries and reduces delay or corruption risks, proving beneficial during the pandemic when physical bank and payment office access was restricted.

Countries like Kenya and Tanzania have successfully integrated mobile money into social protection programmes. Kenya's M-Pesa platform revolutionised cash transfers by enabling direct mobile phone benefit receipt, reducing distribution costs while ensuring quick and secure assistance delivery (Ndung'u, 2018). Mobile money benefit women and rural communities facing the greatest financial exclusion. They provide secure and independent financial management, credit access, and economic participation opportunities, yielding broad benefits including improved household welfare, increased education investments, and enhanced women's economic participation. They also enable farmers and small business owners to access payments, savings, and credit in rural areas with scarce bank branches (Omojolaibi et al., 2019; Suri, 2017).

8.3.5 Implementation Framework and Strategic Considerations

The transformative potential of cryptocurrency and AI must be balanced against ethical considerations to ensure maximum positive impact on marginalised communities. The decentralised nature of blockchain technology necessitates robust data security measures that carefully balance transparency in aid distribution with individual privacy protection. In addition, the perceived anonymity of cryptocurrency transactions raises ethical concerns regarding donor and recipient information confidentiality. This requires prioritisation of sensitive data protection through encryption and stringent protocols to prevent unauthorised access, breaches, and information misappropriation, while also mitigating risks through know your customer (KYC) procedures and compliance measures that encourage responsible and ethical cryptocurrency use for philanthropic purposes. A comprehensive research agenda is essential for deepening understanding and optimising cryptocurrency and AI implementation. It requires rigorous impact evaluations to establish socio-economic effects of cryptocurrency adoption in social protection programmes, assessing cost-effectiveness, transparency impacts, and direct beneficiary benefits. Technological adaptation challenges must be addressed through user interface enhancement for improved accessibility and ensuring cryptocurrency interoperability with existing financial services. Meanwhile, in-depth policy analysis navigates the dynamic regulatory landscapes surrounding emerging technologies, evaluating the efficiency of existing regulatory frameworks and proposing adaptive policies that foster innovation while mitigating risks (Ortega-Rodríguez et al., 2020).

Successful technology adoption requires active collaboration from technology providers, including blockchain platforms, developers, wallet providers, payment processors, and development agencies, extending beyond technical support to strategic collaboration and complex challenge navigation. Given the diverse capabilities and resources of African social protection programmes, technology providers must deliver consistent assistance to ensure seamless integration and operational efficiency. They must also demonstrate compliance expertise through regulatory development awareness and support philanthropic organisations in implementing KYC and anti-money laundering safeguards. Infrastructure development requirements include reliable internet connectivity, digital device accessibility, and comprehensive digital literacy training programmes tailored to different demographic groups. Collaborative frameworks are needed to ensure secure, compliant, and accessible technology use, fostering partnerships and the development of industry standards through multistakeholder engagement.

Clear and comprehensive regulatory frameworks are essential for creating supportive environments for emerging technology adoption within African social protection contexts. These frameworks promote innovation while protecting consumers and ensuring financial stability through flexible, forward-thinking designs that encourage technological advancement and simplify licensing processes and compliance requirements. A structured, phased approach to cryptocurrency and AI integration ensures effective and sustainable rollout. This approach progresses from foundational requirements, including regulatory frameworks, digital literacy programmes, and technological infrastructure development, to medium-term collaborative partnerships and pilot programme implementation. Ultimately, it aims to achieve long-term integration within broader development agendas through adaptable and sustainable transition approaches that address immediate barriers while establishing foundations for transformative, inclusive, and scalable solutions.

8.4 Discussion

This chapter has explored the transformative potential of digital technologies in reimagining social protection systems across Africa, revealing both unprecedented opportunities and critical challenges that will shape the continent's social policy trajectory. The study demonstrates that emerging technologies such as blockchain, AI, and mobile money offer promising pathways to address persistent gaps in coverage, efficiency, and transparency. However, their implementation must be grounded in ethical frameworks and locally owned governance structures. This is crucial to avoid reproducing existing inequalities or creating new forms of exclusion.

8.4.1 Interpreting the Technological Transformation

The convergence of Africa's dynamic development landscape with rapid technological advancement presents a unique historical moment for redefining social protection systems. The research questions posed at the outset reveal three interconnected pathways for advancing social protection in Africa. First is the efficiency paradigm: emerging technologies can address traditional system inefficiencies through enhanced targeting mechanisms, reduced transaction costs, and improved transparency in benefit distribution. Mobile money platforms have already demonstrated their capacity to reach previously excluded populations, while blockchain technology offers unprecedented opportunities for creating tamper-proof benefit registries. Second is the ownership paradigm: sustainable implementation requires moving beyond donor-driven approaches toward locally owned technological ecosystems that reflect African priorities and contexts. This finding challenges conventional development thinking that often prioritises technical solutions over local agency. Third is the governance paradigm: the experiences of Kenya's digital ID initiatives illustrate both the promise and perils of large-scale technological deployment without adequate safeguards, highlighting that technological capability without appropriate governance can reproduce or amplify existing inequalities.

8.4.2 Critical Trade-Offs and Tensions

The study highlights important trade-offs that policymakers must navigate. Efficiency versus inclusion is a key consideration. While AI and algorithmic systems can improve targeting accuracy, they may lead to new forms of digital discrimination if not carefully designed. Innovation versus stability is another critical aspect. Blockchain and cryptocurrency present revolutionary potential for financial inclusion, but their volatility and regulatory uncertainty can pose risks to vulnerable populations who can least afford system failures. Finally, speed versus sustainability is essential to consider. The urgency highlighted by COVID-19 creates pressure for rapid technological deployment; however, sustainable transformation requires careful sequencing and capacity building. These tensions suggest that the choice is not simply between traditional and technological approaches, but rather about how to balance competing priorities within hybrid systems. The mobile money revolution provides important perspectives here, demonstrating that successful technological integration requires deep understanding of local needs, regulatory environments, and cultural contexts.

8.4.3 Policy Implications and Recommendations

Several key policy implications emerge from this analysis. Integrated regulatory frameworks must be developed that can adapt to rapidly evolving technological landscapes while protecting beneficiary rights through multistakeholder collaboration. This requires moving beyond siloed approaches to embrace comprehensive governance structures that span technical, legal, and social dimensions. Investment in digital literacy and institutional capacity represents a critical prerequisite for sustainable technological integration. Infrastructure development must proceed alongside digital literacy programmes, regulatory framework development, and institutional capacity building. This parallel development approach ensures that technological capabilities are matched by human and institutional readiness. Phased implementation approaches should reflect the reality that sustainable transformation requires careful sequencing of interventions and continuous adaptation based on emerging evidence and changing contexts. Rather than pursuing wholesale technological transformation, successful implementation demands iterative approaches that allow for learning and adjustment. Locally owned technological ecosystems must be prioritised to ensure that technological solutions serve African priorities rather than external agendas. The continent's experience with structural dependencies and donor-driven policies highlights the importance of ensuring that technological innovation emerges from and serves local contexts. Ethical frameworks must be established from the design phase of technological interventions to prevent digital discrimination and protect vulnerable populations. The absence of robust data protection frameworks across much of the continent creates significant risks that technological solutions could exacerbate existing exclusions or create new forms of digital discrimination.

8.4.4 Limitations and Constraints

Several limitations constrain the transformative potential of digital technologies in African social protection systems. Infrastructure gaps remain significant barriers, especially in rural areas where connectivity and digital literacy are limited. Regulatory fragmentation across African countries creates challenges for scaling successful interventions. Capacity constraints at institutional levels mean that even well-designed technologies may fail without adequate human resources and governance structures. In addition, the study has primarily focused on successful cases and emerging opportunities, potentially underestimating implementation challenges and unintended consequences. Future research should examine the long-term sustainability of digital social protection systems, comparative analysis of regulatory approaches across African countries, and impact assessments of algorithmic decision-making on vulnerable populations. These research directions would provide crucial evidence for refining technological interventions and governance approaches.

8.5 Conclusion

This chapter contributes to the broader discourse on redefining social policy in contemporary Africa by demonstrating how technological innovation can support more integrated, responsive, and locally owned approaches to social protection. The study reveals that technology alone cannot solve the structural challenges facing African social protection systems, but when combined with appropriate governance frameworks, ethical safeguards, and capacity-building initiatives, digital tools can significantly enhance system effectiveness and reach. Perhaps most importantly, this chapter has shown that the future of social protection in Africa will be shaped not by technological capabilities alone but by the choices made about how these capabilities are governed and deployed. The transformative potential of digital technologies in social protection ultimately depends on their ability to advance human development, reduce inequality, and strengthen social cohesion.

As Africa continues its journey toward more effective and inclusive social protection systems, success will require unprecedented collaboration across sectors, disciplines, and borders. By learning from early experiences, investing in institutional capacity, and maintaining focus on human-oriented outcomes, African countries can harness digital technologies to create social protection systems that truly serve the needs of their most vulnerable populations. The future lies in thoughtfully integrating traditional and technological approaches to create systems that are simultaneously innovative and grounded in local realities, efficient and equitable, responsive and sustainable.

References

Aery, M., & Ram, C. (2017). A review on machine learning: Trends and future prospects. *Research Cell: An International Journal of Engineering Sciences, 25*(63019), 89–96.

Albrecht, C., Duffin, K. M., Hawkins, S., & Morales Rocha, V. M. (2019). The use of cryptocurrencies in the money laundering process. *Journal of Money Laundering Control, 22*(2), 210–216. https://doi.org/10.1108/jmlc-12-2017-0074

Awojobi, O. N., Kwabia, E., & Adeniji, O. A. (2023). *Social protection programmes in mitigating the socio-economic impacts of the Covid-19 pandemic: A comparative study of Ghana, Kenya, and South Africa.*

Baur, D. G., Hong, K., & Lee, A. D. (2018). Bitcoin: Medium of exchange or speculative assets? *Journal of International Financial Markets, Institutions and Money, 54*, 177–189.

Boppiniti, S. T. (2021). Real-time data analytics with AI: Leveraging stream processing for dynamic decision support. *International Journal of Management Education for Sustainable Development, 4*(4).

Buterin, V. (2016). Ethereum: Platform review. *Opportunities and challenges for private and consortium blockchains, 45.*

Cheesman, M. (2022). Self-sovereignty for refugees? The contested horizons of digital identity. *Geopolitics, 27*(1), 134–159. https://doi.org/10.1080/14650045.2020.1823836

Christensen, C. M., McDonald, R., Altman, E. J., & Palmer, J. E. (2018). Disruptive innovation: An intellectual history and directions for future research. *Journal of Management Studies, 55*(7), 1043–1078.

Chu, D. N. (2014). *Bitcoin: A theoretical analysis of money choice* (Doctoral dissertation, University of Puget Sound).

Chu, A. B. (2018). Mobile technology and financial inclusion. In *Handbook of blockchain, digital finance, and inclusion* (Vol. 1, pp. 131–144). Academic Press.

de Luna Martinez, J. (2007). Financial services: Dealing with limited and unequal access. *TRADE*, 155.

Demirgüç-Kunt, A., & Klapper, L. (2013). Measuring financial inclusion: Explaining variation in use of financial services across and within countries. *Brookings Papers on Economic Activity, 2013*(1), 279–340.

Demirgüç-Kunt, A., Klapper, L., Singer, D., & Van Oudheusden, P. (2015). The Global Findex Database 2014: Measuring financial inclusion around the world (Policy Research Working Paper No. 7255). World Bank Group. https://documents.worldbank.org/en/publication/documents-reports/documentdetail/187761468179367706/theglobal-findex-database-2014-measuring-financial-inclusion-around-the-world

Easterly, W., & Pfutze, T. (2008). Where does the money go? Best and worst practices in foreign aid. *Journal of Economic Perspectives, 22*(2), 29–52.

GoodBox. (2019). *How much money goes to charity when you donate?* https://www.goodbox.com/2019/06/how-much-money-goes-to-charity-when-you-donate/

Grohmann, A., & Menkhoff, L. (2021). The relationship between financial literacy and financial inclusion. In *The Routledge handbook of financial literacy* (pp. 517–530). Routledge.

Harðarson, A. (2018). *Can Ripple disrupt the global payments market?* (Doctoral dissertation).

Hoddinott, J. F., Devereux, S., White, P., Klasen, S., Woolard, I., Alderman, H., et al. (2012). *Social protection in West Africa: The status quo, lessons from other regions, implications for research.*

Hughes, N., & Lonie, S. (2007). M-PESA: Mobile money for the "unbanked" turning cellphones into 24-hour tellers in Kenya. *Innovations: Technology, Governance, Globalization, 2*(1–2), 63–81.

Hussain, S. H., Sivakumar, T. B., & Khang, A. (2022). Cryptocurrency methodologies and techniques. In *The data-driven blockchain ecosystem* (pp. 21–29). CRC Press.

Julius Otusanya, O. (2011). Corruption as an obstacle to development in developing countries: A review of literature. *Journal of Money Laundering Control, 14*(4), 387–422.

Kethineni, S., & Cao, Y. (2020). The rise in popularity of cryptocurrency and associated criminal activity. *International Criminal Justice Review, 30*(3), 325–344.

Khan, N., Ahmad, T., & State, R. (2019, October). Feasibility of stellar as a blockchain-based micropayment system. In *International Conference on Smart Blockchain* (pp. 53–65). Springer.

Kizza, J. M. (2024). Blockchains, cryptocurrency, and smart contracts technologies: Security considerations. In *Guide to computer network security* (pp. 575–600). Springer.

Kshetri, N. (2021). Blockchain technology for improving transparency and citizen's trust. In *Advances in Information and Communication: Proceedings of the 2021 Future of Information and Communication Conference (FICC)* (Vol. 1, pp. 716–735). Springer.

Lewis, A. (2018). *The basics of bitcoins and blockchains: An introduction to cryptocurrencies and the technology that powers them.* Mango Media Inc.

Maclean, M., Harvey, C., Yang, R., & Mueller, F. (2021). Elite philanthropy in the United States and United Kingdom in the new age of inequalities. *International Journal of Management Reviews, 23*(3), 330–352.

Martin, A., Markhvida, M., Hallegatte, S., & Walsh, B. (2020). Socio-economic impacts of COVID-19 on household consumption and poverty. *Economics of Disasters and Climate Change, 4*(3), 453–479.

Meer, J. (2023). Are overhead costs a good guide for charitable giving? In *The fundraising reader* (pp. 314–318). Routledge.

Ndung'u, N. (2018). The M-Pesa technological revolution for financial services in Kenya: A platform for financial inclusion. In *Handbook of blockchain, digital finance, and inclusion* (Vol. 1, pp. 37–56). Academic Press.

Odilla, F. (2023). Bots against corruption: Exploring the benefits and limitations of AI-based anti-corruption technology. *Crime, Law and Social Change, 80*(4), 353–396.

Omojolaibi, J. A., Okudo, A. G., & Shojobi, D. A. (2019). Are women financially excluded from formal financial services? Analysis of some selected local government areas in Lagos State, Nigeria. *Journal of Economic and Social Thought, 6*(1), 16–47.

Ortega-Rodríguez, C., Licerán-Gutiérrez, A., & Moreno-Albarracín, A. L. (2020). Transparency as a key element in accountability in non-profit organizations: A systematic literature review. *Sustainability, 12*(14), 5834. https://doi.org/10.3390/su12145834

Sadhya, V., & Sadhya, H. (2018). *Barriers to adoption of blockchain technology.*

Sarmah, S. S. (2018). Understanding blockchain technology. *Computer Science and Engineering, 8*(2), 23–29.

Schwettmann, J. (2020). *Covid-19 and the informal economy. Impact and response strategies in Sub-Saharan Africa.* Friedrich Ebert Stiftung (FES).

Shawa, K. C., Hollingsworth, B., & Zucchelli, E. (2024). Joint effects of ill-health, health shocks and social protection on the intensive margin of labour supply: Evidence from Malawi. *Health Economics Review, 14*(1), 75.

Shifferaw, Y., & Lemma, S. (2021). Limitations of proof of stake algorithm in Blockchain: A review. *Zede Journal, 39*(1), 81–95.

Shin, E. J., Kang, H. G., & Bae, K. (2020). A study on the sustainable development of NPOs with blockchain technology. *Sustainability, 12*(15), 6158.

Suri, T. (2017). Mobile money. *Annual Review of Economics, 9*(1), 497–520.

Tarigan, E. S., & Mawardi, W. (2023). The influence of digital based payment system innovation on financial inclusion and poverty alleviation: An overview of social programs in the Sumatra Region. *Migration Letters, 20*(8), 223–240.

Tern, S. (2021). Survey of smart contract technology and application based on Blockchain. *Open Journal of Applied Sciences, 11*(10), 1135–1148.

The Global Philanthropy Environment Index 2018. (n.d.). Retrieved June 23, 2024, from https://efc.issuelab.org/resources/31188/31188.pdf

Wang, Y., Lucey, B., Vigne, S. A., & Yarovaya, L. (2022). An index of cryptocurrency environmental attention (ICEA). *China Finance Review International, 12*(3), 378–414.

Weichbroth, P., Wereszko, K., Anacka, H., & Kowal, J. (2023). Security of cryptocurrencies: A view on the state-of-the-art research and current developments. *Sensors (Basel), 23*(6), 3155. https://doi.org/10.3390/s23063155

Yadav, S. P. (2022). Blockchain security. In *Blockchain security in cloud computing* (pp. 1–18). https://doi.org/10.1007/978-3-030-70501-5_1

Zhan, Y. (2019). *Local currency networks in rural communities in Africa: A feasibility study of blockchain based community currency* (p. 36). Vrije Universiteit Amsterdam.

Zhauniarovich, Y., Boshmaf, Y., Jawaheri, H. A., & Sabah, M. A. (2019). Characterizing Bitcoin donations to open source software on GitHub. *arXiv preprint arXiv:1907.04002.*

Chapter 9
Exploring the Interplay Among Food and Nutrition Social Policy Actors in Senegal from 1918 to 2021

Alex Nadège Ouedraogo

9.1 Introduction

Food and nutrition are fundamental to the wellbeing of societies hence the need to consider these as salient areas of social policy. Goal 2 of the United Nations' Sustainable Development Goals (SDG) aims to end hunger, achieve food security, improve nutrition and promote sustainable agriculture. Senegal has numerous legal frameworks, including those directly targeting food and nutrition. Though the Senegalese Constitution has provisions that guarantee social protection for citizens, including access to food and nutrition, challenges still exist (Spray, 2018). According to the World Bank, Senegal is among the ten fastest growing countries in the world (Thierry et al., 2019). Yet the food and nutrition question persists. According to the Scaling Up Nutrition Platform, chronic malnutrition in the country increased from 13% in 2014 to 17% in 2017. This is above the 10% standard set by the World Health Organization (WHO).

Given this priority and the need to address food issues is pressing the WHO (2020) figures are of concern. Out of an almost eight billion total world population, 1.9 billion adults are overweight or obese, while 462 million are underweight. Around 45% of deaths among children under 5 years of age are linked to undernutrition. These mostly occur in low- and middle-income countries.

Some scholars such as Masters et al. (2018) and Misselhorn and Hendriks (2017) have highlighted that though food and nutrition in Africa should be a crucial research theme, a vivid lacuna exists. In addition, food and nutrition are minimally prioritized in national and international debates. In Senegal, such gaps are compounded by the monotasking of social policy to social protection, particularly its social assistance dimension, which is widely documented (see Diatta, 2018; Ndiaye et al., 2019).

A. N. Ouedraogo (✉)
Institute of Intercultural and International Studies (InIIS), Faculty 8 (Social Sciences), University of Bremen, Bremen, Germany

E. Ndhlovu, C. Chipenda (eds.), *Redefining the Social Policy Discourse*,
https://doi.org/10.1007/978-3-032-24067-5_9

Against this backdrop that seems prevalent in several African countries and the Global South in general, scholars continue to rather focus on social policy with its productive, redistributive, protective, reproductive and cohesion outcomes (see Adésínà, 2009, 2010, 2011a, 2011b, 2020; Chipenda, 2019; Mkandawire, 2004, 2005, 2015; Tekwa & Adesina, 2018; Yi, 2015).

This article expands the existing literature by specifically focusing on food, nutrition and agriculture, adding figuration, trajectory and international dimensions.

The focus here is on the evolution of food social policy regardless of their appellations, answering questions of who worked on these policies? How are they organized? Under what political circumstances do they function? Nonetheless, the role and steps taken by the Senegalese government seem clear in the literature. A quick overview of all the actors involved in social policy development and the pursuit of food and nutrition security objectives in Senegal reveals that, no matter the labelling and ideology, food provision and production are still a concern yet to be comprehensively addressed. This article will proceed with a three-pronged but intertwined approach based on prime linkages between food-related issues and critical actors of this area of social policy in Senegal.

First is the outline of the conceptual underpinnings of the article. This section is based on the concept of trajectory, drawn from Migdal and Schlichte (2005), complemented by Norbert Elias's figuration, understood as constellations of actors. Second is the adoption of figuration and historical analysis to the interplay of social policy actors in the realm of food and nutrition policy. Through periodization, looking at food-related social policy trajectories, this paper asks: How does the interplay between local stakeholders, national government and international organizations and so forth work? How are these coalitions brought together, and what, beyond monetary interests, binds them? In relation to Senegal, periodization reveals more than trajectories but also answers the question, 'How can the interplay among these social policy actors be designed for improved current and future social policy?' From the perspective of political sociology, the perpetual link between food issues and local and international politics will be acknowledged as a determining vector of people's wellbeing, entangled with the state. This section shows that how social policy is defined, its evolution and contestations, as well as how these link to food and nutrition security have been shifting throughout history and will continue to do so.

The article pays particular attention to historical epochs, including the French colonial era (1918–1960) and the post-colonial era, which is divided into four different periods. Actors varied throughout a history marked by a shift from French imperialism to an independent state but kept similar burdens, for example, the introduction of rice by the French as the staple food to compensate for the reduction in subsistence crop production and primarily to satisfy imperial objectives (see Fanon, 1967). The country has become a food importer and still cannot produce enough local rice to achieve self-sufficiency despite setting goals and funding projects towards this aim.

This article draws on a bibliographical and documentary review and interviews. Case literature and public documents were reviewed along standard procedures drawing on the methodological logics advanced by scholars such as Creswell and Creswell (2018) and Wood et al. (2020). Interviews with heads of public institutions, employees of international organizations employees, as well as beneficiaries of social safety nets were conducted in three different regions (Dakar, Diourbel and Thiès). The field work was conducted within a 3-month research stay in Senegal (January 7, 2019-April 1, 2019).

The benefit of approaching food and nutrition policy based on the two central concepts of figurations and trajectory helps us to transcend Senegal, and the discussion here may be used for a broader context that provides some insights into other African countries and beyond.

9.2 Methodology

The article employs two intertwined conceptual frameworks trajectory and figuration to examine the historical evolution of social policy actors and their interactions in relation to food and nutrition policy in Senegal. The concept of trajectory, introduced by Migdal and Schlichte (2005) and informed by Bayart (1996), highlights the historicity of political institutions, path dependencies and specific political logics, focusing on indigenous roots, local and international dynamics and the influence of historical ruptures on contemporary domination. Figurational sociology, as developed by Norbert Elias, investigates the interdependencies and power dynamics among actors in social policy over time, revealing how these relationships shape policies across historical periods. This comprehensive methodological approach encompasses colonial and post-colonial contexts, linking socio-economic factors to the evolution of food and nutrition policies in Senegal.

To strengthen the methodological rigor, archival sources, including colonial records and policy documents, were analysed thematically to identify patterns in food policy evolution. Qualitative interviews conducted with stakeholders including policymakers, international organization representatives and beneficiaries of social safety nets were transcribed and manually coded. This ensured systematic triangulation of data sources, enhancing reliability and reproducibility (Creswell & Creswell, 2018). For instance, archival data were categorized chronologically, while interview data were grouped thematically to align with key historical periods identified in the trajectory framework.

9.3 Literature Review

9.3.1 Conceptual Underpinnings: Trajectory and Figuration

The article contributes to the dynamics of social policy through intertwined conceptual underpinnings 'trajectory' and figuration. Both fulfil the objective of exploring the historical evolution of the interplay of social policy actors as the basis for understanding and improving current and future social policy relating to food and nutrition as well as other areas of social policy. Trajectory (*trajectoire*) is a unique conceptual framework from Migdal and Schlichte (2005) based on Jean-Francois Bayart's (1996) analysis of the concept. The term 'emphasizes the historicity of political institutions, path dependencies, and specific political logics [...]' (Migdal & Schlichte, 2005: 21). Informed by the view that historical dimensions have been largely relegated in studying non-European societies, Jean-Francois Bayart (2009) utilized the concept in pursuit of understanding to address this lacuna. Notable in trajectory as it relates to these states are their indigenous roots and the complex interaction of local and international forces, institutions and dynamics and the influence of 'historical ruptures, overthrows, continuities and discontinuities in shaping contemporary domination' (Migdal & Schlichte, 2005: 21). In relation to Senegal's food, nutrition and agriculture policy, the concept is pivotal in three ways. It facilitates an understanding of the diversity of states and periods. The concept also includes the influence of external events when applied to state dynamics and provides room for exploring the outcomes of the interaction of multiple actors (e.g. state and non-state actors) interacting in a context of power.

Figurations also feature prominently in this article. The concept tries to capture which constellations of actors were essential for the dynamics of social policy. In other words, figuration shows connections among actors in a field of power, and in this article, exploration is from a long-term perspective. This sociological concept is credited to Norbert Elias (1897–1990) who interchangeably used configuration and figuration to interrogate relationships of interdependence between individuals and society and notions of function, power, coercion, dilemmatic processes, knowledge and social change (Quintaneiro, 2006: 64). Norbert Elias's sociological approach helps to uncover the diversity of actors involved in Senegal's food policy from colonial time to the current phase, interests, agendas and contestations, power gradients and dynamics, figuration dilemmas and social change, and how these have a bearing on the evolution of health policy. Figurational sociology is comprehensively documented (see Elias, 2006, 2008, 2012, 2013).

The literature review contextualizes the conceptual frameworks of trajectory and figuration within existing scholarship. The trajectory framework has been highlighted by Migdal and Schlichte (2005) as a means to address historical dimensions often overlooked in studies of non-European societies. Bayart (2009) further utilizes this framework to fill the gaps in understanding the historical interplay of social actors. The figurational approach, as articulated by Elias (2006, 2008, 2012, 2013), interrogates the interdependence of individuals and society, emphasizing the power relations and contestations within social policy fields. This section also draws on works by scholars such as Bonnecase (2009) and Cooper (1996, 2004) to underline the sociopolitical context of Senegal's food policies, particularly during the colonial and post-colonial periods.

9.4 Results/Findings

The findings reveal a complex interplay of actors and policies across different historical periods in Senegal. In the imperial ruling period (1918–1960), the primary actors included the French colonial administration, Albert Sarraut and the Murids, with policies geared towards profit maximization for the metropolis at the expense of indigenous food security. During WWII and the transition to independence, international scrutiny led to the establishment of the FIDES and the *Mission Anthropologique*, which aimed to address malnutrition and improve living conditions. The post-colonial period (post-1960) is characterized by continuity in reliance on external funding and neocolonial dynamics, with major actors including international stakeholders and local political leaders. This ongoing relationship illustrates how historical legacies shape contemporary food and nutrition policies in Senegal.

9.4.1 Periodization

I distinguish periods roughly according to changes in political power in Senegal during colonial times, starting from 1918 to 1960 as the first category, comprising the post-World War I period and the period from World War II to Senegalese independence.

9.4.1.1 Imperial Ruling (1918–1960)

The paper's departure point is 1918, as important events occurred that year. The Congress of Colonial Agriculture in May 1918 held in Paris (Tourte, 2005: 3), after the World War I, groundnut production represented a crucial turning point for the economy. From exporting about 300,000 tonnes before 1914, exports dropped to 110,000 after the war (Tourte, 2005: 4); this later turned out to be a pivotal year thanks to the development of communication paths, such as railway extensions allowing for the expansion of cultivable zones (Musset, 1930: 554). During the colonial period, making colonies profitable to the metropolis was the sole thrust of colonial policy. This dimension of colonization was ubiquitous to all African colonies (Maddox, 2018; Young, 2018). In Senegal, the Albert Sarraut 'Mise en valeur' Plan was at the heart of the imperial scheme. Sarraut was appointed Minister of the

Colonies six times and is considered one of the pivotal inspirers of colonial policy between the wars (Guy, 1923; Albert Sarraut, 1931).

French economic development portrayed food-related policy as a prerequisite. Yet, the nutritional situation of the native populations remained poor, mainly due to the predominance of capitalist exploitation in the French-colony relationship as summed up in the term 'Mise en valeur'.[1] It is to be noted that the creation of an economy based on commercial agriculture and the application of latent and manifest strategies (see Frankema & van Waijenburg, 2014: 393) for pushing indigenous populations into the new economy was not intended to enhance their food and nutrition security but French capitalist interests. The French colonial administration sought to push peasants to cultivate cash crops, especially groundnuts, to increase their export revenues. The result was widespread neglect of subsistence farming and food and nutrition security and loss of food sovereignty. The French empire specialized its production in each colony and introduced a division of labour in the empire. Therefore, rice produced in Indochina was imported to the Senegalese market to compensate for the decrease in subsistence farming and its detriment to local food production to increase cash crop (groundnut) production.

Industrialization of colonies and broader social protection were not the French colonial masters' interests (see Hugon, 2007: 54; Vitaux, 2018). Albert Sarraut reiterated the need for minimum welfare measures, including reducing food deficiencies of the working force as a means to keep production high (Ndao, 2004: 16). Furthermore, health and malnutrition were already a crucial concern and starkly reflected in the socio-economic inequalities within the territory. For instance, 'depending on economic priorities, urban areas were privileged in the implementation of health and food security measures' (Ndao, 2004: 16). In this regard, areas of economic interest were monitored and taken care of, whereas, in other regions, nutritional deficiencies were rife. This practice created regional development inequalities that Senegal and other former colonies in Africa are grappling with and that are likely to haunt development in the long term. This first period unpacked the constellations of actors being the French empire via the metropolis, Albert Sarraut (the minister of the colonies) and the Murids. Murids are known for their trading and agriculture-centred activities (Monteil, 1962: 98) The 'Murid fact' is also central since it is considered unique in the collaboration between faith and groundnut production (idem: 77). 'Food policies' revolved around Albert Sarraut's advancement of commercial groundnut production and the importation of rice from Indochina to address the reduction in the production of indigenous food crops. Further issues are interrogated in the next section.

[1] The term could be understood as rendering colonies profitable.

9.4.1.2 From World War II Until Independence

The period and dynamics from WWII to independence are expansive; therefore, for illustration, selected key features are brought to the fore. Unlike in the earlier period, rendering colonies profitable as the sole aim of colonial conquest became problematic since there was growing international concern about how empires were treating their colonies. As Bonnecasse explained, 'the immediate post-war period saw an international demand for knowledge of the colonies. In a context of strong questioning of colonization within the new UN organizations, the colonial powers are called upon to produce new information on the living conditions of the colonized populations' (Bonnecase, 2009: 162). At the Hot Springs[2] Conference in 1943 initiated by United States President Franklin D. Roosevelt, representatives of 44 nations gathered to agree on prioritizing food security and the wellbeing of the colonies (FAO, n.d.). This resembles a critical juncture in advancing the living conditions of colonized populations and is particularly true for the food component. An Interim Commission to develop a specific plan for a permanent organization in the field of food and agriculture and to adopt recommendations on the improvement of national diets and so forth was established. However, colonial capitalist interests endured. For example, a gulf emerged between the metropolis and Dakar on the increased use of forced indigenous labour in commercial agriculture, manufacturing and mining in an international context of recognition of the rights of the colonies. An expanded view is provided by other scholars (see Cooper, 1996, 2004: 17; Marseille, 1984).

One of the first considered welfare acts adopted at that time was the creation of the 'Fonds d'Investissement et de Développement Economique et Social' (FIDES) in 1946. The FIDES is often compared to the Colonial Development and Welfare Act of 1940 adopted by the British. It was a huge step as, for the first time, spending was made knowing there would be no profit. Projects were funded to implement water supply, build schools and house workers (Cooper, 2004: 16). A 10-year development plan adopted by France, centrally controlled and representing the main instrument of the development policy of the French Union,[3] the FIDES was jointly funded by the state budget and by territories (Berthélemy, 1980). Its social and economic goals were to increase the population's production capacity and raise living conditions. In 1944, the French started to reconfigure their empire into the French West Africa (AOF)[4] and in 1946 the French Union, where former subjects of the empire became citizens of the French Union. These changes did not imply French goodwill and were marked by the entrenchment of French capitalist interests at the expense of the colonies' wellbeing including food and nutrition security. Problems

[2] Full scanned archival report accessible on https://collections.nlm.nih.gov/bookviewer?PID=nlm:nlmuid-25110080R-bk

[3] The French Union was the new constitutional and therefore legal term gathering French territorial units across the globe previously known as colonies and protectorate countries. Cf. Général Catroux 1953.

[4] The AOF stands for the French acronym 'Afrique Occidentale Française'. The acronym AOF is used in this article.

associated with these changes are documented (see Cooper, 2004: 18; Tourte, 2005: 12).

This political evolution paved the way for the establishment of development goals notable at this stage. This led to a *nutritional scourge (péril alimentaire)* which continued a few years after WWII and shortly before independence. Consequently, the *Mission Anthropologique*, a French West African organization called *Organization of Surveys for the Study of Indigenous Peoples*[5] created in May 1946, targeted malnourishment and nutrition; it aimed at defining the nutritional status of the populations and the deficit of food in the colonies constituting the AOF, as well as identifying growth disorders and the state of malnutrition or undernutrition. This organization was composed of five doctors and four pharmacists from the colonial troop under the lead of Léon Pales and technically assisted by translators and drivers. Other relevant aspects to food and nutrition include the role of the *Organisme de Recherches sur l'Alimentation et la Nutrition Africaines* (ORANA); the Agriculture Department; Muslim brotherhoods (Chafer, 2015; Diop & Diouf, 1992: 70; Grandhomme, 2009: 172; Tamba, 2011); maraboutic Islam (Coulon, 1981: 5; Diouf, 1994; Owen, 2018); and the quest for independence and decolonization (Chafer, 2015; Grandhomme, 2009).

Exhibited by the first period, although based on selected themes of trajectory and figuration, are several pertinent aspects. Social policy in colonized Senegal was not understood as it is done in current social policy—for example, 'the collective public efforts aimed at affecting and protecting the wellbeing of people within a given territory' (Adésínà, 2009: 38). It assumed a colonially engineered process of welfare. The purpose also differed considerably from contemporary aims. Colonial food and nutrition concerns and initiatives were inclined to profit-making as the previous period demonstrated. A primary concern with imperial interests and the absence of a direct focus on improving the wellbeing of the colonized populations are widely documented (see Vitaux, 2018). It is important to understand that what we nowadays understand as social policy, and more particularly as food policy, was, during these two periods, more of a colonial rule addressing nutrition and agriculture issues. With France's control of the AOF and Senegal being central, the resultant contemporary dependency is rooted in colonization. Most African countries continue to exhibit a colonial legacy and are entangled in a relationship of dependency despite indigenizing social policy and decolonial attempts. This is the thrust throughout the next period.

This second period shows that the actors' group slightly changed in relation to the inclusion of colonial powers. France restructured its empire into the French Union, which resulted in a duality of insights from the metropolis and from Dakar. Internationally, concerns about how empires were treating their colonies led to the involvement of representatives of 44 nations, and, later, the UN through FAO. The FIDES in 1946 constituted a turning point in terms of social policy via several

[5]The anthropological mission became, in 1953, the African Food and Nutrition Research Organization.

means and groups, such as the *Mission Anthropologique*, the ORANA, the Agriculture Department and the ongoing presence and influence of Muslim brotherhoods.

9.4.2 The Post-Colonial Period

As the four presidential periods of Senegal under, respectively, Senghor, Diouf, Wade and Sall show, multilateral agreements with international stakeholders, a greater chunk of external funding and control of food policy and other social policies and the predominance of rice consumption are among the major examples of continuities and similarities throughout all the periods and categories. Divergence and breaking away from a period were mainly necessitated by crucial turns in external forces more than local initiatives, particularly in a global context of the rising number of institutional bodies addressing food, nutrition and agriculture issues after independence. Four selected sub-thematic areas are pertinent: an analysis of the dynamics of cooperation and neocolonialism, change in continuity, a break from the past in the so-called *Sopi* (change in Wolof) and more of a renaissance in calling for *Benno Bokk Yakaar* (meaning 'united for the same hope' in Wolof).

9.4.2.1 Between Cooperation and Neocolonialism

Signalling continuity from colonialism, from a political point of view, Léopold Sédar Senghor already represented Senegal as an elected African deputy (Cooper, 2004: 18), giving people in colonial territories a small number of representatives in the Paris legislative elections.[6] He shared a strong state-interventionist understanding with the former colonial power and kept promoting cash crops after independence in 1960, which increasingly engulfed the country in an unsustainable food dependency. This was already the case under colonial rule, showing the continuity of the cash crop option (Ndiaye, 2010). However, during his administration, the government set the price of rice to increase affordability for the majority.

Several public agencies on food and agriculture were created between 1964 and 1976: the *Centres Régionaux d'Assistance au Développement* (CRAD) in 1964 to encourage the creation of cooperatives and the *Office National de Coopération et d'Assistance pour le Développement* (ONCAD) in 1966 to nationalize the marketing functions within this organization, which consolidated the actions of the ONCAD and the CRADs. It held a monopoly on the importation of rice and, from 1975–1976 to 1978–1979, the primary marketing of the main agricultural products (local cereals, groundnuts and cotton). Besides, ONCAD oversaw agricultural input supply and campaign credits (Harré, 1994). Other institutions were set up to provide

[6] Félix Houphouët-Boigny in Ivory Coast as well

technical advice on agriculture to rural populations. Moreover, during his mandates, five Quadrennial Economic and Social Development Plans (QESDPs) were established (Ndao, 2015; Diagne, 2017; Spray, 2018). In 1977, the fifth QESDP was introduced, when the country had to raise its imports of cereals and especially rice. The same year, to avoid such costs, the Senegalese government moved towards a 'voluntarist' policy of replacing imported cereals with local cereals (Benz, 1996).

Official prices were set for production and consumption to favour local cereals. This is how the *Equalization and Stabilization Fund* worked. In 1979, the Fund replaced the ONCAD and was then responsible for ensuring all commercial imports and provisions of rice, while the Office of the Commissioner for Food Security was accountable for the rest of the food aid (idem). In 1980, ONCAD was liquidated and privatized, which 'marked the beginning of the state disengagement from trade in agricultural products and input supply' (Faye et al., 2007: 58). In terms of financial resources need to implement these programmes, available information is very limited.[7] The World Bank produced sectoral distribution of realized investments tables[8] based on the different QESDPs; however, to measure what was allocated to nutrition or food is merged into the sector division.

Overall, this first post-colonial period witnessed a political evolution that had no precedent in having a former colonial state relying on its exceptional good relationship with its former settler in keeping the same economic strategies in farming activities. In addition, after its independence, Senegal has not experienced spectacular industrialization and has remained confined to importing industrial goods (Diagne, 2017: 4). This continuity pervades most African countries well after independence. What this third period reveals is a change in constellation in a new independent state (Senegal) and a former colonial power (France). Nonetheless, both countries kept tight links. This is how, for instance, Senegal continued commercial groundnut production and kept having a comparative cost advantage in groundnut prices, as France set a higher price than that prevailing in the market. In addition, international actors were less prominent than during the late colonial time, as Senghor focused on the implementation of national economic and social development plans.

9.4.2.2 Changing in Continuity

As Prime Minister and, following Senghor's resignation, the new President of the Republic, Abdou Diouf adopted a political stance that was rooted in the Parti Socialiste Sénégalais. Upon serving 2 years of Senghor's mandate, he was re-elected

[7]As Alfred Inis Ndiaye (2010) explained, there is very little information about the funding of the first nutrition policies.

[8]Budgets for the financial resources of the different programmes are not available per se; available information on funding is either digitalized copies of the *Loi de Finances* (Loi no. 77–67 of June 4, 1977) or World Bank memoranda such as the *Senegal: Memorandum economique* of November 5, 1984.

in 1983 and 1988 (Diop & Diouf, 1992). As the second president from 1980 to April 2000, he based his ideology on a liberal economy, although he was left wing. He continued his predecessor's work with three other QESDPs and nutrition programmes until 1995.

His mandate was affected by the SAPs imposed by the International Monetary Fund (IMF) and the World Bank. Several scholars explore the situation of Senegal under SAPs (Dieng, 1996: 49; Dahou & Foucher, 2004: 10; Diagne, 2017; Diop et al., 2000: 163) and other food challenges. For example, a conjectural food deficit due to bad weather and subsequent chronic deficits due to bad harvests worsened the situation. Importation and food aid were adopted (Bonnefond & Couty, 1988: 330–331). In 1982 and 1983, Senegal had known a deficit of 300,000 tonnes of staple food partly covered by emergency food aid (Diagne et al., 2011: 41). In 1984, the state introduced the New Programme Agricole (NPA) that led to the state's disengagement from agricultural value chains, the reduction of subsidies and the abolition of the Programme Agricole, which was the primary source of inputs for farmers in the following years (Dieng & Gueye, 2005: 14: Bonnefond & Couty, 1988: 326–327). Despite problems, the NPA's number one goal was to reach food self-sufficiency (Oya & Cheikh, 2000: 2).

This launched the long period of the state's withdrawal from agricultural economic management and large-scale privatization of state-owned companies (Faye et al., 2007).

Overall, during his tenure (1981–2000), President Abdou Diouf strived to provide his country with effective public services and bring government closer to its citizens going through many different political mechanisms such as unilinear interventionism, liberalization and growing international cooperation (Piveteau, 2005; Diop, 2006). Examples include the Community Nutrition Project (PNC) of 1995 that, however, appeared to be a top-down project of the World Bank and a political opportunity for the Senegalese Government to attract support from the youth (Spray, 2018); and the Conseil National pour la Sécurité Alimentaire (CNSA) created in 1998 under the chairmanship of the Prime Minister. This meant a new constellation of actors in increasing state bodies' roles and international actors. The CNSA led, with UN agencies involved in the humanitarian field, the monitoring of food insecurity and the coordination of technical and financial partner's interventions, signalling the continuity of external domination reoriented the trajectory. The Abdou Diouf presidency also represents the evolving institutionalization of nutrition, particularly from 1990 to 2000, as we have seen with the numerous formalizations of the food and nutrition theme.

9.4.2.3 Sopi: The Call for Change Reached the Presidential Seat

After four decades of Parti Socialiste Sénégalais rule, Abdoulaye Wade was elected with massive support from unemployed youths hungry for change. His period in office from 2000 to 2012 is known for the liberalization of the Senegalese economy. One of the very first measures put in place by the administration was the adoption of a new constitution promulgated in January 2001 (Dahou & Foucher, 2004: 14; Diop, 2006: 114,115), which renewed the presidential supremacy and restored the 5-year term instead of 7. The ruling party had a majority (two-thirds of deputies) in the National Assembly after the elections, which allowed it to have considerable latitude to legislate (Diop, 2006: 108). Therefore, the political opposition could not legally influence major political decisions.

Several key aspects relating to food and nutrition can be noted under his administration: the pursuit of agricultural development through improved and new technologies (like mechanization, irrigation and better-regulated private land ownership), supported by the Agro-Sylvo-Pastoral Orientation Law of 2004 (Coumba Diouf, 2015: 2); the launching of the Accelerated Growth Strategy for the period 2006–2010 to boost an economic growth rate of 7–8% a year to reduce poverty in the country (Diop et al., 2013: 825); the launching of the Back to Agriculture Plan in 2006 with a total estimated budget of 60 million US dollars[9]; the creation of the National Agricultural Insurance Fund of Senegal in 2008 in the form of a semi-public and semi-private joint-stock company and with capital involving the state (IMF, 2019: 67); and the establishment of the Cell for the Fight against Malnutrition (CLM),[10] which was created and placed under the chairmanship of the Prime Minister, leading to the first nutrition policy letter (2001–2014), followed by several nutrition-related policies (Spray, 2018: 2). The Great Agricultural Offensive for Food and Abundance (*Grande Offensive Agricole pour la Nourriture et l'Abondance* (GOANA), launched in 2008 in a political context when wintering[11] began, warned farmers not to acquire inputs on credit, despite the facilitation of their acquisition (Sarrouy, 2010: 27). Against a backdrop of food riots, the Global Hunger Index Scores 2019 (von Grebmer et al., 2019) ranks Senegal 67th out of 117 countries, which means that significant improvement has been made. Many initiatives were adapted between 2002 and 2007, sponsored by the African Development Bank (AfDB), the World Food Programme (WFP) and the United Nations Children's Fund (UNICEF) (World Bank, 2011; Gouvernement République du Sénégal, 2017).

Efforts to address food and nutrition during the tenure of Abdoulaye Wade's administration are commendable. While not trivializing his efforts, the neoliberal ideology is thrust upon the hegemony of the Bretton Woods institutions, external control, dependency, limited opportunities for indigenous policy and increasing vagaries of global economic and political instability. Senegal may be said to be better performing in food security than its West African neighbours (von Grebmer et al., 2019: 55). However, more effort is required to improve the food situation particularly through local investment in agriculture, guaranteeing food availability and stabilizing food prices (Diop, 2010; Owen, 2018: 5; Touré, 2017). This period reveals a similar constellation as the previous one in internationalizing his actions.

[9] Budget estimation in the 2006 introductory note: Plan REVA; *Retour vers l'agriculture*, written by Farba Senghor, Minister of Agriculture, Rural Hydraulics and Food Security

[10] French acronym standing for *Cellule de Lutte contre la Malnutrition*.

[11] Wintering corresponds to the rainy season.

The difference resides in the president's extended power to take decisions, since he and the government have a majority of seats in the National Assembly, which considerably changes decision-making and constitutes a turning point. The trajectory in this period has completely shifted in terms of food policy achievement since many initiatives emerged with few results.

9.4.2.4 Benno Bokk Yakaar: The Promise to the Most Vulnerable

The fourth and current president, Macky Sall, elected in 2012, represents a neoliberal ideology with a totally different approach to governance. His motto 'Benno Bokk Yakaar' means united for the same hope. He is a member of the Alliance pour la République party that he founded in 2008 when he left the Parti Démocratique Sénégalais (PDS) founded in 1974 by his predecessor. The party could be seen as centrist with liberal ideologies. Macky Sall had been former President Wade's mentee and prime minister. He ran for office and won the elections in 2012. Senegal has been facing a double dependence over the years in terms of food and energy (Gassama, n.d.; Système d'information énergétique du Senegal, SIE-Senegal, 2006; Dia et al., 2009). This situation has been reinforced by the decline of the national agricultural production and market imbalances (Gassama, n.d.: 2). This was particularly severe during the previous liberal regime, which focused on short-term issues that brought Senegal prestige but did not address its population's primary needs.

President Sall's main activities concerning Senegal's social developments have been planned with a long-term vision but with short-term means for funding and improving Senegalese everyday life (Ndiaye, 2010:38; Ba, 2015; Lo et al., 2019). Many initiatives and programmes to eradicate poverty, fight against hunger and promote education, as well as grant universal health coverage, are covered under the government's *Plan Sénégal Emergent*, adopted in November 2012, which is the framework for economic and social policy in the medium and long term. In addition, the reform of the third Decentralization Act[12] in 2013 reshaped the territorial action of the state (Sarr, 2015; Owen, 2018: 9).

The government, aiming at rationalizing interventions in the field, set up in 2015 a Single National Register (RNU) intending to register 450,000 households living in extreme poverty (ONU Sénégal, 2018; Ndiaye et al., 2019). To date, 442,019 households have been registered, which represents 98.2% of the 450,000 targeted households (ONU Sénégal, 2018: 74). Eleven national social protection programmes used the RNU data in 2016 (ONU Sénégal, 2018:74; Ndiaye et al., 2019: 3). Local authorities have completed this register through targeting committees[13] that selected

[12] Decree no. 2013–581 of April 29, 2013, establishing the National Steering Committee for the Reform of Decentralization (Act III).

[13] Field activities revealed the composition of these committees. They usually include a religious leader, such as an imam or priest, the village chief, a local authority representative, a teacher and, in some cases, the youth or women representative. Information was collected between January and April 2019.

households in situations of extreme poverty to allow them to access the social safety net. Approximately 99,938 households in 2014 and 197,751 households in 2015 received cash transfer-based support (ONU Sénégal, 2018: 26). However, cash transfers are less useful in transforming the food situation of the beneficiaries. Indeed, interviews[14] revealed that they feed on the moment; however, they do not provide a medium- nor long-term perspective for households to foresee and have a plan B.

School feeding programmes are also evident. Through school canteens, the programme enabled 397,471 pupils from vulnerable families to benefit from school meals in 2016 (ONU Sénégal, 2018: 72). 'However, the withdrawal of technical and financial partners from this program has dramatically reduced its scope to a reduction of 500,000 beneficiaries between 2015 and 2016' (ONU Sénégal, 2018: 72). Students benefited as well from social measures. The government reduced the price of meal vouchers. For instance, the cost of the meal ticket was decreased from 150 to 100 CFA francs for lunch and from 75 to 50 CFA francs for breakfast (Gouvernement de la République du Sénégal, 2018; Diatta, 2018).[15]

Overall, the current government has significantly improved its population's food situation and other aspects of living conditions by increasing public expenditure in social protection. Despite Senegalese efforts to reduce the poverty rate by increasing the economic growth model, promoting decentralization and committing to enhancing both human and economic transformation, the state has insufficient means to reach those goals in the short or middle term. Achieving PSE funding continues to be a major challenge due to insufficient domestic revenues (see OECD, 2019). The desired improvements cannot be sustained due to reliance on external funding compared to limited government expenditure. Cash transfers for food and nutrition security have faltered and do not address the real causes of poverty. A lack of coordination among the different government ministries, multilateral institutions, NGOs and other service providers reduces the effectiveness of social safety nets (ONU Sénégal, 2018: 13; Ndiaye et al., 2019: 1).[16] The interplay of actors has shifted over time as nutrition and food security have taken a more central place on the international organizations' agenda. Over time, both urban and rural populations increasingly suffer when acquiring staple food, although there has been an increasing institutionalization of the food question through legal texts, social protection schemes and so forth. The cash crop dependency forces the state to import a huge quantity of rice and has therefore disrupted local markets (see Sarrouy, 2010: 29). Exploration of the last period reveals a trajectory with previous similar patterns in demonstrating a continuous and increasing role of international actors in shaping

[14] Source: Interview with Family Safety Net beneficiaries in Diourbel on March 26, 2019

[15] Prices in euros at the following rate (1 EUR = 655,957 XOF; 1 XOF = 0,00152449 EUR): 150 XOF = 0.228,674 EUR; 100XOF = 0.152,449 EUR; 75 XOF = 0.114,337 EUR; 50 XOF = 0.0762245 EUR

[16] Not only the literature but also interviews with the director of the national institution (Dakar, March 5, 2019) and the coordinator of the national institution (Dakar, February 28, 2019) emphasized this.

food and social policy. The turning aspect resides in the great expansion of programmes and external funding that led to the current state of food, nutrition and agriculture policies.

9.5 Discussion

The discussion highlights the enduring influence of colonial legacies on Senegal's social policies, particularly in the domain of food security and sovereignty. It critiques the modern concept of social policy by contrasting it with colonial-era welfare practices, which prioritized imperial economic interests over the well-being of the colonized population. This historical context remains relevant today as Senegal continues to depend on former colonial powers and international institutions, a dependency that hinders genuine food sovereignty. Initiatives like *Sopi* and *Benno Bokk Yakaar* emphasize the need for local strategies to break neocolonial influences and foster sustainable social development.

A closer examination of colonial policies reveals structural dependencies that persist to this day. For example, Albert Sarraut's 'Mise en valeur' plan prioritized groundnut exports to France, sidelining subsistence farming and undermining local food sovereignty (Tourte, 2005). These policies entrenched a reliance on imported staples such as rice and external funding for food security programmes. Contemporary challenges, including climate change and global market fluctuations, have only deepened these vulnerabilities. Addressing them requires a thorough understanding of how colonial-era practices shaped present-day dependencies. As Mkandawire (2004) argues in *Social Policy in a Development Context*, decolonizing policy frameworks is essential for fostering adaptive and autonomous governance structures.

To reduce Senegal's dependence on external actors and advance food sovereignty, fostering localized agricultural initiatives is essential. Promoting indigenous crops and traditional farming practices can empower communities and reduce reliance on imported staples like rice. Efforts such as Senegal's Great Agricultural Offensive for Food and Abundance (GOANA) illustrate attempts to boost local production and minimize import dependency (Sarrouy, 2010).

Another crucial step involves strengthening institutional capacity to monitor and implement food security policies effectively. This requires building partnerships among local governments, civil society and private sector actors to ensure diverse representation and sustainable solutions. For example, the establishment of the National Agricultural Insurance Fund in 2008 demonstrates how financial instruments can protect farmers against economic shocks (IMF, 2019). Targeting such mechanisms towards smallholder farmers could help promote equitable development, as Patel (2009) suggests in *What Does Food Sovereignty Look Like?*

Integrating agroecological approaches is also pivotal. These approaches combine environmental sustainability with agricultural productivity, offering long-term solutions to food security challenges. The Great Green Wall initiative, which combats land degradation while enhancing local food production, serves as a valuable model for building resilience.

Finally, insights from other contexts highlight pathways for addressing food security challenges. Rwanda's post-genocide agricultural reforms demonstrate how investments in agroecology and farmer cooperatives can transform food systems (Ansoms et al., 2022). Similarly, Senegal's School Feeding Programme, which provided meals to about 400,000 pupils in 2016, showcases the potential of targeted social protection schemes (ONU Sénégal, 2018). However, scaling these programmes requires sustained funding and improved coordination among stakeholders.

By understanding the historical trajectory and building on successful examples, Senegal can develop governance frameworks that prioritize local agency and resilience. Breaking free from historical dependencies will require reinforcing community-led initiatives and promoting inclusive partnerships, laying the foundation for genuine food sovereignty and sustainable social policies.

9.6 Conclusion

The cardinal argument of this article is that trajectory, periodization and figuration provide valuable lenses to understand the dynamics and design of effective food, nutrition and agriculture policies that improve well-being now and in the future. Figuration highlights the relationships among social policy actors, power dynamics and their influence on food and nutrition policy from colonial to post-colonial times. Trajectory and periodization reveal the critical junctures, continuities and interruptions that have shaped food security policies, along with the factors driving these developments. As food and nutrition security issues intensify and the number of stakeholders increases, it is vital to expand social policy and development agendas beyond national borders.

Lessons from Senegal's historical trajectory and figuration reveal the importance of addressing structural factors that perpetuate food insecurity and marginalization. Current food and nutrition policies rely heavily on formal social protection, particularly cash transfers administered through government and NGO partnerships. While social protection is an essential aspect of social policy, it is insufficient to tackle the underlying causes of poverty, inequality and food insecurity. Senegal, like other African countries, must reimagine social policy by adopting a broader and more transformative vision that addresses structural challenges.

The interplay of local stakeholders, national government and international organizations demonstrates both opportunities and constraints. Senegal's limited resources necessitate reliance on external funding to implement programmes, but this dependence underscores the need to design locally driven, owned and controlled policies for long-term sustainability. Networking and collaboration, when focused on mutually beneficial partnerships, can help bridge gaps in the short term. However, reducing external dependency and dominance should remain a primary objective. This calls for mobilizing domestic resources, fostering economic transformation and designing equitable frameworks that prioritize local agency and innovation.

To enhance local agency and sustainability, restructuring the collaboration between local and international actors is essential. Partnerships should focus on technology transfer, capacity building and resource sharing, reducing dependency while fostering innovation. South-South collaborations with countries like Rwanda or Brazil, which have pioneered transformative agricultural policies, offer valuable lessons. As Frankema and van Waijenburg (2014) argue, local economic resilience can be strengthened through policies that prioritize domestic resource mobilization and equitable development.

In conclusion, Senegal's journey underscores the critical need to address historical legacies while leveraging local innovations to build sustainable food and nutrition policies. By integrating lessons from past trajectories with forward-looking strategies, Senegal can empower local actors, minimize external dependency and chart a path toward genuine food sovereignty. This approach can serve as a model for other African countries, inspiring reforms that promote sustainable and equitable development across the continent.

References

Adésínà, O. J. (2009). Social policy in sub-Saharan Africa: A glance in the rear-view mirror. *International Journal of Social Welfare, 18*(1), 37–51.

Adesina, O. J. (2010). Beyond the social protection paradigm: Social policy in Africa's development. *Canadian Journal of Development Studies, 32*(4), 454–470. https://doi.org/10.1080/02255189.2011.647441

Adésínà, O. J. (2011a). Beyond the social protection paradigm: Social policy in Africa's development. In *Social Protection for Social Justice International Conference, Institute of Development Studies, UK, April 13–15, 2011.*

Adésínà, O. J. (2011b). *Towards a new global anti-poverty agenda: Wider vision, broader instruments.* United Nations Research Institute for Social Development.

Adesina, O. J. (2020). Policy merchandising and social assistance in Africa: Don't call dog monkey for me. *Development and Change*, 1–22. https://doi.org/10.1111/dech.12569

Albert Sarraut, P. G. (1931). Grandeur et servitude coloniales. *Bulletin De L'École Française D'Extrême-Orient, 31*, 557–558.

Ansoms, A., Aoun, E., Chemouni, B., Niyonkuru, R. C., & Williams, T. P. (2022). The politics of policymaking in Rwanda: Adaptation and reform in agriculture, energy, and education. *Journal of Eastern African Studies, 16*(2), 205–227.

Ba, M. (2015). Macky Sall peut-il tenir ses promesses? Jeune Afrique, May 29. https://www.jeuneafrique.com/233242/economie/macky-sall-peut-il-tenir-sespromesses/

Bayart, J. F. (1996). End game south of the Sahara? France's African policy. In C. Alden & J. P. Daloz (Eds.), *Paris, Pretoria and the African continent* (pp. 3–20). Palgrave Macmillan.

Bayart, J. F. (2009). *The state in Africa: The politics of the belly* (2nd ed.). Wiley.

Benz, H. (1996). *Riz local et riz importé en Afrique. Les déterminants de la compétitivé: Les filières riz du Sénégal et de la Guinée face aux exportations de la Thaïlande et des Etats-Unis* (Unpublished PhD thesis, École des Hautes Études en Sciences Sociales).

Berthélemy, J.-C. (1980). L'économie de l'Afrique occidentale française et du Togo, 1946–1960. *Revue française d'histoire d'outre-mer, 67*(248–249), 301–337. https://doi.org/10.3406/outre.1980.2263

Bonnecase, V. (2009). Avoir faim en Afrique occidentale française: Investigations et représentations coloniales (1920–1960). *Revue d'Histoire des Sciences Humaines, 21*(2), 151–174.

Bonnefond, P., & Couty, P. (1988). Sénégal: Passé et avenir d'une crise agricole. *Tiers, 29*(114), 319–340. *Canadian Journal of Development Studies, 32*(4), 1–20.

Chafer, T. (2015). Partir pour mieux rester: Le cas du Sénégal. In B. Salvaing (Ed.), *Pouvoirs anciens, pouvoirs modernes de l'Afrique d'aujourd'hui* (pp. 19–37). Presses universitaires de Rennes. http://books.openedition.org/pur/62347

Chipenda, C. (2019). The youth after land reform in Zimbabwe: Exploring the redistributive and social protection outcomes from a transformative social policy perspective. *Canadian Journal of African Studies*. https://doi.org/10.1080/00083968.2019.1648308

Cooper, F. (1996). *Decolonization and African society: The labor question in French and British Africa*. Cambridge University Press. (French translation: 2004, Paris: Karthala).

Cooper, F. (2004). Development, modernization, and the social sciences in the era of decolonization: The examples of British and French Africa. *Revue d'Histoire des Sciences Humaines, 10*(1), 9–38.

Coulon, C. (1981). Les Marabouts idéologiques. *Politique Africaine, 4*, 111–114.

Coumba Diouf, N. (2015). *Influencer la loi d'orientation agro sylvo pastorale au Sénégal par la concertation paysanne*. International Institute for Environment and Development.

Creswell, J. W., & Creswell, J. D. (2018). *Research design: Qualitative, quantitative, and mixed methods approaches*. Sage.

Dahou, T., & Foucher, V. (2004). Le Sénégal, entre changement politique et révolution passive: 'Sopi' or not 'sopi'? *Politique Africaine, 96*(4), 5–21.

Dia, D., Fall, C. S., Ndour, A., & Sakho-Jimbira, M. S. (2009). *Le Sénégal face à la crise énergétique mondiale: Enjeux de l'émergence de la filière des biocarburants*. Israbame.

Diagne, A. (2017). *Revue des politiques publiques menées au Sénégal sur la période 1960-2016*. [Research report]. ENSAE-Sénégal. ⟨hal-01478937⟩.

Diagne, A., Cabral, F. J., Cissé, F., et al. (2011). *Assessing development strategies to achieve the MDGs in the Republic of Senegal*. UN Department for Social and Economic Affairs.

Diatta, M. (2018, May 29). Augmentation des bourses, baisse du prix du ticket de resto: Le dessert de Macky aux étudiants. *Le Quotidien*. https://www.lequotidien.sn/augmentationdesbourses-baisse-du-prix-du-ticket-de-resto-le-dessert-de-macky-aux-etudiants/

Dieng, A. A. (1996). Le Sénégal au-delà de l'ajustement structurel: Pour une stratégie de développement fondée sur des forces populaires. *Africa Development, 21*(2/3), 43–65.

Dieng, A., & Gueye, A. (2005). *Revue des politiques agricoles au Sénégal: Bilan critique de quarante années de politique céréalière*. http://www.bameinfopol.info/IMG/pdf/Revue_Dieng_et_Gueye.pdf

Diop, M. C. (2006). Le Sénégal à la croisée des chemins. *Politique Africaine, 4*(104), 103–126.

Diop, A. B. (2010). Sénégal: Les mouvements sociaux sous l'alternance. *Alternatives Sud, 17*(4), 139–146.

Diop, S., Thiam, D., Sene, A., Cisse, M., Fall, K., Toure Fall, A. O., Sow, O., & Diakhate, L. (2000). Association drépanocytose-déficit en G-6-PD: Prévalence et influence sur le profil évolutif. *Médecine d'Afrique Noire, 47*(7), 321–326.

Diop, M., Faye, S. S., Hopwood, I., Kinda, O., Lomeña, M., Boumas, G., et al. (2013). SenEval—A decade of advocacy and action for evaluation in Senegal. In J. Rugh & M. Segone (Eds.), *Voluntary organizations for professional evaluation (VOPEs): Learning from Africa, Americas, Asia, Australasia, Europe and Middle East*. UNICEF.

Diop, M. C., & M. Diouf. (1992). L'Administration sénégalaise, les confréries religieuses et les paysanneries. *Afrique et Développement, 17*(2), 65–87.

Diouf, M. (1994). L'échec du modèle démocratique du Sénégal, 1981–1993. *Africa Spectrum, 29*(1), 47–64.

Elias, N. (2006). *The court society: The collected works of Norbert Elias* (Vol. 2). UCD Press.
Elias, N. (2008). *The establishment and the outsiders: The collected works of Norbert Elias* (Vol. 4). UCD Press.
Elias, N. (2012). *On the process of civilisation: The collected works of Norbert Elias* (Vol. 3). UCD Press.
Elias, N. (2013). *Studies on the Germans: The collected works of Norbert Elias* (Vol. 11). UCD Press.
FAO. (n.d.). *The Founding of FAO—The Hot Springs Conference*. http://www.fao.org/3/p4228e/P4228E04.htm
Fanon, F. (1967). Vérités premières à propos du problème colonial. Dans Pour la révolution africaine: Écrits politiques (pp. 141–149). François Maspero.
Faye, J., Ba, C. O., Dieye, P. N., & Dansoko, M. (2009). *Évaluation de la politique de relance du secteur rural au Sénégal. Rapport de mission, September 2009*. IPAR.
Frankema, E., & van Waijenburg, M. (2014). Metropolitan blueprints of colonial taxation? Lessons from fiscal capacity building in British and French Africa, c. 1880–1940. *The Journal of African History, 55*(3), 371–400.
Gassama, K. (n.d.). Sénégal: La spirale de la crise alimentaire mondiale et la montée des périls au pays. *Sud Quotidien*. http://www.hubrural.org/Senegal-La-spirale-de-lacrise.html?lang=fr
Gouvernement de la République du Sénégal. (2018). *Conseil des Ministres du 30 mai 2018*. https://www.sec.gouv.sn/actualit%C3%A9/conseil-des-ministres-du-30-mai-2018
Gouvernement République du Sénégal. (2017). *Programme de Renforcement de la Nutrition (PRN)*. https://www.sec.gouv.sn/programme-de-renforcement-de-la-nutrition-prn
Grandhomme, H. (2009). Connaissance de l'islam et pouvoir colonial: L'exemple de la France au Sénégal, 1936–1957. *French Colonial History, 10*, 171–188.
Guy, C. (1923). La mise en valeur des colonies françaises. *Annales de Géographie, 32*(177), 265–271. https://doi.org/10.3406/geo.1923.9946
Harré, R. (1994). Emotion and memory: The second cognitive revolution. *Philosophy, Psychology and Psychiatry, 37*, 25–40.
Hugon, P. (2007). La politique économique de la France en Afrique. *Politique Africaine, 1*(105), 54–69.
International Monetary Fund. (2019). *Senegal: IMF Country Report No. 19/27*. IMF.
Lo, C., Dia, A. K., Dia, I., et al. (2019). Evaluation of the residual efficacy of indoor residual spraying with bendiocarb (FICAM WP 80) in six health districts in Senegal. *Malaria Journal, 18*, 198. https://doi.org/10.1186/s12936-019-2829-4
Maddox, G. (Ed.). (2018). *Conquest and resistance to colonialism in Africa*. Taylor and Francis.
Marseille, J. (1984). *Empire colonial et capitalisme français: Histoire d'un divorce*. Albin Michel.
Masters, W. A., Rosenblum, N. Z., & Alemu, R. G. (2018). Agricultural transformation, nutrition transition and food policy in Africa: Preston curves reveal new stylised facts. *The Journal of Development Studies, 58*(5), 788–802.
Migdal, J. S., & Schlichte, K. (2005). Rethinking the state. In K. Schlichte (Ed.), *The dynamics of states* (pp. 11–50). Routledge.
Misselhorn, A., & Hendriks, S. L. (2017). A systematic review of sub-national food insecurity research in South Africa: Missed opportunities for policy insights. *PLoS One, 12*(8), e0182399. https://doi.org/10.1371/journal.pone.0182399
Mkandawire, T. (2004). *Social policy in a development context*. Palgrave Macmillan.
Mkandawire, T. (2005). *Targeting and universalism in poverty reduction*. United Nations Research Institute for Social Development.
Mkandawire, T. (2015). Neopatrimonialism and the political economy of economic performance in Africa: Critical reflections. *World Politics, 67*(3), 1–50. https://doi.org/10.1017/S004388711500009X
Monteil, V. (1962). Une confrérie musulmane: Les Mourides du Sénégal. *Archives de sociologie des religions, 14*, 77–102.

Musset, R. (1930). La production et l'exportation des arachides au Sénégal. *Annales De Géographie, 39*(221).

Ndao, M. (2004). *La malnutrition infantile en AOF: Le cas du Sénégal, 1930–1960*. Annales de la Faculté des Lettres et Science Humaines, Université Cheikh Anta Diop.

Ndao, M. (2015). *L'alimentation et la santé des enfants dans le Sénégal colonial, 1905—c.a. 1960*. Harmattan Sénégal.

Ndiaye, A. I. (2010). *Comprendre les facteurs politiques et institutionnels conduisant au changement des politiques de nutrition: Le cas de Sénégal*. World Bank.

Ndiaye, J. L. A., Ndiaye, Y., Ba, M. S., Faye, B., Ndiaye, M., Seck, A., et al. (2019). Seasonal malaria chemoprevention combined with community case management of malaria in children under 10 years of age, over 5 months, in South-east Senegal: A cluster-randomised trial. *PLoS Medicine, 16*(3), e1002762. https://doi.org/10.1371/journal.pmed.1002762

OECD. (2019). Revenue Statistics in Africa 2018: Senegal. http://www.oecd.org/countries/senegal/revenue-statistics-africa-senegal.pdf

ONU Sénégal. (2018). *Bilan Commun de Pays (BCP) du Plan Cadre d'assistance du Système des Nations Unies au Sénégal 2012–2018*. http://onusenegal.org/documents/Rapport_final_du_Bilan_Commun_Pays_CCA_du_Senegalversion_6_mars_2018.pdf

Owen, J. (2018). *BTI 2018 country report: Senegal*. Bertelsmann Stiftung. https://www.btiproject.org/fileadmin/files/BTI/Downloads/Reports/2018/pdf/BTI_2018_Senegal.pdf

Oya, C., & Cheikh, O. B. (2000). Les politiques agricoles 2000–2012: Entre volontarisme et incohérence. In *Sénégal* (pp. 149–178).

Patel, R. (2009). Food sovereignty. *The Journal of Peasant Studies, 36*(3), 663–706. https://doi.org/10.1080/03066150903143079

Piveteau, A. (2005). Décentralisation et développement local au Sénégal: Chronique d'un couple hypothétique. *Revue Tiers Monde, 1*(181), 71–93.

Quintaneiro, T. (2006). The concept of figuration or configuration in Norbert Elias' sociological theory. *Teoria and Sociedade, 12*(1), 54–69.

Sarr, I. (2015). *Acte III de la Décentralisation: Une réforme, des questions*. Les cahiers de l'alternance, No 18.

Sarrouy, C. (2010). *Insécurité Alimentaire au Sénégal: L'agroécologie comme réponse à la sous-alimentation et à la dégradation de l'environnement dans un pays en développement* (Unpublished Master Thesis, Université Libre de Bruxelles). http://memenvi.ulb.ac.be/Memoires_en_pdf/MFE_09_10/MFE_Sarrouy_09_10.pdf

Spray, A. L. (2018). *Evolution of nutrition policy in Senegal*. World Bank. http://documents.worldbank.org/curated/en/946271537165819880/pdf/EvolutionofNutrition-PolicyinSenegal.pdf

Système d'information énergétique du Sénégal. (2006). *Un outil d'aide à la décision. Rapport—2006*. https://ec.eurpa.eu/energy/intelligent/projects/sites/ieeprojects/files/projects/documents/sieafrique_phase_ii_sie_senegal_2006_fr.pdf

Tamba, M. (2011). *Mutations politiques au Sénégal: Bilan de cinquante ans d'indépendance, (1960–2010)*. Konrad-Adenauer Foundation.

Tekwa, N., & Adesina, J. (2018). Gender, poverty and inequality in the aftermath of Zimbabwe's land reform: A transformative social policy perspective. *Journal of International Women's Studies, 19*(5), 45–62.

Thierry, B., J. Hurley, L. Martin, and D. McGrenra. (2019). République du Sénégal. Programme d'options stratégiques pour le pays. 2019–2024. Document no. EB 2019/126/R.18. Rome: International Fund for Agricultural Development.

Touré, I. (2017). Jeunesse, mobilisations sociales et citoyenneté en Afrique de l'Ouest: Étude comparée des mouvements de contestation "Y'en a marre" au Sénégal et "Balai citoyen" au Burkina Faso. *Africa Development, 42*(2), 57–82.

Tourte, R. (2005). La période coloniale et les grands moments des jardins d'essais: 1885/189 1914/1918. In *Histoire de la recherche agricole en Afrique tropicale francophone* (Vol. iv). Organisation des Nations Unies pour l'alimentation et l'agriculture.

Vitaux, A. (2018). *Aspects économiques et douaniers du système colonial français et ses particularismes en AOF et en Côte d'Ivoire*. Association pour l'Histoire de l'Administration des Douanes. https://histoire-de-la-douane.org/aspects-economiques-et-douaniers-du-systeme-colonial-francais-etses-particularismes-en-aof-et-en-cote-divoire-2/

von Grebmer, K., Bernstein, J., Patterson, F., Wiemers, M., Ní Chéilleachair, R., & Foley, C. (2019). *2019 Global Hunger Index: The challenge of hunger and climate change*. Welthungerhilfe.

Wood, L. M., Sebar, B., & Vecchio, N. (2020). Application of rigour and credibility in qualitative document analysis: Lessons learnt from a case study. *The Qualitative Report, 25*(2), 457–470.

World Bank. (2011). *Senegal—Nutrition Enhance Second Phase Project: Partners to support joint mission has implementation 20–24 June*. World Bank Group. http://documents.worldbank.org/curated/en/395811468296713803/SenegalProgrammederenforcement-de-la-nutrition-Phase-2-mission-conjointe-des-partenaires-d-appui-a-la-mise-eneuvre-20-24-juin-2011

World Health Organization. (2020). Malnutrition Key Facts. https://www.who.int/news-room/fact-sheets/detail/malnutrition.

Yi, I. (2015). *New challenges for and new directions in social policy*. United Nations Department of Economic and Social Affairs.

Young, C. (2018). The heritage of colonialism. In J. W. Harberson & D. Ro (Eds.), *Africa in world politics: Constructing political and economic order* (6th ed.). Routledge. https://doi.org/10.4324/9780429495472

Chapter 10
Agricultural Market Participation and Welfare in Rural Zimbabwe: Land Reform as a Social Policy Instrument

Clement Chipenda and Jimi Adesina

10.1 Introduction

Land and agrarian reforms have until recently been an overlooked vector of social policy (Chipenda, 2024; Tom, 2024). This is changing with the recognition that they are a policy instrument with functional equivalents to other conventional forms of social policy. As Adesina (2015: 113) has argued, land and agrarian reforms can be viewed as a social policy instrument, and social policy frameworks can be used to evaluate the outcomes of redistributive land reform in enhancing human welfare and wellbeing. These assertions have come in a context where little attention has been attached to the concomitant transformative social policy dimensions of land and agrarian reforms, an important dimension that some scholars are robustly exploring (Tom, 2024; Chipenda, 2019, 2024; Tekwa, 2023). Principal to this chapter is that land and agrarian reforms are social policy instruments with immense capacity to transform people's wellbeing through dislodging the structural causes of poverty, inequality, and marginalisation. Land and agrarian reform and the associated lessons of development and social policy can be assessed from diverse conceptual, ideological, and epistemological perspectives.

This article is anchored on a political economy approach which is integrated with the transformative social policy (TSP) approach. The TSP, credited to the contributions of various scholars, has at its core the desire to improve social policy and development thinking, design, and delivery through norms of equality and solidarity, multiple instruments that are relevant to Africa, and broader transformative outcomes (Mkandawire, 2004, 2007; Adesina, 2015, 2020). This is distinct from development and social policy informed by the neoliberal philosophy that is

C. Chipenda (✉) · J. Adesina
South African Research Chair in Social Policy, College of Graduate Studies, University of South Africa, Pretoria, South Africa
e-mail: chipec@unisa.ac.za; adesij@unisa.ac.za

E. Ndhlovu, C. Chipenda (eds.), *Redefining the Social Policy Discourse*,
https://doi.org/10.1007/978-3-032-24067-5_10

dominating Africa, and that has for long been negating transformative, inclusive, innovative, and sustainable socioeconomic wellbeing (Tom, 2024). The TSP can be seen as facilitating the understanding and adoption of land and agrarian reforms as social policy tools in Africa and other parts of the Global South (Chipenda, 2024; Tom, 2024).

The African continent has for a long time endured legacies of colonial land dispossession, with land reform emerging as a critical mechanism for addressing historical injustices and restructuring agricultural systems (Moyo, 2011a, 2011b). Zimbabwe presents a particularly compelling case study of radical land redistribution, embodying the intersections of historical dynamics, social justice, and agricultural transformation. Against the backdrop of protracted colonial land appropriation that systematically marginalised indigenous populations, the country's approach to land reform represents a bold and unprecedented intervention in addressing structural inequities (Moyo, 2013; Chipenda, 2024). It is against this background that this chapter looks at Zimbabwe as a case study, following the Fast Track Land Reform Programme (FTLRP).

Zimbabwe's FTLRP, which was officially initiated by the state in 2000, represents one of Africa's most significant agrarian transformations of the twenty-first century (Moyo & Chambati, 2013). The FTLRP can be seen as a comprehensive land redistribution initiative that fundamentally restructured the country's agricultural landscape as it sought to address historical inequities rooted in colonial dispossession (Tom & Banda, 2023). The significance of the FTLRP lies in the manner in which it was implemented and its contemporary outcomes. Polemical and divisive scholarly debates have for decades been crystallised around its complex dynamics, focused on social justice, illegal land dispossession, human rights abuses, and violence (which accompanied its implementation), among other issues (Cliffe et al., 2011). Despite being highly controversial due to the radical and unprecedented way it was executed, the programme resulted in a substantial amount of land acquisition and the resettlement of thousands of households within a very short period. The FTLRP succeeded in resettling approximately 182,000 beneficiaries under the small-scale A1 model and 22,000 under the medium-scale A2 model, with 13 million hectares of land being acquired (Matondi, 2020). The programme and the evolving legal and policy framework led to significant outcomes, including the emergence of a diverse agrarian class, a transition from a bimodal to a trimodal agrarian structure, and a shift in the land tenure regime from private property rights to state-allocated land-use rights (Moyo, 2011a, 2013).

It is now 24 years after Zimbabwe officially embarked on this radical and unprecedented FTLRP; yet, there remains limited empirical evidence on how the reconfigured agrarian structure influences patterns of market participation across different farming categories. The post-2017 political transition witnessed in the country, marked by President Robert Mugabe's departure after 37 years and the ascension of Emmerson Mnangagwa's 'New Dispensation' or 'Second Republic', introduced substantial policy shifts in the agrarian landscape. This new era embraced a notably neoliberal economic trajectory, diverging from previous redistributive and black empowerment policies, which were championed by the previous administration

(Mkodzongi, 2022). Key policy frameworks including the Transitional Stabilisation Programme (2018–2020), National Development Strategy (2020–2025), and Zimbabwe National Agriculture Policy Framework (2019–2030) have been instrumental in emphasising market-oriented agricultural transformation, which has in turn created new dynamics in rural economies (Musungwini et al., 2024; Chipenda, 2024).

This study employs an integrated theoretical framework that synthesises the political economy approach with the TSP framework to address a critical research gap. Despite substantial scholarly attention to the immediate outcomes of land redistribution, there remains a significant lacuna in systematic research examining how the new agrarian structure influences agricultural market participation and, consequently, rural household welfare. This knowledge gap is particularly compelling given the current policy environment's dual emphasis on market-led growth and social protection.

Focusing on Goromonzi and Zvimba districts in rural Zimbabwe, the study engages with two strategically selected localities that represent the diverse agro-ecological zones and farming systems within Zimbabwe's dynamic agricultural landscape. These districts have undergone substantial transformation through the FTLRP and subsequent policy shifts, thereby offering rich and diverse contexts for examining market participation patterns and welfare outcomes. The chapter systematically explores three key analytical dimensions: first, the extent and comparative patterns of market participation among resettled A1 (small-scale) and A2 (medium-scale) farmers; second, the critical determinants of market participation across these distinct farming categories; and third, the relationship between market participation and household welfare outcomes. Interwoven throughout the discussion are the institutional and structural factors that either enable or constrain market participation, providing a comprehensive analytical lens.

Through this comprehensive analysis grounded in an integrated political economy and TSP framework, the article makes a contribution to our understanding of how structural transformation in land ownership patterns, coupled with recent policy shifts, influences agricultural commercialisation and rural welfare in post-land-reform Zimbabwe. The timing of this research is particularly opportune, coinciding with both the 24-year milestone of the FTLRP and the implementation of new market-oriented agricultural policies. This temporal context provides an opportunity to assess the long-term implications of radical land reform on rural markets and livelihoods, especially given the increasing state pressure for productivity among resettled farmers (see Chipenda, 2024). In the subsequent section, we will look at the theoretical framework that underpins this critical examination of Zimbabwe's agricultural transformation.

10.2 Research Methods and Study Area

This study adopted a pragmatic research paradigm with a sequential explanatory mixed-methods design, integrating both qualitative and quantitative approaches to examine agricultural market participation and welfare outcomes in Goromonzi and Zvimba districts. This philosophical stance was considered particularly valuable for examining the multidimensional aspects of agricultural transformation and rural development (see Gillespie et al., 2024), while facilitating a comprehensive understanding of complex phenomena (see Shannon-Baker, 2016; Johnson & Onwuegbuzie, 2004; Foster 2023). The research began with quantitative research followed by qualitative inquiry, which was seen as offering advantages in capturing diverse perspectives, explaining underlying trends, identifying causal processes, and enriching theoretical insights (Caesar, 2024; Gerzso & Riedl, 2024). This approach proved particularly valuable as initial quantitative findings informed subsequent qualitative interviews, enabling effective triangulation and richer contextual understanding (Creswell, 2012, 2014; Chali et al., 2022).

The study population comprised A1 and A2 farmers in Goromonzi and Zvimba districts. Through a combined sampling approach utilising both purposive and random techniques, 200 resettled farmers were selected across both districts, in addition to 100 communal area farmers (for comparative purposes). The purposive sampling component, which Babbie (2012) notes is used to choose a sample based on the researcher's judgement of comprehensive understanding, ensured balanced representation across farming categories and geographical locations. Data collection employed multiple instruments to ensure methodological rigour and comprehensiveness. A detailed survey questionnaire incorporating over 1500 variables was administered to the 300 farmers, covering farmer profiles, agricultural production activities, household demographics, farming activities, asset accumulation, and social dynamics. This quantitative tool was complemented by in-depth interviews with 20 key informants purposively selected from government officials, resettled farmers, and local authorities, providing crucial institutional and policy perspectives.

The analytical approach maximised the strengths of both methodologies. Quantitative data was processed using SPSS Statistics software, enabling robust statistical analysis of market participation patterns and welfare outcomes. Qualitative data underwent systematic thematic analysis, with careful categorisation and coding of emerging themes. This dual analytical approach ensured statistical accuracy while providing a rich contextual understanding, with qualitative findings helping to explain statistical patterns.

Ethical protocols were rigorously observed throughout the research process. The study obtained formal ethical clearance from the university ethics committee and adhered to established institutional research guidelines (see Calia et al., 2023). Necessary authorisations were secured from government ministries, and local authorities facilitated community entry protocols. Informed consent was obtained from all participants, with clear explanations about the study's purpose, methods, and implications. Participant welfare was prioritised through robust measures ensuring privacy, confidentiality, and anonymity.

The study sites, Goromonzi District in Mashonaland East Province and Zvimba District in Mashonaland West Province, represent diverse agro-ecological zones and farming systems within Zimbabwe's agricultural landscape. Both districts feature significant populations of resettled A1 and A2 farmers as well as communal farmers from old native reserves, making them ideal for comparative analysis of market participation patterns and welfare outcomes in post-land-reform contexts.

10.3 Findings

10.3.1 Land Utilisation and Agricultural Production Patterns

Drawing on the integrated theoretical framework, the study examined land utilisation and agricultural production patterns as key mechanisms for understanding market participation across different tenure regimes in the study sites. The analysis focused on unpacking the interactions between land allocation, class formation, power relations, and resource access in shaping agricultural production and market engagement. The FTLRP's differentiated land allocation model created distinct class categories within the agrarian structure, with A1 farmers receiving smaller parcels (5–50 ha) and A2 farmers allocated larger commercial holdings (50–2000+ ha). This initial resource endowment differentiation provided a useful lens for investigating how structural inequalities and institutional arrangements influence agricultural strategies and market participation. Employing Bernstein and Oya's (2014) political economy framework, the study explored fundamental questions of 'who owns what?' through land access patterns and 'who does what?' through production strategies across farming categories. This approach allowed for an examination of how power relations and institutional contexts shape farmers' capacities to navigate resource constraints and opportunities.

The analysis engaged land-use efficiency, crop diversification strategies, and the balance between subsistence and commercial production. By considering how different farming categories develop agricultural strategies within the transformed agrarian landscape, the research provided insight into the broader social and economic transformations initiated by the land reform process. This methodological approach aligned with the TSP's emphasis on understanding structural changes as dynamic processes of social and economic restructuring. It moved beyond simplistic assessments of land redistribution to explore how institutional power, resource allocation, and market participation interact to potentially create pathways for social and economic transformation in post-land-reform Zimbabwe.

10.3.1.1 Land Utilisation

Land utilisation, as highlighted in the preceding section, remains a critical aspect that has been at the centre of the FTLRP. It has often served as a key metric for evaluating the programme's success, although this has been subject to contestation. As an initial step toward understanding market participation patterns and household welfare outcomes, it was imperative to examine land utilisation trajectories across the two tenure regimes. Figure 10.1 below summarises land allocation size and cropped area between 2019 and 2022 in Zvimba and Goromonzi. The empirical evidence reveals significant disparities between allocated land size and actual cropped area across both A1 and A2 farming categories. A2 farmers, despite having substantial land allocations of 8006 ha consistently across the study period, exhibited relatively low land utilisation rates, with cropped areas ranging from 1202 ha in 2019 to 1720 ha in 2022. This represented a gradual increase in land utilisation from 15% to 21.5% over the 4-year period. This suggested modest improvements in their capacity to bring more land under cultivation, though a significant portion remained underutilised. The Agricultural Extension Officer (AEO) in Goromonzi District, gave an explanation for these dynamics when he said:

> Over the past few years there has been a modest increase in agricultural production here in Goromonzi. The challenge is that a lot of the land is lying fallow. The increase in output which we have seen lately is due to government support and policy. We have the *Pfumvudza/Intwasa* (Climate-Proofed Presidential Inputs Scheme) for small-scale farmers and the Command (National Enhanced Productivity Scheme) for the A2 farmers. The level of financial and input assistance is at another level. Most farmers get it…our Ministry invests so much in these programmes and the results can be seen, farmers are cultivating diverse crops such as maize, sunflower, small grains, and soya beans…While the numbers of those getting help have increased the size of the assistance is limited hence expansion of cultivated fields remains constrained. (Interview, 05/03/24)

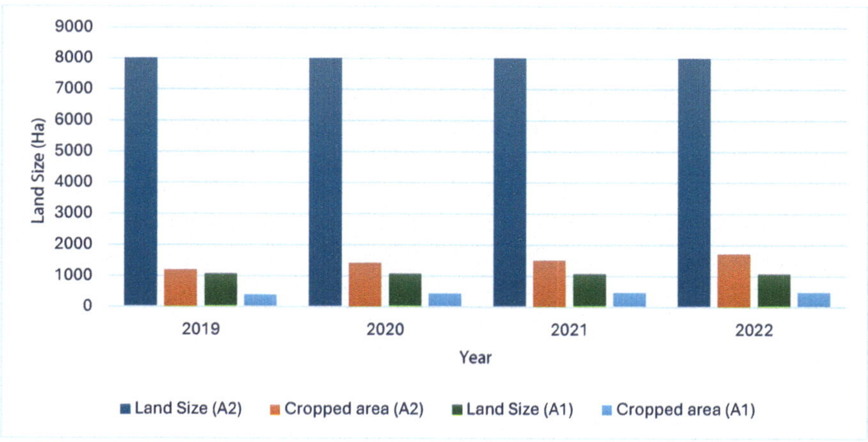

Fig. 10.1 Land size and utilisation. (Source: Author Fieldwork 2022)

In both study sites at the time of the study, it was observed that in the ongoing season, farmers had received substantial support in the form of agricultural inputs. This confirmed what the AEO had highlighted. Maize production through state inputs for A1 farmers was supported by an input package of 3 kg seed, 50 kg basal, and 50 kg topdressing fertilisers, which was expected to cover at least a 0.125-ha plot. For respondents who were interviewed, this support was '...very important; it ensures that on every farm there is a piece of land that is cultivated. This ensures a measure of food security, at least in the months immediately after the harvest...' (Interview, 09/03/24). This was seen as being a critical aspect of agrarian welfare dynamics as noted by Chipenda (2024), and it was also a contributor to the recent increase in land utilisation witnessed in recent years.

Figure 10.1 above highlights a notable contrast between A1 and A2 farmers in the study, showing that A1 farmers operated on smaller allocated plots averaging 1.068 ha. Despite this limitation, they exhibited higher land utilisation efficiency, increasing their cropped area from 399 ha in 2019 to 479 ha in 2022. This improvement reflects a rise in land utilisation from 37.4% to 44.8%. When discussing this trend, both A1 farmers and the AEO in Zvimba attributed the higher proportional utilisation of land by A1 farmers to government initiatives that promote intensive farming practices. In addition, the pressing need to maximise returns from limited land holdings was highlighted as playing a critical role, as farmers strive for food security while also responding to state mandates for productivity amid threats of eviction and the mandatory request for all farmers in the different tenure regimes to submit annual agricultural production reports.

For the study, the analysis of land utilisation patterns is vital for understanding the political economy of market participation within different tenure regimes in the resettlement areas. The evolving capacity for production, particularly among A2 farmers, is evident in their increased cropped area from 1202 to 1720 ha over the study period. This growth can be linked to improved access to resources, inputs, and agricultural expertise, which are crucial for navigating the power dynamics and class formations present in the agricultural sector. The persisting gap between allocated and utilised land, particularly in the A2 tenure regime, was seen as being a major source of concern. It was seen as revealing some of the significant structural constraints that hinder full land utilisation potential. A respondent gave insight into this by indicating that:

> Yes, there is a big gap between the land we have been allocated and what we can actually use. This shows the real troubles we face. We struggle with not having enough money to use all the land and we cannot find trustworthy and committed partners to work with in joint ventures...We have challenges accessing high paying markets even those from overseas and most of us do not have the right political connections and those who do are better off. These are some of the things that keep us from using our farms to their full potential. (Interview, 06/03/24)

The constraints which are highlighted that A2 farmers face were seen as not only limiting agricultural productivity but also restricting their market participation capabilities, ultimately impacting household welfare outcomes. As the study examined land utilisation, production dynamics, and the subsequent income generation through the integrated theoretical framework lens, it was able to gain a deeper understanding of how emergent class relations and access to resources are shaping agricultural production and market engagement in the resettlement areas. Linked to land utilisation were production and income generation dynamics, which are looked at below.

10.3.1.2 Agricultural Production

For the study, the analysis of agricultural production patterns was seen as revealing complex power relations, institutional arrangements, and structural conditions that shape the differentiated market participation capacities of A1 and A2 farmers. This analysis was considered to be crucial for understanding how historical land ownership patterns, contemporary resource access, and institutional support mechanisms are interacting to influence marketable surplus generation across different tenure regimes. The agricultural production landscape across Goromonzi and Zvimba districts was seen to be reflecting both the persistence of historical agrarian structures and emerging forms of agricultural capitalism. Resettled farmers were seen to have developed diversified cropping portfolios that span a comprehensive range of 13 crops. This included cereals such as maize, millet, sorghum, and *rapoko*; oilseeds including sunflower and soya beans; various legumes comprising groundnuts, sugar beans, cowpeas, and roundnuts; and high-value crops such as tobacco and potatoes. This was different from the cohort of communal area farmers who reported that they did not have a comprehensive portfolio of crops compared to their counterparts in the resettlement areas. Most of them could be seen mainly concentrating on the production of food crops.

For the resettled farmers, intercropping was seen as being another important strategy contributing to crop diversification. As one respondent indicated:

> … we are doing a lot of intercropping in response to the excessive dry spells which we have experienced of late. We also grow a lot of crops, and this helps us to stay safe from the risks of poor harvests or not harvesting at all. Food security is very important and so is having something to sell to get money for the household expenses. (Interview, 10/03/24)

The diversification strategy witnessed in the study sites was seen as representing not merely risk management but a reflection of the dynamics between food sovereignty imperatives, market integration pressures, differential access to productive resources, and power relations within agricultural value chains. The 2019–2021 production trajectories were clear in revealing patterns of accumulation and market integration across the different tenure regimes. Among the A2 farmers, production patterns to some extent reflected historical advantages in resource access and superior integration with formal market institutions. Their maize production increased substantially from 5,253,300 kg to 6,201,000 kg, representing an 18% growth. When it came to tobacco production, however, they showed only modest growth, with a 1% increase from 148,165 kg to 149,760 kg. These patterns and responses during interviews indicated that they had greater capacity to mobilise labour and capital, alongside enhanced bargaining power in value chains compared to A1 and their communal area counterparts. The production scale was interpreted as reflecting both historical advantages and contemporary institutional support mechanisms that favour larger-scale producers.

On the other hand, A1 farmers could be seen exhibiting emerging patterns of commercial integration through different pathways. Their maize production grew from 1,402,100 kg to 1,522,800 kg, showing an 8.6% increase. Of importance was that they achieved remarkable growth in tobacco production, with a 57% increase from 45,275 kg to 72,055 kg. These patterns indicated growing integration into contract farming arrangements and the emergence of new forms of agricultural entrepreneurship, as these farmers strategically navigate institutional barriers and adapt to market opportunities despite resource constraints. The differentiated production patterns were seen to be revealing how institutional arrangements and power relations mediate market participation through multiple mechanisms. Resource access and control continue to be influenced by historical patterns of distribution as they favour A2 farmers. Institutional frameworks also typically favour these larger-scale producers in input markets. Differential access to credit and technical support was reportedly fundamentally shaping production choices and market possibilities.

Major takeaways from the study, when looked at from a production perspective, were that differentiated production patterns in maize and tobacco show the varying capacities of the resettled farmers to utilise resources and access markets. It also signals the current dynamics that are contributing to the enhancement of their productive capacities. The ability of A2 farmers to maintain large-scale cereal production while A1 farmers show dramatic growth in high-value crop production was seen to be suggestive of diverse pathways to market participation and income generation. These patterns have important implications for understanding how the resettled farmers in the different tenure regimes circumnavigate the transition from subsistence to commercial agriculture, and ultimately, how production decisions influence market participation and welfare outcomes.

10.3.2 Income Generation and Wellbeing

The examination of agricultural commercialisation patterns in the study sites was believed to be important for a better understanding of market participation and household welfare. To this end, an analysis of marketed agricultural output and income generation was undertaken. These metrics were seen as useful in providing important information on the market integration levels of the resettled farmers and their capacity to generate income, across the different tenure regimes and geographical locations. For this article, the focus of the output analysis was on maize, tobacco, groundnuts, and sugar beans. Findings revealed significant differentiation in commercialisation patterns across the tenure regimes in both districts between 2018 and 2021. Maize marketing data demonstrated A2 farmer's market dominance in

Goromonzi, where marketed volumes increased from 2,559,000 kg (US$787,150) to 3,897,900 kg (US$972,880). Their Zvimba counterparts increased from 1,239,400 kg (US$317,610) to 1,626,000 kg (US$425,560). A1 farmers, though operating at smaller scales, showed steady growth, with Goromonzi farmers increasing from 790,300 kg (US$214,669) to 1,007,500 kg (US$246,825). In Zvimba it was from 264,500 kg (US$61,780) to 280,000 kg (US$67,520).

In the tobacco sector, an interesting pattern emerged where Goromonzi's A1 farmers demonstrated remarkable growth from 41,800 kg (US$154,125) in 2018 to 72,055 kg (US$224,193) in 2021. During this period, A2 farmers experienced a slight decline from 162,275 kg (US$485,065) to 149,760 kg (US$526,879). Zvimba showed similar trends but at lower volumes, with A1 farmers increasing from 14,600 kg (US$39,970) to 21,500 kg (US$65,000), and A2 farmers showing minimal growth from 49,900 kg (US$148,880) to 52,600 kg (US$163,500). Groundnut commercialisation exhibited regional disparities, with Goromonzi's A1 farmers increasing sales from 35,200 kg (US$11,053) to 45,500 kg (US$15,486). A2 farmers, meanwhile, maintained stable volumes of around 42,000 kg (US$13,460). Zvimba showed significantly lower participation, particularly among A2 farmers who marketed only 1000–1500 kg (US$455–875). Sugar bean marketing revealed strong commercialisation in Goromonzi in the period under review, where A2 farmers increased sales from 97,940 kg (US$91,912) to 119,650 kg (US$113,244), and A1 farmers from 14,450 kg (US$13,430) to 19,500 kg (US$18,455). Zvimba's engagement was lower but showed growth, with A2 farmers increasing from 15,300 kg (US$13,090) to 19,000 kg (US$16,840), and A1 farmers from 8350 kg (US$7480) to 10,600 kg (US$9500).

The findings in both districts on marketed output and income generation patterns were seen as revealing dynamic regional and categorical differences in market participation. The important takeaways were that from 2018–2019 to 2020–2021, agricultural output in Goromonzi increased by 42%, leading to a 20% rise in income. This was considered indicative of the economic potential of access to land, as beneficiaries' engagement in agricultural activities was seen to be positively impacting income generation. In contrast, Zvimba saw a 26% increase in agricultural output during the same period, and this also led to a higher income increase of 27%. These dynamics were reflected in the gross income realised by the farmers. In Zvimba, the gross income of A1 farmers increased significantly from US$255,230 in the 2019–2020 season to US$317,740 in 2020–2021. A2 farmers in Zvimba also saw growth, with their income rising from US$558,860 to US$691,925. In Goromonzi, A1 farmers experienced a similar trend, with income rising from US$474,329 to US$568,808. A2 farmers in Goromonzi could be seen as having excelled, with their gross income climbing from US$1,512,864 to US$1,803,136 during the same period.

Respondents noted that the steady increase in output, product sales, and income reflected a positive trajectory that was indicative of improved livelihoods. They emphasised that the levels of output, sales, and income achieved in the resettlement areas were not comparable to those in communal areas, where farmers faced significant constraints that limited their production capacity and, by extension, income generation. This substantial income growth was said to have translated into tangible improvements in household welfare across both districts. Survey data revealed that resettled farmers invested their increased earnings in various aspects of household development. Many households reported significant improvements in housing quality, with 85% of respondents reportedly having upgraded from traditional to modern housing structures. Educational investments were particularly significant, with 80% of households reporting that they were managing to send their children and relatives to better schools, in addition to supporting higher education pursuits.

Healthcare access was said to have improved. Despite the challenges in health infrastructure in both districts, 65% of households reported having a better ability to afford medical care and essential medicines. In the past 5 years, 76% of households reported having invested in productive assets such as agricultural equipment and livestock. For them this was considered to be an important and critical investment which was seen as creating a positive cycle of enhanced production capacity and income generation. Dietary diversity also improved significantly, with 82% of households reporting increased consumption of varied food groups. This was considered to be indicative of better nutrition outcomes. An important observation that was made was that these welfare improvements were particularly distinct among households that were generating income and had demonstrated consistent market participation. For the study, from an integrated theoretical approach, this was seen as suggesting a strong correlation between market engagement, income stability, and enhanced living standards, resulting in improved welfare and wellbeing. In the next section, we look at market participation dynamics.

10.3.3 Market Participation

> As farmers, the market is essential for us. It affects everything we do with our crops and livestock. It shows whether our hard work pays off and decides if we get what we deserve. How well we do in the market can really shape our lives and those of our families in the months and year ahead. (Interview, 06/03/24)

The quote above summarises the importance that farmers in the study sites have for the market. Agricultural market participation in the study sites was seen as exhibiting interesting dynamics, as it was characterised by diverse marketing channels and institutional arrangements. Findings revealed the existence of a sophisticated market ecosystem spanning formal and informal channels across local, national, and international value chains. The market architecture demonstrated significant variation in crop-specific marketing channels, with diverse patterns emerging based on commodity types, institutional requirements, and geographical considerations. Maize marketing was seen as predominantly occurring through state marketing boards and urban markets, driven by price incentives and proximity considerations. Tobacco marketing, constrained by regulatory requirements, was seen as being channelled through formal auction floors, while legumes (groundnuts and sugar beans) utilising more flexible, direct-to-consumer channels, including farm-gate sales and urban markets. Perishable horticultural products could be seen as following localised marketing patterns, primarily through farm-gate sales and nearby urban markets. The reason for this was highlighted by a respondent who pointed out that product shelf life, transportation costs, and customer availability were the main considerations. She had said:

> We sell our fresh vegetables right from the farm and to nearby markets because they do not last long. It is all about getting them to customers quickly, as transportation costs can be high, and we need to find buyers closer to home. (Interview, 02/03/24)

The evidence suggested that farmers exercise considerable agency in market channel selection, with decisions influenced by several factors. These include price differentials, market access costs, institutional requirements, and household resource capabilities. This market participation is embedded within broader political economy dynamics, where state intervention (particularly in staple crops) coexists with market-based mechanisms in other commodity chains. The emerging market structure was seen as revealing interesting dynamics between formal institutional arrangements and informal market mechanisms, with farmers strategically navigating these channels to optimise returns. When probed on probable explanations for some of these dynamics, respondents indicated that with their experience, they have learnt to exercise considerable agency in selecting market channels, with their decisions being primarily influenced by price differentials, the costs of market access, and the need to optimise financial returns. They also indicated that they recognise the complex political economy dynamics affecting their market choices. This was taken as an illustration and explanation of how they are strategically navigating between formal and informal channels, ensuring that they get the best return for their products. Critical was their emphasis on the importance of adaptability in response to changing market conditions and external factors, and the need to always be aware of potential fraudsters and unscrupulous agents and at times seemingly credible businesses who pretended to offer legitimate business deals but would end up cheating farmers of their produce.

Despite the challenges that the farmers highlighted, a significant finding in the study was the emergence of entrepreneurial behaviour among farming households. Interactions with respondents revealed that the farmers view markets not merely as exchange platforms but as vehicles for wealth accumulation and economic advancement. This transformation is evident in both districts, and it aligns with national policy objectives of transitioning from subsistence to commercial agriculture. Pathways to commercialisation, however, were seen to be differing significantly between the A1 and A2 farmers. A major takeaway and observation from the study was that the market ecosystem's evolution has demonstrated the dynamic interrelationships between agricultural production, market institutions, and rural transformation processes. This was interpreted as suggesting that market participation serves as both a driver and an outcome of agricultural commercialisation in the country's contemporary rural context.

10.4 Discussion

The examination of market participation and welfare outcomes in Goromonzi and Zvimba districts reveals a landscape of agricultural transformation, analysed through the combined lenses of integrated political economy and the TSP frameworks. Utilising Bernstein and Oya's (2014) foundational questions: 'Who owns what?', 'Who does what?', 'Who gets what?', and 'What do they do with it?', the study uncovers the multifaceted dynamics of agrarian change following the FTLRP. The findings can be seen as challenging traditional narratives about agricultural productivity, revealing that market participation is shaped by a range of institutional frameworks and structural conditions. It is clear that the relationship between land size and market performance is not straightforward; rather, it involves a complex nexus of entrepreneurial strategies, institutional support, and the ability of farmers to adapt to changing circumstances. A1 farmers, in particular, exemplify this complexity. Despite their smaller landholdings, they are demonstrating impressive market integration capabilities. This points out to the transformative potential of strategic institutional interventions. This success aligns with the TSP framework's emphasis on reshaping economic opportunities beyond simple redistribution, providing a counternarrative to neoliberal perspectives that prioritise market-driven solutions. Instead, it shows that targeted social policies can significantly enhance productive capacities in the agricultural sector. In contrast, the experiences of A2 farmers reveal ongoing structural constraints within Zimbabwe's agricultural landscape. Their modest production growth and lower land utilisation rates highlight the necessity of viewing market participation as a multidimensional process, deeply influenced by historical power relations, access to resources, and institutional support. For the A2 farmers, institutional barriers, particularly the lack of access to credit and support services, indicates the lingering effects of capital flight following the FTLRP. These constraints are indicative of the power dynamics at play, showing how historical inequalities continue to influence agricultural market participation.

The study engages with household welfare dynamics. It provides evidence of the transformative effects of land reform when effectively linked to market participation. The notable increases in agricultural output of 42% in Goromonzi and 27% in Zvimba, along with corresponding income growth, substantiate the TSP framework's assertion that land reform can reshape economic and social structures beyond mere redistribution. The research critically examines state interventions such as Pfumvudza/Intwasa and Command Agriculture, which serve as important mechanisms for facilitating market participation. These initiatives embody the TSP approach's vision of social policy as a transformative tool that transcends minimal safety nets. Their effectiveness, however, is shown as being often hampered by resource limitations, reflecting the persistent challenges faced in implementing comprehensive agricultural reforms. The higher incomes of resettled households compared to their non-resettled counterparts affirm the multifunctional nature of social policy, as articulated by scholars like Adesina (2015) and Mkandawire (2004).

This study shows how access to land, combined with strategic market engagement, can simultaneously address various dimensions of welfare, including production, reproduction, protection, and social cohesion.

By integrating the political economy and TSP perspectives, the study offers an important analytical lens that transcends conventional market-driven development approaches. It reveals how farmers, especially A1 farmers, are forging adaptive strategies within the dynamics of neoliberal economic policies and agrarian welfare initiatives. The study and findings made are important in enhancing our understanding of structural transformation within Zimbabwe's agricultural sector. It demonstrates that radical land reform, coupled with evolving institutional arrangements and farmer strategies, can open pathways for meaningful rural welfare improvements. Nonetheless, the research acknowledges the enduring challenges rooted in structural conditions and power dynamics. The findings emphasise the necessity for comprehensive, holistic approaches to agricultural development that prioritise social solidarity, equity, and the active involvement of historically marginalised communities. By questioning conventional narratives and providing deep-seated empirical insight, this study contributes a critical perspective on the ongoing process of agricultural transformation in post-land-reform Zimbabwe.

10.5 Conclusion

This article has been important in bringing to the spotlight the evolving nature of Zimbabwe's agrarian transformation and its implications for future agricultural development policy. The findings point to the need to reconceptualise how we understand the relationship between the resettled farmers and market participation in post-land-reform Zimbabwe. An important aspect worth noting is smallholder market participation and its relationship to land size, where there has been a challenging of traditional assumptions about economies of scale in agriculture. The experiences of A1 and A2 farmers in Goromonzi and Zvimba suggest that successful agricultural commercialisation pathways are more closely tied to farmer's adaptive capacities and institutional support systems than to land size alone.

From the study findings, it can be seen as critical for there to be agricultural support policies in place that recognise the diverse capabilities and constraints of farmers in the different tenure regimes and that provide them with the requisite support. The study's findings on output growth and income improvements point to the potential for targeted interventions that could further enhance market participation across both the A1 and A2 sectors. Realising this potential will, however, require addressing the persistent challenges of resource access, particularly the critical need for improved credit systems and market infrastructure.

The broader implications of this research can be seen as extending beyond Zimbabwe. They potentially contribute to debates about land reform, market participation, and rural development in Global South contexts. The study demonstrates how radical land reform policies can, over time, generate new patterns of agricultural commercialisation and rural welfare enhancement, even within constrained institutional environments. These outcomes suggest the need for policy frameworks that can better respond to the differentiated needs of emerging farmer categories while addressing structural inequalities in agricultural markets.

Future research directions are important. They can explore how evolving market dynamics are reshaping Zimbabwe's agrarian structure. Focus needs to be on the role of emerging technologies and alternative marketing channels in bridging current institutional gaps. Such investigations would complement this study's findings on the nexus between land reform, market participation, and rural welfare enhancement in post-land-reform contexts.

References

Adesina, J. (2015). Return to a wider vision of social policy: Rereading theory and history. *South African Review of Sociology, 46*(3), 99–119.

Adesina, J. (2020). Policy merchandising and social assistance in Africa: Don't call dog monkey for me. *Development and Change, 51*(2), 561–582.

Babbie, E. R. (2012). *The practice of social research (13th ed.).* Belmont, CA: Wadsworth, Cengage Learning.

Bernstein, H., & Oya, C. (2014). *Rural futures: How much should markets rule? IIED Working Paper*. IIED.

Caesar, L. D. (2024). Justifying an application of the explanatory sequential mixed methods research design in maritime research. *Australian Journal of Maritime & Ocean Affairs*, 1–20. https://doi.org/10.1080/18366503.2024.2335713

Calia, C., Chakrabarti, A., Sarabwe, E., & Chiumento, A. (2023). Maximising impactful, locally relevant global mental health research conducted in low- and middle-income country settings: Ethical considerations. *Wellcome Open Res, 7*(240), 1–21. https://doi.org/10.12688/wellcomeopenres.18269.2

Chali, M. T., Eshete, S. K., & Debela, K. L. (2022). Learning how research design methods work: A review of Creswell's research design: Qualitative, quantitative and mixed methods approaches. *The Qualitative Report, 27*(12), 2956–2960.

Chipenda, C. (2019). The youth after land reform in Zimbabwe: Exploring the redistributive and social protection outcomes from a transformative social policy perspective. *Canadian Journal of African Studies, 54*(3), 497–518. https://doi.org/10.1080/00083968.2019.1648308

Chipenda, C. (2024). A transformative social policy perspective on land and agrarian reform in Zimbabwe. *Africa Spectrum, 59*(1), 89–118. https://doi.org/10.1177/00020397241234024

Cliffe, L., Alexander, J., Cousins, B., & Gaidzanwa, R. (2011). An overview of Fast Track Land Reform in Zimbabwe: Editorial introduction. *The Journal of Peasant Studies, 38*(5), 907–938. https://doi.org/10.1080/03066150.2011.643387

Creswell, J. (2012). *Educational research: Planning, conducting and evaluating quantitative and qualitative research*. University of Nebraska Press.

Creswell, J. (2014). *Research design: Qualitative, quantitative, and mixed method*. SAGE.

Foster, C. (2023). Methodological pragmatism in educational research: from qualitative-quantitative to exploratory-confirmatory distinctions. *International Journal of Research & Method in Education, 47*(1), 4–19.

Gerzso, T., & Riedl, R. (2024). The potential of mixed methods for qualitative research. In J. Cyr & S. W. Goodman (Eds.), *Doing good qualitative research* (pp. 72–84). Oxford University Press.

Gillespie, A., Glăveanu, V., & de Saint Laurent, C. (2024). *Pragmatism and methodology: Doing research that matters with mixed methods*. Cambridge University Press.

Johnson, R. B., & Onwuegbuzie, A. J. (2004). Mixed methods research: a research paradigm whose time has come. Educational Researcher, 33(7), 14–26.

Matondi, P. (2020). *Rural farm size regulations: Analysis of the 2020 statutory instrument of the Government of Zimbabwe*. Policy Analysis Working Paper. Retrieved 09–25, 2024, from https://zimbabweland.files.wordpress.com/2020/03/farm-size-regulations-an-analysis-prosper-b-matondi-050320_-003.pdf

Mkandawire, T. (2004). Introduction: Social policy in a development context. In T. Mkandawire (Ed.), *Social policy in a development context* (pp. 1–33). UNRISD/Palgrave Macmillan.

Mkandawire, T. (2007). Transformative social policy and innovation in developing countries. *The European Journal of Development Research, 19*(1), 13–29.

Mkodzongi, G. (2022). Zimbabwe's changing agrarian sector in perspective. In G. Mkodzongi (Ed.), *The future of Zimbabwe's agrarian sector: Land issues in a time of political transition* (pp. 1–13). Routledge.

Moyo, S. (2011a). Three decades of agrarian reform in Zimbabwe. *Journal of Peasant Studies, 38*(3), 493–531.

Moyo, S. (2011b). Changing agrarian relations after redistributive land reform in Zimbabwe. *The Journal of Peasant Studies, 38*(5), 939–966.

Moyo, S. (2013). Land reform and redistribution in Zimbabwe since 1980. In S. Moyo & W. Chambati (Eds.), *Agrarian reform in Zimbabwe: Beyond white-settler capitalism* (pp. 29–77). CODESRIA.

Moyo, S., & Chambati, W. (2013). Introduction: Roots of the Fast Track Land Reform in Zimbabwe. In *Land and agrarian reform in Zimbabwe: Beyond white-settler capitalism* (pp. 1–28). CODESRIA.

Musungwini, S., Madongonda, Y., & Hogo, H. (2024). Contemporary agriculture marketing strategies for smallholder farmers in a developing context: Echoes from Zimbabwe. In J. Garwi, R. Masengu, & O. Chiwaridzo (Eds.), *Sustainable practices for agriculture and marketing convergence* (pp. 200–225). IGI Global.

Shannon-Baker, P. (2016). Making paradigms meaningful in mixed methods research. *Journal of Mixed Methods Research, 10*(4), 319–334.

Tekwa, N. (2023). "That woman is a 'Farmer": Gender and the changing character of commercial agriculture in Zimbabwe. *Feminist Africa, 4*(1), 103–130.

Tom, T. (2024). Decades of land and agrarian reform in Zimbabwe. A review of transformative social policy lessons for other African countries. *The African Review*. https://doi.org/10.1163/1821889x-bja10143

Tom, T., & Banda, R. (2023). Women farm workers in Zimbabwe: The social policy outcomes two decades after the transformative Fast Track Land Reform. *Feminist Africa, 4*(1), 75–102.

Part III
Cross-Cutting Themes (Civil Society, Informal Economies and Gender)

Chapter 11
Civil Society as a Catalyst for Inclusive Governance: Examining the Role of Non-state Actors in Shaping Social Policies in South Africa

Otshepeng Mazibuko

11.1 Introduction and Background

Civil society has played an important role since its inception in South Africa, with organisations like the United Democratic Front (UDF) and trade unions mobilizing against the apartheid state (Voster, 2015). Civil society's engagement in post-apartheid transformation enabled and shaped the social policies of the democratic state during the transition period (Habib & Kotzé, 2003; Lehman, 2008). This period saw health policies notably influenced by organisations like the Treatment Action Campaign (TAC) (Voster, 2015). The democratic dispensation highlighted the importance of civil society organisations, making them key actors in crucial areas like healthcare, education, and social protection. Environmental and social movements like Abahlali baseMjondolo, which advocated for housing rights, exemplify this engagement (Mdlalose, 2014; Zikode, 2008).

This chapter examines civil society's evolving role in influencing policy-making processes within South Africa's democratic context, specifically in healthcare, education, and social protection. It explores how CSOs shape social policies to achieve more inclusive and effective outcomes, despite challenges posed by political dynamics and resource limitations. The chapter addresses key questions: How does the state engage with CSOs within South Africa's socio-political landscape? What specific roles do these organisations play in shaping critical social sectors? What obstacles do CSOs face such as limited political will and financial constraints and what strategies do, they employ to overcome these barriers and drive policy change?

O. Mazibuko (✉)
Department of Anthropology, Archaeology and Development Studies, University of Pretoria, Pretoria, South Africa

E. Ndhlovu, C. Chipenda (eds.), *Redefining the Social Policy Discourse*,
https://doi.org/10.1007/978-3-032-24067-5_11

11.2 Methodology

The chapter deploys a qualitative study approach, putting a focus on civil society organisation engagement in the development of policy (Colebatch, 2009; Smit, 2003). Key sectors abovementioned, especially healthcare, education, and social protection were explored in-depth. This methodology demonstrates how CSOs shape policies in these areas (Colebatch, 2009). The key areas are chosen due to the CSO's significant role they have played in influencing policy outcomes. The study explores the Treatment Action Campaign (TAC), which is the organisation that advocates for HIV/AIDS patients to be provided with antiretroviral (ARV) drugs. Such an important step taken in South Africa's public health policy, TAC influenced the expansion of accessible life-saving treatments through the government, ensuring that their efforts are recognized in the healthcare sector (Sadovnik, 2017).

Equal Education, as CSO enabled and led campaigns that sought to improve the quality of education, by rigorously improving the school infrastructure and sanitation. The campaigns leveraged both the engaged communities and government, making Equal Education a pivotal stakeholder that has worked in confronting critical unavoidable challenges in the education sector (Jansen, 1995; Spreen & Vally, 2006). Further, enhancing educational outcomes, in marginalized, remote, and under-resources, through meaningful dialogues and collective aims (Draga, 2017).

Social justice plays a crucial role in advocating for policies that promote equity and protect vulnerable communities. One such organisation is Black Sash, which is committed to advancing social protection policies and ensuring justice for all. Black Sash is an organisation advocating for social protection policies and dedicated to social justice (Benjamin, 2004). The CSO focused on protecting vulnerable population groups and ensuring that the social support systems are equitable to everyone, and implementation of such initiatives. This has depicted its instrumental role in trying to improve social status (Mitchels & De Graaf, 2010). The chosen case studies demonstrated how effective and impactful are the CSOs in society, in shaping policies in the stated sectors, and others that are important to the communities (Habib, 2005; Ross, 2009). The CSO's involvement in policy-making processes is imperative in adopting a comprehensive understanding of involved other key stakeholders, in driving a collective call to action (Cwaile, 2023).

The chapter provides an analysis of the assessed degree of alignment by the integration of governmental and civil society-produced briefs, in the adaptation of the recommendations made by CSOs and the subsequent policy decisions taken by the government (Fitzgerald, 2012; Sadovnik, 2017). This type of dissemination tool also brings into light the CSO interventions, as they often respond to immediate political and legislative developments (Colebatch, 2009; Smit, 2003; Wach & Ward, 2013). CSOs document their exclusive activities in these reports, by offering and proposing alternatives and improvements to government policies as society advocates. The basis of such engagements with policymakers from advocacy agendas is probed by the public's perspective and research findings that these reports contain. Through the analysis of these documents, the chapter assesses the presented evidence that CSOs use to frame policy issues, and further do policy change recommendations. It also explores how the public domain uses these reports as advocacy tools and is adopted by policymakers as direct communication.

The chapter uses content analysis to explore themes, and language, that are being systematically employed, for further recommendations within these documents (Fitzgerald, 2012). It allows for a thorough identification of key issues raised by CSOs, including their observed policy preferences and positionality. This helps to understand how they respond to governmental actions and both internal and external shifts. Further, the chapter weighs the degree to which CSOs' input is being incorporated into policy frameworks, by tracing their presence and contributions to government documents (Fitzgerald, 2012; Wach & Ward, 2013).

Civil society's influence on policy development is traced through several critical indicators (Smit, 2003). By examining where and how CSOs make recommendations that appear in government policy documents, the chapter assesses civil society's direct impact on policy formulation. Beyond specific policy changes, the analysis examines language and framing within government documents. CSO advocacy can have subtler but equally significant impacts on policy narratives, leading government to adopt new terminology that frames education as a human right rather than merely a service. This analysis reveals broader governance dynamics where CSOs directly impact and influence policy outcomes. Through document analysis, the chapter traces civil society's policy influence, showing how CSOs shape policy agendas and debates during the inception process while holding government accountable to public interests. The emphasis on advocacy through documentation and archival work, written reports, and briefs forms the foundation for CSO engagement with the state.

11.3 Theoretical Framework

Antonio Gramsci's theory of civil society and participatory governance concepts are key theoretical perspectives that offer essential insights, clarifying dynamic interactions between the state and civil society organisations (CSOs). The frameworks enable the CSOs to contextualize their influence on governance processes, especially in shaping social policy (McSweeney, 2014).

11.3.1 Gramsci's Theory of Civil Society as a Counter-Hegemonic Force

Gramsci's theory of civil society suggests that counter-hegemony forces exert influence over the dominant state powers, engaging in an ongoing ideological struggle, with the CSOs operating as a primary for this context, their influence shapes the direction of policy advocacy and social justice initiatives within the community (Bates, 1975; Im, 1991). According to Gramsci's view, the political society that encompasses direct control and exhibits measures through laws is referred to as the state, whereas the establishment of non-governmental organisations and social institutions are civil society, that is responsible for exerted influence by consent and following cultural hegemony (Gramsci, 1971; McSweeney, 2014). A key role played in reshaping the dominant and at most times indirectly oppressing ideologies promoted by the state is perceptible through the involvement of civil society (Habib, 2005; Ngxabi, 2016).

Supporting underrepresented and vulnerable communities, CSOs can be considered the counter-hegemonic actors in this context. Further, challenging the power of the state, advocating for democratic changes, and opposing policies that continuously enable inequality (Ngxabi, 2016). This view is contextually relevant in South Africa, where a need to push for more social policies that are equitable has to be continued, taking into account the historic role of CSOs in challenging apartheid-era policies and continuing their advocacy in the post-apartheid era.

Gramsci's idea of hegemony accentuates gaining the consent of the society governed, through the promotion of definite ideologies as common sense, which the state maintains control over (Bates, 1975; Im, 1991). Also, as a potential, the hegemony can be disrupted by CSOs when mobilizing civil society to support policy change, through introducing different discourses. For example, the Treatment Action Campaign emerged as a counter-hegemonic force during the HIV/AIDS crisis in South Africa, confronting the government's reluctance to provide antiretroviral treatment (Heywood, 2009; Ndinda et al., 2015). Public demonstrations, media campaigns, and a combination of legal challenges to shift public opinion and exert influence on government policy were effectively used by TAC and contested the state's approach to addressing healthcare governance (Ndinda et al., 2015). Equally, an inevitable consequence of underfunded, under-resourced schools of post-apartheid was confronting common sense, which Equal Education acted as a counter-hegemonic force throughout. Equal Education was able to improve school infrastructure, targeting the rural and underprivileged areas, and shaping policies that the CSO did through mobilization, research, and legal challenges (Draga, 2017).

11.3.2 Participatory Governance and the Role of CSOs in Policy-Making

Participatory governance as a concept merely opposes the state by moving beyond the idea of civil society (Speer, 2012). Instead, policy formulation and governance through an emphasis on fostered collaboration between the state and civil society (Edwards, 2001; Speer, 2012). This concept of participatory governance perceives that CSOs' active engagement allows for more inclusive and democratic structures in decision-making processes (Edwards, 2001; Osmani, 2007). Civil society in participatory governance is an integral player in shaping policy through dialogue activities, negotiations, and collaboration with the state, not only demonstrating as a counter-hegemonic force (McSweeney, 2014; Speer, 2012). This concept is seen as a response to put forth and dismiss the limitation of traditional, top-down governance, which is supposedly viewed to be too centralized and exclusive, hence the alluded critics (Edwards, 2001).

The National Economic Development and Labour Council (NEDLAC) is one of the examples where institutionalized forms of participatory governance, business, civil society, and other entities gather in a collective approach to negotiate policies related to economic and social development, which can benefit the majority of South Africans (Ngxabi, 2016). Participatory governance is demonstrated through institutionalization and involvement of civil society in policy-making processes. Empirical evidence on the role of civil society in participatory governance is argued for in evident South African case studies. For instance, advocacy for social protection policies has been notable by the engagement made by Black Sash in publicizing participatory governance. It has proved to be successful in the improvement of the social grant system, monitoring of grant payment through consistent social audits, and efficient service delivery through enhancement collaboration with the government (Tshangana, 2013).

Advancement of safety and sanitation in remote and informal settlements had worked by the involvement of The Social Justice Coalition (SJC). In Cape Town, the SJC has significantly contributed to shaping policies in the municipalities, by encouraging the involvement of communities in policy advocacy and monitoring service delivery (Overy, 2013). As for Equal Education, they strategized and employed a participatory governance approach in policy advocacy for better school infrastructure, by engaging students, parents, and educators in the process (Draga, 2017). This is depicted in the research outputs and policy recommendations they came up with, which were integrated in government deliberations, and demonstrated the potential for collaboration policy-making between the state and civil society.

There are critiques and challenges posed by opponents to this approach of Gramsci's counter-hegemony theory and participatory governance, which is supposed to prescribe useful frameworks to better understand civil society's role (McSweeney, 2014; Overy, 2013). Co-optation of civil society by the state can often take place, especially where participatory governance is concerned. Excessive reliance on government funding and accessibility may lead to CSOs losing their autonomy structure and capacity to confront the same state policies effectively. Other unfortunate instances involve power imbalances that participatory governance suffers from, evidently where policy processes are dominated by state actors, leading to CSOs resisting to exert their influence. Incapabilities of being under-resourced CSOs often excluded them from significant participation. As such, Gramsci's theory puts forth ideological struggles poised by civil society's role, which sometimes are fragmented. Competing for influence and sometimes having conflicting goals as CSOs weaken the main purpose of collective power in the process of shaping policies in South Africa (Mavee, 2022).

Different perspectives in understanding the role of CSOs in governance seem possible through reference and possible adaptation and practice of Gramsci's concept of civil society as a counter-hegemonic force and participatory governance framework. The examined theories could be put into practice in measuring the extent of civil society's role in shaping and implementing social policy through advocacy, contestation, and collaboration. This is evident in the act of confronting and working with the state in the promotion of inclusive governance and social justice, in which the integration of Gramsci's counter-hegemony and the participatory governance model accentuates the dual role of civil society.

Advocacy coalitions concept is a crucial tool to be used in understanding, collaborating, and amplifying the civil society organisations (CSOs) exerted influence on policy-making processes (Schlager, 1995). This concept is simply a diverse network of organisations and individuals who share collective values, beliefs, and goals in addressing a certain policy issue (Jenkins-Smith et al., 2018). These integrations of coalitions leverage their collective influence on decision-makers, from pushing collectively in pooling resources through policy changes and coordinated strategies.

Paul Sabatier and Hank-Smith through the Advocacy Coalition Framework (ACF) jointly developed, drawing on how such coalitions operate through the stipulated exploration of the theoretical lens (Jenkins-Smith et al., 2018; Sabatier & Weible, 2019). This framework puts forth an argument that the interactions between competing coalitions over time shape the policy change, which sets a path on how a particular issue should be confronted, by incorporating a unified set of core beliefs (Jenkins-Smith et al., 2018; Sabatier & Weible, 2019). Contextually, it is observed that coalitions that are formed by CSOs bring forth a cohesive policy message and unify their voices through strengthened and exerted influence (Mavee, 2022). Moreover, in putting into place public support, expertise, and funding, they pool such resources to ensure that their advocacy efforts are coordinated to be aligned and strategically times and that the lobbying, protests, and media campaigns are relevant at that time (Moyo, 2018).

11.3.3 The Importance of Advocacy Coalitions in Civil Society

It is known that the struggle with resources, political accessibility, and a certain extent of influence brings about a positive and meaningful policy change, advocacy coalitions are important for CSOs. CSOs can bring in their respective political leverage, which in most cases is effective with policy-making institutions, and is then important in the derivations of coalitions. In dissecting the complexity of socio-political contexts in South Africa, coalition-building is relevant in ensuring that the significant power of the state is retained in policy formulation, especially in instances where formal decision-making platforms are limited or for smaller or less influential organisations (Mavee, 2022).

The power of collective civil society efforts in South Africa is highlighted in evident notable examples of advocacy coalitions that drive the process of shaping public policy. The mentioned case studies demonstrated how CSOs in critical sectors such as health, education, and social protection have used coalition-building to confront these key societal issues. Firstly, coalitions foster accountability by enabling CSOs to work together in holding government accountable; it contributes to the sustainability of social change by combining resources. Coalitions serve as powerful vehicles for influencing policy, and lastly, they help ensure that the voices of these communities are heard.

11.3.4 The Treatment Action Campaign (TAC) and the AIDS Coalition

Treatment Action Campaign (TAC) is marked as one of the most major advocacy coalitions formed in South Africa's history, in the fight against HIV/AIDS in the late 1990s and early 2000s. TAC, alongside other CSOs, encouraged labour unions such as the Congress of South African Trade Unions (COSATU) and international partners, in making demand for access to antiretroviral (ARV) treatment for people living with HIV/AIDS, led to the mobilized powerful coalition to exist.

The formation of the coalition demonstrated its significant role in government policy under then-President Thabo Mbeki, who opposed the widespread provision of ARVs (Sabi & Rieker, 2017). In contributing to the enhancement of the public health system, the led rollout of ARV treatment through a TAC-led coalition confronted the government's attitude, by uniting around a collective goal, and ultimately ensuring access to life-saving medication to the communities (Friedman & Mottiar, 2005; Nattrass, 2004). The political resistance and transformative policy changes were shown through the combined efforts of diverse organisations, perceiving a positive side of the coalition's successes (Heywood, 2009; Martinez & Kohler, 2016).

11.4 Case Studies on Accountability, Sustainability, and Policy Influence

Equal Education (EE) in the education sector advocated greatly for improved school infrastructure, especially in less or under-resourced marginalized areas. This joint demonstrated how the possibilities of having the voices of marginalized communities amplified, which depicts the importance of the core role of advocacy coalitions. Collectively by the vigorous involvement and participation of student organisations, parent groups, legal advocates, and other civil society actors, the advocacy coalition group brought forward a demand for non-negotiable minimum norms and standards for school infrastructure.

In the most marginalized, remote, and underprivileged areas, where evident unhygienic sanitation, water, and electricity are experienced by students, in improving school facilities, through compelling the involvement of government, the coalition's coordinated efforts expanded to public mobilization, research, and litigation in ensuring the success of such (Draga, 2017). Driving this initiative, the South African government obliged and agreed to adopt the set minimum standards for school infrastructure, by enforcing and securing a legal victory, through the EE coalition in 2013. This achievement underscored the power of sustained collective action in overcoming bureaucratic inertia and resource constraints in the education sector (Chisholm, 2005; Spreen & Vally, 2006). Therefore, sustained collective action in overcoming bureaucratic lethargy and limited resources in the education sector is featured and shown in the coalition's advocacy efforts.

11.4.1 Black Sash and the Coalition for Social Protection

In the scope of social protection, a key player in advocating for the rights and benefits of social grant beneficiaries has been the Black Sash. Integrated coalitions that seek to confront ineptitudes and selfish corruption activities within the South African Social Security (SASSA) were mused by Black Sash and other CSOs. The exposure of irregularities in payments of social grants and instilled protection of vulnerable beneficiaries' rights was enabled by the coalition of social protection (Seekings & Nattrass, 2008).

Notably, the collective coalition in the Constitutional Court, summoning and declaring that SASSA's inception of contracting private payment service providers was invalid as a result of irregularities and breaches of social protection policy, was instrumental. Such collective actions by civil society actors demonstrated the capability to mobilize legal, social, and political resources, to protect the rights of vulnerable South Africans who are highly dependent on social grants (Kenny, 2015).

11.4.2 The Structure and Dynamics of Advocacy Coalitions

Complementary skills and resources that grassroots organisations possess are standard embedment that stems advocacy coalition, to drive community support and provide hand-on-the-ground perceptions into policy issues (BARASA, 2019). Also, collected and analysed data, that form a directive of evidence and policy expertise in supporting advocacy efforts, brings in another key actor of think tanks and research institutions. Referral to the enforcement of legal obligations and confrontation of unjust and biased policies are what legal advocacy groups litigate on. Inclusive public opinion and coverage that instigate investigative journalism affirm media organisations to amplify coalition messages (Mitchels & De Graaf, 2010). Another important and overlooked role of labour unions is how their role is vital in mobilizing mass membership organisations that should be responsible for the provision of political clout and a broad support base.

The success of advocacy coalitions is entirely dependent on and driven by coordination, which is a very critical aspect. An alignment of diverse interests and capacities of member organisations for coalition leaders to enable, that is to ensure that the shared vision is manageable through encountered internal dynamics. Clear goals that are established, maintained open communication, and developed strategic advocacy plans are key points in achieving successful coalitions, that can leverage the strength of each member organisation.

11.4.3 Challenges of Advocacy Coalitions

There are several challenges that at some times hinder the success of advocacy coalitions, which are perceived as powerful vehicles for policy change. This is because of the integration of varying priorities, resources, and strategies that now have to find a common ground and embracement of diversity, ensuring that the encountered difference does not overcome the main purpose of collating and managing the difference by compromising and maintaining unity, especially to coalition members that have competing hidden agendas. In other cases, the dissolvement of coalitions happens when their immediate goals have been achieved, which motivated the formation of policy issues. The precariousness is likely to happen where the long-term sustainability of coalitions is threatened by funding and political momentum decline. In maintaining the status quo, well-organized coalitions run the risk of facing significant opposition from state actors that hold power, or the emergence of vested interests. The remedies in preventing these challenges can be political compromise and sustained advocacy that could overcome this resistance.

The role of advocacy coalitions is very important in exerting civil society's influence on public policy. The ability to pool resources by coalitions, augment policy messages, and unite diverse actors through collective goals and objectives can break political and institutional barriers (Bhuiyan et al., 2023). The Treatment Action Campaign, Equal Education, and Black Sash prove the ability and success of coalitions in powerful collective action to shape social policy. However, in ensuring that concrete and sustained effective coalitions are established, strategic planning, coordination, and navigation around complex existing political environments are required (Sotirov & Memmler, 2012; Wampler, 2021).

11.4.3.1 Overview of the Role of Non-state Actors in Social Policy Across Different Regions

Impressive impactful developments have been seen through education and health policies, which were influenced by civil society organisations in Brazil and India and have set a global example of how grassroots mobilization holds significant influence in government policy. In Brazil, the advocated accessible education and health services for the under-resourced and disadvantaged communities have been instrumental through the efforts of the Movement of the Landless Rural Workers (MST) and other various health-focused organisations (Carter, 2010, 2015; Mohanty et al., 2011). Inclusive and impactful rural education policies and land reforms have been successfully lobbied for by the MST, resulting in the establishment of schools that develop and integrate the curricula around agrarian education, for the benefit of the local needs (Carter, 2015).

Looking at the health sector and its policy, the creation of the Unified Health Systems (SUS) by the Brazilian civil society had demonstrated a vital role for a powerful sustained public advocacy. CSOs ascertained universal health access, by adopting and embedding a participatory approach in health governance, through establishments and partnerships with community activists, and health professionals. Also, in India, the bridging of educational policy gaps and health service delivery, accentuating the data-driven approaches that will enable and drive the shaping of inclusive policies to confront the needs of both India's rural and urban populations, were functioned by the CSOs like Pratham in education and the Public Health Foundation of India (PHFI) (Das, 2022; Dwivedi, 2007). Such instances demonstrate the championed equity and access by challenging policy norms and collaborating with government bodies, which civil society organisations can do.

In Africa, active civil society engagement has been witnessed in Kenya and Nigeria, which sought to confront their respective socio-political contexts, which varied in focus and effectiveness. In Kenya, the vibrancy and active participation of CSOs significantly shaped policies around social justice and public accountability, which were active in electoral advocacy and constitutional reform. Elimu Yetu coalition, which is an education-focused organisation, depicted its vital role in advancing education access, and free primary education policies through the act of both advocacy and policy intervention (Coalition, 2009). In the Nigerian civil society's landscape, maternal health and HIV/AIDS programs showed a significant improvement through organisations like the Nigeria Network of NGOs (NNNGO) (Olawoore & Kamruzzaman, 2019). Also, an interactive and active engagement has demonstrated strong CSO involvement for the balanced and controlled electoral reform and anti-corruption efforts.

Relatively, inspecting both Kenya and Nigeria's robust and vigorous CSO engagement in social policy, there is an important contrast that makes each case unique (Ajulor, 2018; Coalition, 2009; Nyamu-Musembi & Musyoki, 2004). Kenya's civil society advocates for good governance through thorough accountability and inclusivity in policy, while navigation through a complex landscape of resistance by the state is where Nigeria's CSOs maintained focus, stressing grassroots mobilization and the need for unbiased legal involvement (Dan-Azumi, 2018; Elueze, 2023; Matelski et al., 2018). The examples that have been briefly outlined showcase a crucial intermediary in social policy formation, in which the civil society functions and pilot strategies to drive inclusive governance and society equity are attributed to the local political climates.

11.4.3.2 Existing Research on South Africa's Civil Society Influence on Policy Development

Scholarship on civil society organisations in South Africa, in the crucial domains of health, education, and social protection, ponders on dense yet impactful relationships between CSOs and the state (Cwaile, 2023). A case study of the Treatment Action Campaign (TAC) has been frequently referenced by scholars, in how policy change can be influenced by civil society, following the TAC successfully pushed for the inclusion of antiretroviral drugs for HIV/AIDS treatment in the public health system in the early 2000s (Friedman & Mottiar, 2005).

Studies show that the approach used by TAC for combined grassroots mobilization, legal challenges, and advocacy initially was resistant to widespread ARV distribution, to put exhibit pressure on the government. The adaptation of this strategy mobilized society toward health equity and increased transparency and accountability within the health sector, as the public became aware of HIV/AIDS, which eventually influenced health policy (Friedman & Mottiar, 2005; Habib, 2005). Nonetheless, an indication made by these studies is that heavy reliance and dependency on international support, availability of resources, and the political will drive the success, and other factors contribute over time and indirectly and directly influence the degree of sustained which CSOs can maintain (Heywood, 2009).

In the education space, studies demonstrate how Equal Education set a significant milestone in civil society's policy influence through their campaign for improved infrastructure standards (Jansen, 1995). This led to the establishment of the Norms and Standards for Public School Infrastructure in 2013 through Equal Education, commanding deserved minimum infrastructure standards for South African schools, in under-resourced areas. This made a shift in government policy, pointing out that one of the factors that contribute to the success of quality education is safe and adequate facilities for all key actors, especially pupils (Jansen, 1995; Spreen & Vally, 2006). However, notable challenges made by scholars imply that they have hindered the full realization of these goals, where the implementation phase is supposed to take place, encountered limitations in resources, bureaucratic delays, and inconsistent state commitment.

Mixed outcomes have been experienced in social protection. Although Black Sash has been instrumental and at the forefront of improving the administration of social grants, limitations that they face are connected to the capacity of administration and the processes the government adapts for contraction (Overy, 2013). The state's operational and political dynamics are often constrained, as the available literature portrays the nexus picture of civil society's policy, that is it always substantial (Orvis, 2001; Overy, 2013). The mentioned constraints demonstrate the urgency and immediate action by CSOs in ensuring that policy reforms are achieved for their intended objectives, by adapting and employing versatile strategies, for the benefit of these crucial sectors.

11.5 Findings

11.5.1 The Treatment Action Campaign (TAC) in Healthcare Policy

TAC has proven to possess the critical skills that potentially influence public health policy, particularly at times of HIV/AIDS treatment access. Its initial aim was to be at the forefront of advocating for health services accessibility, especially when the government was resisting the provision of antiretroviral drugs (ARVs) for people living with the virus, despite the escalating prevalence (Cwaile, 2023; Friedman & Mottiar, 2005; Overy, 2013; Smit, 2003). Public campaigns and international partnerships, combined with legal legislation, led TAC's advocacy to emerge as a powerful force for reforming health policy.

As such, South Africa's appraised constitutional documents which in the blueprint maintain that the right to health by all citizens should be provided, TAC was able to use it head-on and challenge the policies the government came up with. A remarkable victory by TAC was when the 2002 lawsuit against the Department of Health was won, which had demands including immediate consideration of Nevirapine in the reduction of mother-to-child transmission to HIV-positive pregnant women (Friedman & Mottiar, 2005). It was witnessed by the public when TAC was favoured by the Constitutional Court in fulfilling the government's obligation to its respective citizens and applauded the CSOs for observing legal mechanisms and using them rightfully to hold the state accountable.

The growing momentum of grassroots mobilizations and public awareness and involvement campaigns was leveraged by TAC to gain support. In maintaining and upholding the dignity and equity of a human being, the TAC efforts underscored the human rights dimension of HIV/AIDS treatment consistently and accessible (Friedman & Mottiar, 2005; Gómez, 2018). Involvement and awareness campaigns catalysed treatment rollout, and the government was resistant. The integration of consulting legal teams and mobilizing the public through the campaigns resulted in the government's decision to be influenced by the CSOs, for nationwide accessibility of the ARV program (Habib & Kotzé, 2003). This bold action salvaged millions of lives of people to not succumb to the virus and placed South Africa as the eventual global leader in the treatment of HIV/AIDS (Nattrass, 2004). This demonstrates how critical consultations are in all dimensions, when used strategically, to substantially shift policies that were working against the people, to now benefit them, particularly in healthcare.

11.5.2 Equal Education (EE) in Education Policy

Remote and marginalized communities always face hurdles to proper infrastructure, especially at schools. The formation of Equal Education, established in 2008, came as a CSO to confront those inequalities in the education sectors in South Africa, ensuring that even in the rural areas, proper infrastructure can be accessed by school pupils. The CSO in pushing for equitable education policies used evidence-based advocacy and legal action (Jansen, 1995). The CSO influenced the policy in establishing a conducive space for learning for all safe students, by ensuring that there are Norms and Standards for Public School Infrastructure for adoption and further implementation.

The studies conducted by EE focused on advocating for infrastructure standards, actively engaging the communities in their inputs and priorities, and observing the challenges faced by students, particularly in rural and remote areas, which potentially impacted their learning outcomes. These efforts align with the findings of (Jansen, 1995; Spreen & Vally, 2006) who highlight the importance of infrastructure in shaping educational experiences. Some of the calamitous conditions students unfortunately had to bear were overcrowded, unsafe, and unsanitary facilities, which the CSO gathered as evidence, in building a strong case (Spreen & Vally, 2006). The evidence data collected was a base for petitioning policymakers to develop infrastructure standards.

EE resorted to a lawsuit when the petitions were not producing any significant changes anticipated. The inception of the Norms and Standards for Public School Infrastructure was won by EE in 2013, against a legal battle with the Department of Basic Education. The policy documentation demanded access to water and sanitation, provision of electricity, and safe building in school facilities, which were mandated in the blueprint. In turning ligation into a catalyst, the CSO obliged the government to perform its rightful functions and deliver on its commitment to tangible infrastructure improvements, which was a remarkable victory in the education realm (Chisholm, 2005). During the inception and adoption of the Norms and Standards, EE then monitored the progress of the government's compliance with the policy, and the implementation of infrastructure improvement had taken place. This demonstrated the commitment that the CSO role comes with, particularly in being an accountability partner. Further, illustrates how the potential of CSO to see the execution of mandates and sustained impact that is documented in the policy.

11.5.3 Black Sash in Social Protection Policy

The administration of social grants, economic inclusion, and social justice has been advocated by Black Sash, which has been an instrumental CSO. The unfortunate incidents that the vulnerable population had to bear due to the inconsistent distribution of social grants have been attributed to corruption, inefficiency, and accountability, which Black Sash has uncovered (Benjamin, 2004). The practice for fair and transparent administration of social grants has improved, which behind the scenes has been long campaigned by Black Sash within the South African Social Security Agency (SASSA). The CSO highlighted the urgent necessity for reliable and efficient service delivery to protect the rights of the beneficiaries, as millions of South Africans highly depend on the grants for survival.

A private company responsible for administering and distributing social grants, Cash Paymaster Services (CPS), contracted by SASSA, was critiqued by Black Sash. The issue of data privacy, sharing, and unauthorized deductions, allegedly done by the contractors, posed a concern to Black Sash, which then the CSO filed a case in 2017 with the Constitutional Court. The court established and concluded that the contract did not follow the standardized proper procurement procedures and was unlawful in how it was awarded, then ordered SASSA to find a different payment system. The importance of transparency and accountability in social protection programs was highlighted in the case. This led to the restructuring of administered grants, which highlighted the importance and potential that Black Sash's advocacy as a CSO had, and that they were vocal representatives for vulnerable populations and exerted influence on policy change, through legal channels.

CSOs have played a crucial role in South Africa, particularly in advocating and confronting the challenges that continuously incur in the most critical sectors such as healthcare and education. Its active participation has been vibrant in influencing significant changes in the policy. Public health policy has shown improvements through the Treatment Action Campaign (TAC) advocacy. Its feasible measures in taking legal action and engaging the public were influential in combating the resistance from the South African government to the inclusion and distribution of antiretroviral (ARV). This depicts the capabilities and capacity of the CSOs to play an important role in advocating for expanded access to life-saving treatment for people living with HIV/AIDS, by reforming and mandating urgent public health needs in the policy frameworks. TAC used a model in rights-based approaches to draw attention to the right to health in the Constitution, in holding the government accountable for ensuring essential health interventions.

Improved school infrastructure in public schools in under-resourced areas was successfully advocated and campaigned by Equal Education (EE), within the education spectrum. The establishment of regulations relating to the Norms and Standards for Public School Infrastructure policy, which culminated EE's activism to demanding minimum standards for public schools. This included school facilities, the provision and access to clean water and functioning hygienic sanitation facilities, and safe buildings for conducive teaching and learning. Rural and township areas were able to benefit from the substantial improvements in school infrastructure, which halved the impeded quality of education, overcrowding which affects the quality of teacher-to-learner relationships, and shabby facilities. Extensive research, public engagement and advocacy, and legal routes exerted influence on EE's approach to push for the reform of policies to enhance educational equity and close the evident gap that harms a safe learning environment for students.

With all the alluded successes CSOs have achieved, extensive challenges remain within the cycle of CSOs, which in most cases affect their effectiveness. Observed are government reluctance, political obstacles and internal instabilities, and lack of political will to endorse the implementation of policies. For example, from the mentioned challenges encountered from the political view, TAC efforts evident and succeeded in ARV treatment advocacy, from government resistance that potentially delayed the implementation and impacted the policy. This indicates how state actors within political systems tend to be hesitant in establishing or integrating with CSOs to prioritize certain policies, which shows the complexities underlying advocacy. Also, the government may face challenges in the allocation of resources and expertise for successful implementation of the new policies, due to the intensified lack of state capacity.

Ensuring the sustainability of advocacy efforts is largely dependent on the resources, which can pose a constraint, as CSOs rely on volunteer support and sourcing external funding in maintaining operations. A notable example, TAC which sourced and benefited from international funding could enlarge the scale-advocacy campaigns, but as other organisations longevity of its respective efforts depends on access to resources, which is rare and unlikely. Also, funding and maintenance of grassroots campaigns were unstable and Equal Education unfortunately faced the challenges across the diverse regions of South Africa. The advocacy impacts from resource limitation ponders the success of the CSO, which in unfortunate cases results in the dissolvement of coalitions and organisations, or shifted focus on the core objectives when the funding decreases. With support from state actors, stable commitment in the political space, a capacitated state, and resource security, this highlights the need for a supportive environment to continuously enable CSOs to shape South African social policies effectively and efficiently and to be impactful to the society.

11.6 Discussion

A growing phenomenon of good governance and policy inclusivity demonstrates the exerted influence by the establishment and formation of civil society organisations to align and address public challenges in introducing and practice of participatory governance in South Africa. As such, a notable illustration of policy influence during policy processes fundamentally set to put forth the complex societal needs, to which the non-state actors are at the forefront. The centralized governance models were challenged by the civil society organisations (CSOs), in holding them accountable for their actions and chronicling their focus on social equity by adaptation and reconsideration to practical realities in policy formulation. For instance, an outcome in a more inclusive and transparent governance landscape in South Africa was impacted by TAC's role in the confrontation of public health needs on the health policy, but most importantly the inclusivity of the public in the decision-making processes.

The uncertainty from the state on including non-state actors and the CSO advocacy pushing for social accountability reflects the dense power dynamics between the government and civil society. The initial resistance to CSO demands for ARVs shows that there is tension and unexplainable coexistence between state autonomy and the pressure the public brings. With the encountered ambivalence from the state, the government gave in, cooperated, and became more responsive, which CSOs persisted in the advocacy, to navigate around these existing dynamics. A strategic balance between the authority of government and civil society as the custodian for public inclusivity depicts working interactions, where each unique influence contributes to the successful reformation of the policy outcomes. This establishment interaction expands on the values of openness toward CSO integration by the state, where engagement of the public is concerned and vested interest and welfare are potentially implicated, input to the design of the policy.

Preparing for the future, CSOs are seen as key non-state actors that exhibit an important role in inclusive governance frameworks, regardless of the ongoing challenges that are politically motivated and limitations in resources, and are highly expected to salvage the situation. With the rise of dense societal issues, these recognized partnerships between non-state actors and the state are a need, to also navigate technological advancement that is rapidly taking over. A collaboration where both institutions tabulate their traits and desired objectives toward establishing a common social objective is directive for CSOs to gain confidence to negotiate in intensive policy-making spaces in structured access. The growth of technological advancement and transition to digital advocacy tools endows and equips the CSOs to be more direct and efficient in the process of engaging policymakers and the public. CSOs can bypass the primitive barriers that make them perceived as adaptable and showcase their ability to reach wider audiences and mobilize online support, by echoing influencing policy agendas to good governance practices.

There are commonalities to South Africa's experiences with countries like Kenya, in how constitutional reforms were reshaped and how CSOs were influential in the process. Also, how social justice and electoral transparency were executed in Nigeria, similarly to South Africa. These proportional examples make a provision of possible strategies that CSOs in South Africa could adopt, considering the valuable insights. This could be possible by the formation of dynamic advocacy coalitions to strengthen their vocal presence, as witnessed in Kenya, and the development of vigorous legal frameworks for accountability in Nigeria. Therefore, these evident and successful practices from other countries put South Africa in a position to transform and be adaptable to best practices based on successful stories, which could build a long-term capacity in confronting societal issues within the South African socio-political context. The need for South African organisations to seek and leverage collaborative networks underscores the significance of the adaptability aspect and coalition-building for CSOs across Africa, enhancing them to be influential and sustainable in the inception.

As mentioned, the introduction and integration of technology tools in the realm of advocacy have a significant impact in creating greater transparency in government processes, allowing South African CSOs to enable a vibrant interaction with the public and policymakers in one space. These enable the CSOs to align in real time the mobilization of communities, gathering of public opinions, and holding government accountable. An evident example is the TAC's outreach, in which they used digital tools, to increase awareness of HIV/AIDS and also apply pressure on the government publicly for the demand for ARV treatments. Social media campaigns came in as efficient digital tools for Equal Education, to raise public awareness of the unpleasant and inadequate school infrastructure, and further develop a motion for policy change. The gap between geographical and social divides will be likely closed by the expanded influence of the CSOs, attributed to the internet accessibility expansion, and have a varied cross-section of society gathered support.

As notable as the positive impact, the technological advancement may sound, some challenges will need to be considered. These are the potential of misinformation, technological skills, and cybersecurity. This compiles the CSOs to take responsibility for the navigation around the digital landscape and establish internal capacity, which could be effective for online engagement. Also, the analysis of government regulatory frameworks may evolve and adapt, and authorities for balanced freedom of expression with governance priorities are closely monitored through digital activism. South Africa should remain flexible for their CSOs to strategize for relevant and secured information sharing, through the investment of digital literacy, taking into cognizance the abiding regulations. In forestalling the possibilities that CSOs can encounter concerning digital activism, as Nigeria was imposed to, South Africa could derive a proactive response plan to such anticipated challenges. In doing so, with the exerted influence in shaping social policy amidst evolving technological and political landscapes, CSOs in South Africa will be equipped to secure their substantial role at the forefront of participatory governance in the future.

11.7 Conclusion

In conclusion, the essential sectors like healthcare, education, and social protection have benefited from the influential role of civil society organisations, despite the encountered constraints within South Africa's social policy development. With victories attained from improving the HIV/AIDS treatment policies, advocating and improving the standards of school infrastructure, and monitoring the administration of fair and equitable social grant distribution by the Treatment Action Campaign (TAC), Equal Education (EE), and Black Sash, respectively, their vibrant influence has then been affected by political, financial, and institutional challenges. CSOs constantly have to manoeuvre around the encountered challenges through the underlying complexity landscape, to see the profound contributions that will bring about a positive impact to vulnerable communities, by the revamped practical policies. The contrast between the existing limitations and the intended influence puts forth the necessity for CSO to effectively shape social policy by adopting targeted strategies.

11.8 Recommendations

The study then puts forth feasible recommendations intended for strengthening the potential collaboration between the state and the CSOs to develop channels that are more structured, engagement that is accessible, and transparency, which are practical steps. The state is then able to establish accommodative platforms that allow CSOs to be more influence policies proactively, which would first need the CSO participation to be institutionalized in policy-making by the inception of councils and formal frameworks. In relevant, capacity-building efforts should prioritized by the CSOs which will halve the dependency on donor aiding that is external and restrictive, which will then be trained and establish cross-sector partnerships and securing funding sources. The existing documented resources, epistemology, and platforms that international and private sector links have can assist CSOs to maintain and sustain their advocacy efforts amongst government support that is constrained. Also, in enabling more collaborative and responsive policy-making, in structured feedback mechanisms between the state and CSOs, the process should be encompassed by values of trust, transparency, and accountability.

An empowered civil society sector can exhibit sustainable inclusive governance, which is imperative for the democratic health of South Africa. In ensuring that the policies are responsive and prioritize the needs of diverse communities, with a particular focus on marginalized populations, the CSO has attributes to hold the state accountable when it is not performing accordingly. This encourages the public to be more involved and actively engage in immediate issues and practical realities, which fosters CSO processes to be effective. This could be attained by having access to an enabling environment that could facilitate an efficient collaboration from both the CSOs and the state, and the community at large, for civil societies to execute this role diligently, with encompassed legitimacy and grassroots support.

It is affirmed by the undertaken study that long-term governance and social development by civil society is essential in South Africa. In having practical equitable policies that speak and advocate for the marginalized South African distress and complex socio-economic challenges, the vigorous involvement of civil society is crucial for ensuring initiatives that the government comes up with are directly prioritized and aligned with the practical realities of the public. This can be done by offering necessary support that will encourage the ongoing open and participatory governance model that the CSOs have to be involved in, to steer and support social development and comprehensive social policies that speak to the people. Also, this will enable their immediate impact, which will positionally instil foundational pillars from the framework that is more participative, transparent, and accountable.

Moreover, investigations and assessments of CSO functions go beyond their usual approach in increasing their impact and influence over policies. For instance, CSOs such as TAC and Black Sash made use of legal action to confront and challenge unfair, unjust policies and actions that the state was posing to society, and it demonstrated how the legal weapon used by CSOs can be effective when rightfully applied, and for purposeful reasons. This method not only highlights important issues but also establishes legal precedents that impact policy frameworks, even when direct policy cooperation is not possible. When CSOs involve the judiciary, they have an additional way to impact policy, strengthening their position as a watchdog of government authority and an advocate for social equality, especially in cases concerning human rights.

CSOs in the context of South Africa need to adapt to the changing political landscapes, which hold the most influence on the policy domain and political environment. In times of difficulty in state-CSO cooperation, these groups must function as watchdogs, highlighting policy deficiencies through campaigns, reports, and public events. For example, during periods of political tension, organisations in the fields of education and healthcare have rallied public backing to ensure that policy matters remain a topic of national discussion. This flexible approach helps CSOs deal with changing levels of government transparency, making sure important social issues are still recognized and that public opinion can influence policymakers indirectly.

References

Ajulor, O. V. (2018). *Civil Society Organizations (CSOs) participation in policy implementation on poverty reduction: A comparative analysis of Nigeria and Ghana.*

BARASA, T. (2019). *The contribution of non-state actors to inclusive public policy process in Kenya.*

Bates, T. R. (1975). Gramsci and the theory of hegemony. *Journal of the History of Ideas, 36*(2), 351–366.

Benjamin, E. (2004). *An historical analysis of aspects of the Black Sash, 1955–2001.* University of Stellenbosch.

Bhuiyan, M. I. H., Rumi, M. H., Mesbah, T. B., & Rahman, D. (2023). Examining the role of Civil Society Organizations (CSOs) in promoting participatory local governance at the grassroots in Bangladesh: A case study of Jhilongjha Union in Cox's Bazar District. *Journal of Governance and Social Policy, 4*(2), 227–244.

Carter, M. (2010). The landless rural workers' movement and democracy in Brazil. *Latin American Research Review, 45*(S1), 186–217.

Carter, M. (2015). *Challenging social inequality: The landless rural workers movement and agrarian reform in Brazil.* Duke University Press.

Chisholm, L. (2005). The state of South Africa's schools. In *State of the nation: South Africa 2004–2005* (pp. 201–226).

Coalition, E. Y. (2009). *Country case study: Kenya.*

Colebatch, H. K. (2009). Governance as a conceptual development in the analysis of policy. *Critical Policy Studies, 3*(1), 58–67.

Cwaile, M. S. (2023). *The influence of civil society organisations in the new governance framework of democratic South Africa.*

Dan-Azumi, J. D. (2018). CSOs advocacy and the legislative process in Nigeria: A case study of the National Health Act, 2014. *JL Pol'y & Globalization, 74*, 69.

Das, N. (2022). How important is the non-governmental sector for providing public health services in India? *Development in Practice, 32*(5), 624–634.

Draga, L. (2017). Infrastructure and equipment. In *Basic education rights handbook: Education rights in South Africa.*

Dwivedi, S. (2007). Mobilising civil society campaigning at the grassroots. In *Citizen participation and democratic governance, in our hands* (p. 87).

Edwards, M. (2001). Participatory governance into the future: Roles of the government and community sectors. *Australian Journal of Public Administration, 60*(3), 78–88.
Elueze, C. T. (2023). *Civil society advocacy and women participation in politics: Implications for 2023 elections in Nigeria.*
Fitzgerald, T. (2012). Documents and documentary analysis. In *Research methods in educational leadership and management* (pp. 296–308).
Friedman, S., & Mottiar, S. (2005). A rewarding engagement? The treatment action campaign and the politics of HIV/AIDS. *Politics & Society, 33*(4), 511–565.
Gómez, E. J. (2018). Civil society in global health policymaking: A critical review. *Globalization and Health, 14*, 1–11.
Gramsci, A. (1971). *Selections from the prison notebooks.* Routledge.
Habib, A. (2005). State-civil society relations in post-apartheid South Africa. *Social Research: An International Quarterly, 72*(3), 671–692.
Habib, A., & Kotzé, H. (2003). Civil society, governance and development in an era of globalization: The South African case. In *Governance in the new South Africa: The challenges of globalization* (pp. 246–270).
Heywood, M. (2009). South Africa's treatment action campaign: Combining law and social mobilization to realize the right to health. *Journal of Human Rights Practice, 1*(1), 14–36.
Im, H. B. (1991). Hegemony and counter-hegemony in Gramsci. *Asian Perspective, 15*(1), 123–156.
Jansen, J. D. (1995). Effective schools? *Comparative Education, 31*(2), 181–200.
Jenkins-Smith, H. C., Nohrstedt, D., Weible, C. M., & Ingold, K. (2018). The advocacy coalition framework: An overview of the research program. In *Theories of the policy process* (pp. 135–171).
Kenny, B. (2015). Retail, the service worker, and the polity: Attaching labour and consumption. *Critical Arts, 29*(2), 199–217.
Lehman, B. (2008). The emergence of civil society organizations in South Africa. *Journal of Public Affairs: An International Journal, 8*(12), 115–127.
Martinez, M. G., & Kohler, J. C. (2016). Civil society participation in the health system: The case of Brazil's Health Councils. *Globalization and Health, 12*, 1–12.
Matelski, M., Otundo, B., Zijlstra, S. G, Dekker, M., van Kempen, L. A. C. M., Nangulu, A., & Spierenburg, M. J. (2018). *Civil society engagement with land rights advocacy in Kenya: What roles to play? Literature review.*
Mavee, S. E. A. (2022). The relevance of civil society participation in democratic governance. *Administratio Publica, 30*(3), 171–191.
McSweeney, J. (2014). The absence of class: Critical development, NGOs and the misuse of Gramsci's concept of counter-hegemony. *Progress in Development Studies, 14*(3), 275–285.
Mdlalose, B. (2014). The rise and fall of Abahlali baseMjondolo, a South African social movement. *Politikon, 41*(3), 345–353.
Mitchels, A., & De Graaf, L. (2010). Examining citizen participation: Local participatory policymaking and democracy. *Local Government Studies, 36*(4), 477–491.
Mohanty, R., Thompson, L., Coelho, V., & S. (2011). Mobilizing the state? Social mobilization and state interaction in India, Brazil, and South Africa. *IDS Working Papers, 359*, 1–39.
Moyo, Q. (2018). *Civil society participation in the Southern African Development Community (SADC) policy formulation and implementation process.*
Nattrass, N. (2004). *The moral economy of AIDS in South Africa.*
Ndinda, C., Hongoro, C., Chilwane, D., & Mokmane, Z. (2015). Civil society activism for the prevention and treatment of non-communicable diseases in South Africa: Implications for policy and practice. *Millennium Development Goals (MDGs) in Retrospect: Africa's Development Beyond, 2015*, 191–207.
Ngxabi, S. (2016). *The role of social dialogue (civil society participation) in policy decision-making in South Africa: The case study of NEDLAC.*

Nyamu-Musembi, C., & Musyoki, S. (2004). *Kenyan civil society perspectives on rights, rights-based approaches to development, and participation.*

Olawoore, B., & Kamruzzaman, P. (2019). NGOs, partnership and accountability—A case study of ActionAid and its local NGO partners in Nigeria. *Forum for Development Studies, 46*, 451–471.

Orvis, S. (2001). Civil society in Africa or African civil society. *Journal of Asian and African Studies, 36*(1), 17–38.

Osmani, S. (2007). Participatory governance for efficiency and equity: An overview of issues and evidence. Background paper prepared for the *World Public Sector Report*, Issue.

Overy, N. (2013). The social justice coalition and access to basic sanitation in informal settlements in Cape Town, South Africa. *International Budget Partnership Impact Case Study*, 11.

Ross, F. (2009). *Raw life, new hope: Decency, housing and everyday life in a post-apartheid community.*

Sabatier, P. A., & Weible, C. M. (2019). The advocacy coalition framework: Innovations and clarifications. In *Theories of the policy process* (2nd ed., pp. 189–220). Routledge.

Sabi, S. C., & Rieker, M. (2017). The role of civil society in health policy making in South Africa: A review of the strategies adopted by the Treatment Action Campaign. *African Journal of AIDS Research, 16*(1), 57–64.

Sadovnik, A. R. (2017). Qualitative research and public policy. In *Handbook of public policy analysis* (pp. 443–454). Routledge.

Schlager, E. (1995). Policy making and collective action: Defining coalitions within the advocacy coalition framework. *Policy Sciences, 28*(3), 243–270.

Seekings, J., & Nattrass, N. (2008). *Class, race, and inequality in South Africa.*

Smit, B. (2003). Can qualitative research inform policy implementation? Evidence and arguments from a developing country context. *Forum Qualitative Sozialforschung/Forum: Qualitative Social Research, 4.*

Sotirov, M., & Memmler, M. (2012). The Advocacy Coalition Framework in natural resource policy studies—Recent experiences and further prospects. *Forest Policy and Economics, 16*, 51–64.

Speer, J. (2012). Participatory governance reform: A good strategy for increasing government responsiveness and improving public services. *World Development, 40*(12), 2379–2398.

Spreen, C. A., & Vally, S. (2006). Education rights, education policies and inequality in South Africa. *International Journal of Educational Development, 26*(4), 353–362.

Tshangana, A. (2013). The impact of litigation by the legal resources centre for adequate classroom infrastructure in South Africa. *International Budget Partnership Impact Case Study*, Issue.

Voster, J. M. (2015). The possible contribution of civil society in the moral edification of South African society: The example of the 'United Democratic Front' and the 'Treatment Action Campaign' (1983–2014). *HTS Teologiese Studies/Theological Studies, 71*(3).

Wach, B., & Ward, R. (2013). *Learning about qualitative document analysis.*

Wampler, B. (2021). Entering the state: Civil society activism and participatory governance in Brazil. *Political Studies, 60*(2), 341–362.

Zikode, S. B. (2008). The greatest threat to future stability in our country is the greatest strength of the Abahlali baseMjondolo Movement (SA) (Shackdwellers). *Journal of Asian and African Studies, 43*(1), 113–117.

Chapter 12
The Unseen Architects: Civil Society's Role in Social Policy Innovation Across Africa

Takudzwa Musekiwa

12.1 Introduction

Civil society organisations (CSOs) in Africa are crucial in shaping social policies and addressing various issues. The essence of local CSOs' involvement in shaping social policy across Africa has particularly focused on creating an agenda for action that maintains equitable and sustainable solutions to social challenges, often reflecting the priorities and values of their communities. This chapter focuses CSOs interventions in the areas of healthcare, education, human rights, and governance. During Africa's HIV/AIDS crisis between 1990 and 2009, which was one of the most devastating public health challenges in modern history (Devereux & Sabates-Wheeler, 2021), CSOs played an instrumental role in shaping social policies, contributing to a 25% reduction in HIV incidence in 22 African countries from 1990 to 2009 (Wamai, 2014). However, CSOs have not only been active in advocating for social policies to do with healthcare, education, human rights, and governance but also social security, housing, labour and employment, child welfare, land and resource rights, and more recently, climate change. Some of these CSOs have built on pre--colonial civic associations that mobilised against colonialism and continue to influence post-colonial policy processes (Saruchera & Odhiambo, 2004). In Mozambique, for example, CSOs strongly influenced the environmental policy during the development of mining and petroleum laws in 2019 by providing expertise, setting agendas, representing communities, and monitoring processes (Sanches & Júlio, 2021). Additionally, across Africa, CSOs have been increasingly involved in gender-related policy-making, with South African women's organisations showing greater

T. Musekiwa (✉)
Centre for the Advancement of Scholarship, University of Pretoria, Pretoria, South Africa
e-mail: takudzwa.musekiwa@up.ac.za

E. Ndhlovu, C. Chipenda (eds.), *Redefining the Social Policy Discourse*,
https://doi.org/10.1007/978-3-032-24067-5_12

visibility and impact compared to their Nigerian counterparts (Adeogun, 2019). These examples demonstrate the growing relevance of CSOs in shaping social policies across Africa. Chikoto-Schultz and Uzochukwu (2016), in their study examining social policy process and non-state actors' involvement in Zimbabwe and Nigeria, found that although political elites continue to dominate the policy process, non-state actors increasingly permeate the process through various formal and informal strategies. Atukunda et al. (2015) make a similar observation about CSOs and healthcare policy in Uganda, pointing out that CSOs had played a key role in promoting the introduction and use of misoprostol for postpartum haemorrhage, despite a lack of government support for the initiative.

Most studies on African social policy by scholars such as Huijbregts and van Ginneken (2023) and Hickey and Bukenya (2020) emphasise the roles of state institutions and international organisations such as the United Nations (UN) and World Bank while overlooking the contributions of local CSOs. This creates an incomplete understanding of the broader social policy-making ecosystem, which this chapter hopes to address by highlighting the role of CSOs in shaping social policies across Africa. The chapter explores CSO contributions to social policy development and implementation across Africa. Addressing these gaps can provide a more holistic understanding of the roles and contributions of civil society in shaping sustainable and inclusive social policies across Africa. The chapter also discusses the conflicts between CSOs and governments in Africa and limitations that have emerged in Africa's CSO interventions.

12.2 Methodology

The chapter employs a multiple case study approach to examine diverse examples of CSOs across African countries. This approach allows for an analysis of how CSOs influence social policy in varying socio-political and economic contexts. This qualitative case study methodology enables a rich, contextual exploration of CSO contributions, addressing the research gap in understanding their nuanced roles in social policy-making in post-colonial Africa. Data was collected from policy documents, government reports, CSO publications, academic articles, books, and media coverage of CSO activities. Historical records were also reviewed to trace the evolution of CSO engagement in social policy-making. Cases were selected based on the documented or identified impact of CSO initiatives, with a focus on organisations that represent marginalised or vulnerable populations. While case studies provide in-depth insights, their findings may not be universally applicable. In this chapter, this is mitigated by selecting a diverse range of cases for broader relevance and addressed by triangulating data from multiple sources, including independent reports and policy documents.

12.3 Literature Review

Some scholars, such as Allen (1997), have argued that the concept of civil society is problematic and ideologically loaded. They argue that it is the proponents of liberal democratic reform, rather than the African polities themselves, who need this concept (Allen, 1997:333). While rooted in Western origins, the concept of civil society has been adapted to the African contexts, inspiring action and shaping local meanings in the negotiation between citizens, states, and markets (Lewis, 2002). In the context of this chapter, civil society refers to the sphere of organised, voluntary collective action that exists outside the state. It encompasses a broad range of non-state actors and organisations representing various groups' interests, values, and identities.

Historically, civil society actors increased their participation in social policy-making processes in Africa in the 1970s following Uhuru (Smith, 2019). However, their influence on social policy in different countries was affected by various factors, including the presence of participative norms, governing structures, and regime changes (Smith, 2019). Today, CSOs play a crucial role in democratic consolidation and development across different countries (Botchway, 2019). In Africa, the weakening of state authority has created opportunities for societal actors to enhance their role in political change (Bratton, 1989). CSOs' power is strengthened when they join epistemic networks, form broader coalitions, secure resources, and frame issues in alignment with national policies and political priorities (Smith, 2019). In South Africa and Uganda, CSOs have influenced government policy and legislation implementation, although their impact on policy outcomes has been limited (Robinson & Friedman, 2007). In the Southern African Development Community (SADC), CSOs have been involved in HIV/AIDS policy implementation, primarily through service delivery. However, their participation in decision-making has been restricted by member states (Godsäter & Söderbaum, 2017). CSOs have advocated for land and resource rights across sub-Saharan Africa, building on pre-colonial civic associations. In South Africa, the Treatment Action Campaign successfully shaped HIV/AIDS policy through various initiatives, including legal action against drug companies (Sabi & Rieker, 2017). Despite some challenges, CSOs remain crucial in extending citizen participation and enhancing pluralism in African social policy. CSOs in Africa are crucial in shaping social policies and advocating for citizens' rights. These organisations have roots in pre-colonial African associations and have evolved to address modern challenges (Saruchera & Odhiambo, 2004). CSOs across Africa engage in various activities, including policy advocacy, accountability efforts, and social transformation initiatives. However, the concept of civil society in Africa is complex, with some scholars arguing that it can be influenced by powerful actors, including Global North governments and donors, potentially compromising its role as a force for challenging the status quo (Hearn, 2001). Despite these challenges, the concept of civil society remains useful for analysing contemporary African politics and inspiring action (Lewis, 2002). CSOs continue to adapt international concepts to local contexts, navigating the delicate balance between citizens, states, and markets in their pursuit of social change and good governance (Lewis, 2002).

12.4 CSOs and Social Policy in Africa

12.4.1 CSOs and Healthcare Policy

When it comes to shaping healthcare policy in Africa, particularly in response to the HIV/AIDS crisis, CSOs have been influential in developing social policy that considers marginalised groups, provides expertise, and enhances policy representation and legitimacy. In sub-Saharan Africa, through coalitions of CSOs such as the AIDS and Rights Alliance for Southern Africa (ARASA), CSOs shaped the implementation of HIV/AIDS prevention and treatment programs, contributing to a 25% reduction in HIV incidence in 22 African countries between 1990 and 2009 (Wamai, 2014). Recent developments highlight the ongoing efforts of these organisations. ARASA continues to promote a human rights approach to HIV/AIDS, sexual and reproductive health and rights (SRHR), and tuberculosis (TB) across 18 countries in Southern and East Africa. Their work focuses on training and capacity strengthening of people living with HIV and other key populations, aiming to promote evidence-based policy and community advocacy. In the case of South Africa, the Treatment Action Campaign (TAC) founded in 1998, successfully advocated for improved HIV/AIDS health service delivery, shaping current policies with legal action against drug companies (Sabi & Rieker, 2017). Apart from HIV/AIDS crisis, CSO in Africa have been actively involved in shaping health care policy. Across the continent, CSOs have also influenced policy changes, as demonstrated by their role in promoting misoprostol for postpartum haemorrhage in Uganda, supporting its registration, guideline development, and distribution (Atukunda et al., 2015). TAC remains active in monitoring, advocacy, and campaigning within the healthcare system to ensure that all public healthcare users can access quality and dignified healthcare services. Their efforts include engaging in discussions on teenage pregnancy, TB, STIs, HIV prevention, and the importance of adherence to PrEP.

Kenyan CSOs have made significant strides in health advocacy. The Health NGOs Network (HENNET), a coalition of over 100 CSOs operating across all 47 counties, has been at the forefront of promoting health initiatives. During the 2024 vaccine stockout, HENNET emerged as a key advocate, championing the urgent need to replenish vaccine supplies. Drawing on insights from SMART advocacy training, HENNET strategically amplified the voices of caregivers and children in underserved communities missing essential vaccinations. Additionally, the Kakamega County Maternal, Newborn, and Child Health CSOs Alliance, established in 2016, coordinates advocacy efforts to improve access to healthcare services for women and children. This alliance exemplifies how localised CSO collaborations can effectively address specific health challenges within communities. However, challenges remain, including organisational limitations and political constraints. These barriers highlight the need for more supportive environments and stronger institutional frameworks to enable CSOs to sustain and expand their impact.

12.4.2 CSOs, Education, and Gender Rights

Since the enactment of Africa Union's Maputo Protocol in 2003, African CSOs have increasingly played an important role in influencing education and gender rights. CSOs such as the African Women's Development and Communication Network (FEMNET), which coordinates campaigns to promote women's rights in 37 African countries, continue to influence government policies on education, gender equality, and economic rights. One of the more tangible policy changes influenced by FEMNET includes the stricter laws against Female Genital Mutilation (FMG) in Kenya (Ahere, 2019). In 2024, FEMNET launched the Inclusive Dialogue Mechanism to strengthen EU-Kenya civil society engagement, aiming to enhance collaborative efforts in promoting gender equality and women's rights (FEMNET, 2024). Another pan-African CSO, Women in Law and Development in Africa (WiLDAF), has supported legal reforms in the various African countries through partner NGOs whilst also providing legal aid to women across several African countries. A study comparative by Adeogun (2019) on CSOs and policy-making on gender issues in South Africa and Nigeria found that women's organisations, in particular, have become more visible in decision-making processes compared to other CSOs in the two countries, impacting policies related to women's issues.

In Tanzania, WiLDAF has been at the forefront of advocating for women's political inclusion, legal reform, economic empowerment, and education, aiming to bridge the gender gap and empower women and girls to shape their futures. Some CSOs have also been instrumental in addressing sexual diversity struggles, lobbying for the decriminalisation of same-sex sexualities, and developing transnational networks to support LGBT rights (Currier & Cruz, 2014). In Côte d'Ivoire, women's rights groups organised a march in Grand-Bassam in December 2024, culminating 16 days of activism against femicide. The event highlighted the ongoing prevalence of gender-based violence and called for legal recognition of femicide with harsher penalties, underscoring the critical role of CSOs in advocating for women's safety and rights.

In Egypt, the Egyptian Centre for Women's Rights (ECWR) founded in 1996, focuses on advocating for women's legal rights, combating gender-based violence, and promoting women's political participation. Similarly, the Zimbabwe Women's Resource Centre and Network (ZWRCN) aims to promote gender equality and women's empowerment through information dissemination, advocacy, and capacity building. While the Association Tunisienne des Femmes Démocrates (ATFD), a CSO dedicated to defending women's rights, promoting gender equality, and advocating for legal reforms, has made significant strides enhancing women's status in Tunisia. However, media and gender equality CSOs face challenges such as financial constraints, legal restrictions, and cultural prejudices, which limit their effectiveness (Kassa & Sarikakis, 2021). Implementing Millennium Development Goals (MDGs) related to gender equality and education in Kenya and South Africa has raised questions about different government officials' ownership and interpretation of these goals (Unterhalter & North, 2011). These cases highlight the complex role of CSOs in shaping gender and education policies in Africa.

12.4.3 CSOs, Human Rights, and Anti-corruption

In the context of combating corruption and protecting human rights, CSOs have played a significant role despite facing numerous challenges. These organisations have been instrumental in detecting human rights abuses and advocating for solutions, often more effectively than governments (Welch, 1995). Pan-African CSOs such as Afrobarometer, African Network on Environmental and Economic Justice (ANEEJ), and Southern Africa Litigation Centre (SALC) have pushed for new anti-corruption laws and institutions while pushing anti-corruption litigation. In post-colonial Africa, where transitional democracies have ineffective formal accountability mechanisms, these CSOs have emerged as crucial actors in promoting political reforms and developing accountability systems (Chalmers & Setiyono, 2012). Working independently and in collaboration with local and international stakeholders, these CSOs have been able to influence policy on these issues. In May 2023, Afrobarometer conducted a 4-day workshop in Accra, Ghana, aimed at equipping CSOs and journalists with the necessary skills to use public attitude data effectively, thereby enhancing their capacity to advocate against corruption and for human rights.

Another example is the campaign by the African Centre for Justice and Peace Studies (ACJPS) to end impunity in Gambia under the Presidency of Yahya Jammeh which brought international attention to human rights violations under Yahya Jammeh's regime (Huijbregts & van Ginneken, 2023). The ACJPS's effort led to key recommendations for justice that Adama Barrow, Jammeh's successor, later implemented. Another example is the End SARS movement in Nigeria, a campaign led by Feminist Coalition a CSO and activists, that forced the Nigerian government to disband the SARS police unit in 2020 (Huijbregts & van Ginneken, 2023). The success of the Feminist Coalition shows how influential CSOs are in the region's social policy. In South Africa, the AmaBhungane Centre for Investigative Journalism exposed the Gupta family's influence over state institutions, leading to investigations, prosecutions, and reforms against state capture (Ahere, 2019). In July 2024, at least 42 youths in Uganda were charged for their involvement in a banned anti-corruption protest held in Kampala. The demonstrators marched through the capital's streets, shouting anti-corruption slogans and holding placards condemning corrupt lawmakers. This incident underscores the challenges faced by anti-corruption activists in the region. However, the impact of CSOs on reducing corruption in Africa remains limited, suggesting a need for a deeper understanding of corruption's underlying causes. Despite these challenges, CSOs continue to be vital partners in the fight against corruption and the protection of human rights in Africa. In November 2024, the United Nations Office on Drugs and Crime (UNODC) reported efforts in the Democratic Republic of Congo to empower voices against corruption, highlighting the importance of whistle-blower protection laws to safeguard those exposing corruption. In Ethiopia, CSOs continue contributing to human rights protection, mass atrocities prevention, and civilian protection in conflict-affected areas, complementing government efforts (Chukwuemeka, 2018). Despite facing challenges from anti-imperialist rhetoric and government attacks, CSOs continue investigating policies, exposing corruption, and advocating for vulnerable communities (Hearn, 2001). In financial transparency, CSOs have demanded beneficial ownership disclosure to curb Illicit Financial Flows (IFFs) and promote accountability. While beneficial ownership provisions are still nascent in African countries, various policy frameworks are being explored to implement them, including anti-money laundering initiatives and transparency efforts in the extractives sector (Jalipa & Danzi, 2020). These efforts underscore the importance of CSOs in fostering positive transformation and universal protection of human rights across the continent. In July 2023, Amnesty International called upon West and Central African states to stop persecuting human rights defenders who expose corruption, bribery, and abuse of office, emphasising the need to protect and support these individuals.

12.4.4 Conflicts Between CSOs and Governments in Africa

Conflicts between civil society organisations (CSOs) and governments also affect how much CSOs can influence social policies. These conflicts emerge due to competition for resources (class formation), diverging objectives, and governments' (mis)understanding of CSO roles (Campbell, 1996). Wood (2016) argues that these tensions are exacerbated by Western donors' actions, which are often perceived as neo-colonial by several African governments. Some states view CSOs as a source of social conflict. As Marchetti and Tocci (2011) point out, in ethno-political conflicts, civil society organisations, which they termed "conflict society organisations" (CoSOs), play a crucial role as both violators and promoters of human rights. The impact of civil society on conflicts is determined by factors such as political identities, frameworks of action, and political opportunity structures (Marchetti & Tocci, 2011). Depending on these factors, civil society actors can fuel conflict, maintain the status quo, or contribute to social harmony (Marchetti & Tocci, 2011). Hence, civil society is not homogenous, with some CSOs causing volatility in Africa. Resolving these conflicts between the government and civil society requires building trust, mutuality, and cross-sectoral partnerships. Continental Union organisations such as the African Union's (AU) mediation can potentially improve CSO-state relations. However, institutional and policy barriers may hinder this role.

12.4.5 *Limitations That Have Emerged in Africa's CSO Interventions*

Despite their potential to improve social policies, citizen participation, and pluralism, research by Yeboah and Ayelazuno (2021), Hickey and Bukenya (2020), and Devereux and Sabates-Wheeler (2021) suggests that CSOs face limitations. These limitations include political constraints, financial and resource limitations, limited access to decision-making, fragmentation and lack of CSO coordination, capacity and expertise deficits, public perception and legitimacy issues, cultural and social barriers, and restrictive legal frameworks (Robinson & Friedman, 2007). Political constraints include a repressive political environment, with some African governments such as Zimbabwe, Ethiopia, Kenya, and Egypt viewing CSOs as adversaries, restricting CSO operations through stringent regulations, outright bans, and state violence. Ethiopia's 2009 Charities and Societies Proclamation provides just one example of restrictive legislation used to limit CSOs' operation and influence (Ahere, 2019). Political constraints also include a lack of political will, with governments deliberately ignoring CSO policy advocacy efforts, particularly when these social policies threaten political elites' interests or expose corruption (Ahere, 2019). The government's co-option of CSOs is a significant constraint that dilutes their independence and effectiveness.

Financial and resource limitations include CSOs' donor dependence. With many organisations heavily relying on foreign funding, their interventions often focus on donor priorities rather than local needs (Yeboah & Ayelazuno, 2021). Additionally, constrained domestic funding makes it difficult for CSOs to sustain their long-term initiatives and even their ability to respond quickly to emerging crises (Devereux & Sabates-Wheeler, 2021). Limited access to decision-making for CSOs includes exclusion from policy processes by state actors. Sometimes, when invited, it is for tokenism, with their input ignored in the final decision. Fragmentation and lack of coordination are limitations for CSOs (Yeboah & Ayelazuno, 2021). These include their overlapping mandates and competition for funding, which leads to duplication of funding instead of collaboration. The fragmentation of CSOs, particularly divergent ideologies and priorities, limits their collective advocacy on social policy. Due to capacity and expertise deficits, CSOs across the continent struggle to assess the impact of their advocacy and initiatives, limiting their credibility and effectiveness (Ahere, 2019). Some CSOs are sometimes viewed as elitist and out of touch with grassroots realities, leading to distrust from communities and negatively affecting their public perception and legitimacy. Again, CSOs heavy reliance on foreign donor funding also creates scepticism about their independence and motives (Hickey & Bukenya, 2020). Cultural and social barriers, such as traditional norms and values, also limit CSO interventions, particularly in areas such as LGBTQ+ rights advocacy. Restrictive laws targeting CSOs, such as funding caps and registration barriers, have also limited CSOs' interventions and ability to operate effectively. State actors often restrict CSO involvement in decision-making and policy formulation, relegating them primarily to service delivery roles (Godsäter & Söderbaum, 2017). The effectiveness of CSOs is further hampered by knowledge gaps regarding successful intervention mechanisms and insufficient attention to political incentives underlying government failures (Devarajan et al., 2014). Moreover, donor support for African CSOs may inadvertently contribute to maintaining the status quo rather than challenging it. In some cases, well-funded CSOs intervening in national development strategies act more as consensus-builders than as forces for change (Hearn, 2001). These limitations highlight the complex challenges CSOs face in effectively influencing social policy across various African contexts.

12.4.6 How Can Regional or Pan-African Collaborations Among CSOs Enhance Their Effectiveness?

Regional and Pan-African collaborations will most likely enhance their effectiveness through various means in shaping and influencing social policy across the continent. Complementarity in collaborations allows CSOs to capitalise on diversity and reshape the roles of CSOs as complementary rather than contradictory (van Wessel et al., 2020). One example is regional peacebuilding efforts by the East and Horn of Africa Human Rights Defenders Project (EHAHRDP), a CSO in Africa's Great Lakes Region, addressing conflicts with regional dimensions (Kamatsiko, 2017). EHAHRDP has helped build resilience against the oppression of CSOs with cross-border collaborations of CSOs in repressive environments with CSOs in countries with more enabling environments. These collaborations and networks have protected CSOs in repressive environments, providing solidarity and amplifying advocacy when other CSOs are targeted, offering safety in numbers (Kamatsiko, 2017).

Capacity building and training collaborations can enhance the effectiveness of African CSOs. The success of FEMNET in training partner CSO activists on gender advocacy is just one example of how regional and pan-African collaboration CSOs can enhance effectiveness. Partnering with local or grassroots CSOs when conducting research and data collection creates a more comprehensive understanding of issues in their local context and better informs the advocacy and interventions (van Wessel et al., 2020). The work of Afrobarometer, which partners with local CSOs to produce reports, surveys, and case studies on pressing issues, is another example of how collaboration enhances effectiveness.

12.4.7 Key Capacity-Building Needs for CSOs in Africa

Research indicates that capacity building is crucial for African civil society organisations (CSOs) to enhance their policy advocacy and implementation roles. Key needs include resource development, policy analysis, and budget advocacy skills (Despard et al., 2018; Walker, 2016). CSOs require support to shift health funding from aid to domestic sources and improve social accountability (Walker, 2016). Capacity constraints in poverty reduction strategies include limitations in poverty impact analysis, monitoring and evaluation, and public expenditure management (Ikhide & Obadan, 2011). Institutional factors such as policymakers' non-receptivity to policy analysis and politicisation of policy-making affect capacity-building efforts (Wubneh, 2003). While international funding and prior capacity-building assistance can reduce needs, many organisations still require significant support (Despard et al., 2018). Long-term investment in indigenous NGOs, rather than relying on international NGOs, is recommended for sustainable capacity development (Walker, 2016). Careful consideration of gestation periods, program integration, and institutional settings is crucial when designing capacity-building projects in Africa (Wubneh, 2003).

12.4.8 What Partnerships in Africa Can Strengthen CSO Effectiveness in Social Policy?

Partnerships between civil society organisations (CSOs) and African governments can enhance CSO effectiveness in social policy implementation. Important factors for successful partnerships include clear objectives, defined roles, and the capacity to fulfil responsibilities (Devarajan et al., 2014). However, challenges persist in North-South donor-funded partnerships, particularly in the financial and management control systems of Northern organisations (Ashman, 2001). In South Africa, a pluralist approach involving state-CSO partnerships has the potential to address developmental challenges, but discrepancies between policy and practice raise concerns about the model's efficacy (Patel, 2012). Public–private partnerships (PPPs) have emerged as a strategy for poverty reduction and local development. However, they can also lead to increased state control over CSOs, as seen in the case of the Agri-FoodBank in South Africa (Warshawsky, 2012). While civil society engagement shows promise in addressing government failures in Africa, there are significant knowledge gaps regarding effective interventions. Donor support for CSOs should focus on strengthening political incentives and improving government functioning rather than bypassing the state and incorporate rigorous evaluations to determine impact mechanisms (Devarajan et al., 2014). To strengthen CSO effectiveness, partnerships should focus on collaborative accountability, proactive change strategies, and addressing governance issues in welfare service delivery.

12.4.9 *What Policy or Institutional Changes Are Necessary to Create a More Supportive Environment for CSO Engagement?*

Creating a supportive environment for CSO engagement requires several policies and institutional changes. Establishing new global mechanisms and advisory services for Bank-CSO engagement can promote mutual understanding and cooperation (Monico et al., 2005). Donors can help defend CSOs' operating space and support local networks and international contacts (Claessen & de Lange, 2016). Developing guidelines for best practices and implementing integrated learning programs for staff, governments, and CSOs can improve engagement effectiveness (Monico et al., 2005). Universities should create research and educational environments responsive to community needs and foster partnerships with external communities (Holland, 2008). Multisectoral nutrition governance systems should involve CSOs at national and subnational levels to ensure community ownership and diverse representation (Busse et al., 2020). Investing in civil society, improving accountability mechanisms, and creating platforms for CSO engagement at all levels are crucial for enhancing CSO participation in policy influencing and governance systems (Busse et al., 2020; Monico et al., 2005).

12.5 Discussion: The Evolving Role of CSO in Shaping Social Policy

The findings have revealed the multifaceted and dynamic role of civil society organisations (CSOs) in shaping social policies across Africa. It shows how CSOs have become critical players in pursuing sustainable and inclusive development by influencing policy processes. However, their effectiveness depends on navigating complex political, social, and economic landscapes.

One of the key findings is the impact of CSOs in influencing social policy on pressing social issues. The case of the Treatment Action Campaign (TAC) in South Africa provides an example of how grassroots mobilisation by CSOs can drive significant social policy changes, such as improved access to HIV/AIDS treatment. The case of the African Women's Development and Communication Network (FEMNET) in Kenya, which has contributed to stricter laws against female genital mutilation and enhanced gender equality, is also another example. These examples show the potential of CSOs when they engage with local communities and government. However, the restrictive environments characterised by political hostility and resource constraints often affect CSO successes.

While foreign donor funding enables CSOs to operate effectively, donor priorities and local needs are not always aligned. Hence, new financing structures prioritising financial sustainability, such as domestic philanthropy and public–private partnerships, are needed. This can help CSOs retain greater autonomy over their initiatives grounded in the community's priorities and not only donor priorities. As the cases show, fragmentation within CSOs is a significant challenge. This fragmentation, characterised by conflicting agendas, overlapping mandates, and lack of coordination, often limits CSOs' impact. The establishment of regional and pan-African collaborations has helped address some of these issues. By creating platforms for shared resources and knowledge exchange, CSOs have resolved some of these issues. CSOs such as Afrobarometer and FEMNET show the potential of collaborative efforts with positive results for social policy across the continent.

The cases also show a strained relationship between CSOs and governments. Governments sometimes view CSOs suspiciously as adversaries rather than collaborators. Measures such as restrictive legislation and co-option of CSOs show this mistrust. Hence, building trust and fostering dialogue between CSOs and governments is crucial for creating an enabling environment for civil society to influence social policy across Africa. As Africa continues to face challenges such as climate change, economic inequality, and governance deficits, CSOs role in shaping social policy is indispensable.

12.6 Conclusion

CSOs are essential actors in Africa's social policy, and they address gaps left by the government and international agencies such as the UN. This chapter demonstrated CSOs' essential contributions to sectors such as healthcare, education, gender rights, human rights, and anti-corruption. Through case studies, the chapter showed how CSOs have influenced social policies across the continent, from advocating for HIV/AIDS treatment in South Africa to advancing gender rights in Kenya and exposing corruption in Nigeria. CSOs ability to organise communities, act for marginalised voices, and provide support services highlights their transformative potential in shaping equitable and inclusive social policies.

Despite the success of CSOs in shaping social policy in Africa, it has not been without challenges. Political repression, restrictive legislation, dependency on foreign funding, and fragmented coordination often impede their operations. These challenges show a need for institutional reforms and capacity-building initiatives. In order for CSOs to overcome some of these challenges and fully realise their potential, a more enabling environment is vital. An environment values CSOs' contributions and provides mechanisms for meaningful participation in policy-making processes.

The chapter highlighted that regional and Pan-African collaborations such as ARASA offer promising pathways to strengthen the impact of CSOs when it comes to shaping social policy. These initiatives such as cross-border coalitions, capacity-building programs, and partnerships with grassroots organisations create a multiplier effect, strengthening the collective capability of CSOs. Additionally, creating trust and dialogue between CSOs, governments, and international donors is crucial for CSOs influence. To improve social policy, governments must move beyond viewing CSOs as adversaries but rather embrace them as partners in addressing pressing social and developmental issues.

Going forwards, the success of CSOs in shaping social policy across Africa will largely depend on their ability to adapt to changing socio-political contexts while leveraging emerging opportunities to overcome systemic challenges. These measures include adopting innovative strategies to secure financial freedom from foreign donors, improving their advocacy through collaborative data-driven approaches, and deepening their connections with local communities. It also requires a commitment to long-term capacity building and training, ensuring that African CSOs are not only resilient but can also lead transformative change. Ultimately, the role of CSOs extends beyond simple service delivery or policy advocacy. CSOs are architects of social transformation and custodians of civic space in Africa. As Africa confronts challenges ranging from economic inequalities to climate change, the inclusion of CSOs in shaping social policy will be essential to achieving sustainable development and social justice. Hence, concerted efforts by all stakeholders to recognise, strengthen, and integrate the work of CSOs into governance frameworks are necessary. By doing so, Africa can harness the full potential of its CSOs to build a more inclusive, equitable, and resilient future.

References

Adeogun, T. (2019). Civil society organizations and policy making on gender issues: South Africa and Nigeria in comparative perspective. In O. Tella (Ed.), *Nigeria-South Africa relations and regional hegemonic competence* (Advances in African Economic, Social and Political Development) (pp. 179–194). Springer. https://doi.org/10.1007/978-3-030-00081-3_11

Ahere, J. (2019). Civil society and democratic governance in Africa: Challenges and opportunities. *Frontiers in Political Science*.

Allen, C. (1997). Who needs civil society? *Review of African Political Economy, 24*(73), 329–337. https://doi.org/10.1080/03056249708704266

Ashman, D. (2001). Strengthening North-South partnerships for sustainable development. *Nonprofit and Voluntary Sector Quarterly, 30*, 74–98.

Atukunda, E. C., Brhlikova, P., Agaba, A. G., & Pollock, A. M. (2015). Civil Society Organizations and medicines policy change: A case study of registration, procurement, distribution and use of misoprostol in Uganda. *Social Science & Medicine, 130*, 242–249.

Botchway, T. P. (2019). Understanding the dynamics and operations of civil society in the 21st century: A literature review. *Journal of Politics and Law, 12*(1), 108–121.

Bratton, M. (1989). Beyond the state: Civil society and associational life in Africa. *World Politics, 41*(3), 407–430. https://doi.org/10.2307/2010506

Busse, H., Covic, N., Aakesson, A., & Jogo, W. (2020). What is the role of civil society in multisectoral nutrition governance systems? A multicountry review. *Food and Nutrition Bulletin, 41*, 244–260.

Campbell, W. (1996). *The potential for donor mediation in NGO-state relations: An Ethiopian case study* (No. 33). Institute of Development Studies.

Chalmers, I., & Setiyono, B. (2012). the struggle against corruption during the democratic transition: Theorising the emergent role of CSOs. *Development and Society, 41*, 77–102.

Chikoto-Schultz, G., & Uzochukwu, K. (2016). Governing civil society in Nigeria and Zimbabwe: A question of policy process and non-state actors' involvement. *Nonprofit Policy Forum, 7*(2), 137–170. https://doi.org/10.1515/npf-2015-0051

Chukwuemeka, B. E. (2018). The role of CSOs in promoting human rights protection, mass atrocities prevention, and civilian protection in armed conflicts. In *Regionalism and human protection*. Brill | Nijhoff. https://doi.org/10.1163/9789004363212_011

Claessen, A., & de Lange, P. (2016). Lessons for supporting policy influencing in restrictive environments. *Development in Practice, 26*(5), 544–554. https://doi.org/10.1080/09614524.2016.1188881

Currier, A., & Cruz, J. (2014). Civil society and sexual struggles in Africa. In E. Obadare (Ed.), *The handbook of civil society in Africa. Nonprofit and civil society studies* (Vol. 20). Springer. https://doi.org/10.1007/978-1-4614-8262-8_20

Despard, M. R., Ansong, D., Nafziger-Mayegun, R. N., & Adjabeng, B. K. (2018). Predictors of capacity-building needs among nongovernmental organizations in sub-Saharan Africa. *Journal of Community Practice, 26*, 204–224.

Devarajan, S., Khemani, S., & Walton, M. (2014). Can civil society overcome government failure in Africa. *World Bank Research Observer, 29*, 20–47.

Devereux, S., & Sabates-Wheeler, R. (2021). Social protection in Africa: Evidence, politics, and rights. In A. H. Dawson & R. A. Macmillan (Eds.), *Global social policy: Themes and issues* (pp. 123–147). Routledge.

FEMNET. (2024, December 20). FEMNET 2024 highlights. FEMNET. https://www.femnet.org/2024/12/femnet-2024-highlights

Godsäter, A., & Söderbaum, F. (2017). Civil society participation in regional social policy: The case of HIV/AIDS in the Southern African Development Community (SADC). *Global Social Policy, 17*, 119–136.

Hearn, J. (2001). The 'uses and abuses' of civil society in Africa. *Review of African Political Economy, 28*(87), 43–53. https://doi.org/10.1080/03056240108704502

Hickey, S., & Bukenya, B. (2020). The politics of promoting social protection in Africa: The case of Uganda's National Social Protection Policy. In S. Hickey, T. B. Adato, & J. Seekings (Eds.), *The politics of social protection in Africa* (pp. 47–67). Oxford University Press.

Holland, B.R. (2008). *Creating a supportive environment for community-university engagement: Conceptual frameworks*.

Huijbregts, M., & van Ginneken, W. (2023). Institutional dynamics in social protection systems in Africa: The role of civil society in promoting inclusion. *Journal of Social Policy Studies, 52*(1), 35–50.

Ikhide, S., & Obadan, M. I. (2011). The next generation of poverty reduction strategy papers in Africa: What is the role for capacity building? *Poverty & Public Policy, 3*, 1–27.

Jalipa, R., & Danzi, E. (2020). The case for beneficial ownership disclosure: A discussion paper on the policy frameworks promoting beneficial ownership transparency in Africa (August 2020). *Tax Justice Network Africa*. Available at SSRN: https://ssrn.com/abstract=3754859

Kamatsiko, V. V. (2017). Civil society organizations, Africa's Great Lakes Region conflict, and attempts at regional peacebuilding. *Journal of Civil Society, 13*, 54–70.

Kassa, B. E., & Sarikakis, K. (2021). Shrinking communicative space for media and gender equality civil society organizations. *Feminist Media Studies, 22*(7), 1745–1762. https://doi.org/10.1080/14680777.2021.1917640

Lewis, D. (2002). Civil society in African contexts: Reflections on the usefulness of a concept. *Development and Change, 33*, 569–586.

Marchetti, R., & Tocci, N. (2011). *Civil society, conflicts, and the politicization of human rights* (pp. 1–265). United Nations University Press.

Monico, C., Gears, C., Mandell, C. R., Thindwa, J., Garrison, J. W., & Reuben, W. (2005). *Issues and options for improving engagement between the World Bank and civil society organizations*.

Patel, L. (2012). Developmental social policy, social welfare services and the non-profit sector in South Africa. *Social Policy & Administration, 46*, 603–618.

Robinson, M., & Friedman, S. (2007). Civil society, democratization, and foreign aid: Civic engagement and public policy in South Africa and Uganda. *Democratization, 14*(4), 643–668. https://doi.org/10.1080/13510340701398329

Sabi, S. C., & Rieker, M. (2017). The role of civil society in health policy making in South Africa: A review of the strategies adopted by the Treatment Action Campaign. *African Journal of AIDS Research, 16*, 57–64.

Sanches, E. R., & Júlio, N. M. (2021). Shifting power dynamics? The role of civil society organizations in the policymaking for Mozambique's extractive sector. *Forum for Development Studies, 48*, 571–592.

Saruchera, M., & Odhiambo, M. O. (2004). Civil society and social movements: Advocacy for land and resource rights in Africa. In *Policy brief: Debating land reform and rural development* (Vol. 12). School of Government, University of the Western Cape.

Smith, S. L. (2019). Factoring civil society actors into health policy processes in low- and middle-income countries: A review of research articles, 2007–16. *Health Policy and Planning, 34*, 67–77. https://doi.org/10.1093/heapol/czy109

Unterhalter, E., & North, A. (2011). Girls' schooling, gender equity, and the global education and development agenda: Conceptual disconnections, political struggles, and the difficulties of practice. *Feminist Formations, 23*(3), 1–22.

van Wessel, M., Naz, F., & Sahoo, S. (2020). Complementarities in CSO collaborations: How working with diversity produces advantages. *VOLUNTAS: International Journal of Voluntary and Nonprofit Organizations, 32*, 717–730.

Walker, J. (2016). Achieving health SDG 3 in Africa through NGO capacity building—Insights from the Gates Foundation Investment in Partnership in Advocacy for Child and Family Health (PACFaH) Project. *African Journal of Reproductive Health, 20*(3), 55–61.

Wamai, R. (2014). Civil society's response to the HIV/AIDS crisis in Africa. In E. Obadare (Ed.), *The handbook of civil society in Africa. Nonprofit and civil society studies* (Vol. 20). Springer. https://doi.org/10.1007/978-1-4614-8262-8_11

Warshawsky, D. N. (2012). State, civil society, and the limits of NGO institutionalization in post-apartheid South Africa. *African Geographical Review, 32*(1), 1–13. https://doi.org/10.1080/19376812.2012.715549

Welch, C. E. (1995). *Protecting human rights in Africa: Roles and strategies of nongovernmental organizations*. University of Pennsylvania Press.

Wood, J. (2016). Unintended consequences: DAC governments and shrinking civil society space in Kenya. *Development in Practice, 26*(5), 532–543. https://doi.org/10.1080/09614524.2016.1188882

Wubneh, M. (2003). Building capacity in Africa: The impact of institutional, policy and resource factors. *African Development Review, 15*, 165–198.

Yeboah, F. K., & Ayelazuno, J. A. (2021). NGOs, state policies, and social welfare: Civil society and Ghana's Livelihood Empowerment Against Poverty (LEAP) program. *African Affairs, 120*(479), 72–92. https://doi.org/10.1093/afraf/adaa001

Chapter 13
Informal Economies and Transformative Social Policy in Sub-Saharan Africa: Bridging Critical Gaps

Neo Mokone and Emmanuel Ndhlovu

13.1 Introduction

The informal economy plays a central role in Sub-Saharan Africa, contributing significantly to employment and livelihoods, particularly in areas where formal job opportunities are scarce. Within contemporary Africa, there is a convergence of global disruptions and local development imperatives. The informal economy emerges as both a critical component of economic resilience and a significant blind spot in traditional social policy approaches. This is especially important as the continent faces accelerating progress toward the Sustainable Development Goals (SDGs) while preparing for post-2030 development frameworks, as has been argued by Ndhlovu and Chipenda (see Chap. 1).

According to the International Labour Organization (ILO, 2018), the informal economy accounts for a substantial share of employment in Sub-Saharan Africa, with Central, Eastern, and Western Africa reporting informal sector employment rates of 80% or more, far surpassing levels observed in industrialized regions, which include countries in North America, Europe, and parts of East Asia. While other scholars, such as Meagher (2022), suggest slightly lower figures, the ILO data is considered the most consistent and widely accepted for the region. This substantial informal participation directly impacts multiple SDGs simultaneously, especially SDG 1 (No Poverty), SDG 8 (Decent Work and Economic Growth), and SDG 10 (Reduced Inequalities).

N. Mokone (✉)
Centre on African Philanthropy and Social Investment (CAPSI), Wits Business School, University of the Witwatersrand, Johannesburg, South Africa
e-mail: neo.mokone@wits.ac.za

E. Ndhlovu
School of Public Management, Governance and Public Policy, College of Business and Economics, University of Johannesburg, Auckland Park, South Africa

E. Ndhlovu, C. Chipenda (eds.), *Redefining the Social Policy Discourse*,
https://doi.org/10.1007/978-3-032-24067-5_13

Despite this substantial contribution, informal workers often face precarious working conditions marked by low wages, lack of job security, and exclusion from formal social protection systems. This exclusion exemplifies the limitations of traditional social policy approaches exposed by contemporary global challenges, necessitating more integrated and adaptive frameworks. In many parts of the region, informal work is not a voluntary choice but rather a survival strategy in the face of high unemployment and inadequate formal job opportunities (Williams, 2023). Consequently, informal workers are particularly vulnerable to economic shocks, health crises, and poverty.

The persistence of informality is influenced by multiple factors, including high poverty levels, regulatory burdens, and the absence of effective labor market policies that encourage formal employment (Meagher, 2022). These challenges are compounded by structural constraints that have historically limited African states' policy autonomy, complicated by persistent donor-driven agendas that often prioritize external preferences over locally defined priorities (see Chipenda and Tom, Chap. 2). Formalization—the process of transitioning informal work into the formal sector—has proven challenging due to significant structural barriers, including high costs associated with formal business registration, a lack of trust in government institutions, and concerns about losing economic flexibility. This mistrust reflects broader challenges of states becoming more accountable to external forces than to their citizens. Informal workers, particularly those in small-scale businesses, fear that formalization will lead to higher taxes, stricter regulations, and reduced autonomy (Noyoo & Boon, 2018). These challenges are compounded by a lack of accessible social protection schemes tailored to the unique characteristics of informal work, such as fluctuating incomes and a lack of formal contracts.

Despite these obstacles, the informal economy remains vital for both economic stability and poverty alleviation in Sub-Saharan Africa. Informal businesses, ranging from street vending to small-scale agriculture, contribute significantly to regional GDP and provide livelihoods for millions (Vorng, 2024). However, their economic potential is often undermined by the lack of formal social protection, leaving informal workers vulnerable to income insecurity, health issues, and old-age poverty. The COVID-19 pandemic has exposed these vulnerabilities, with informal workers bearing disproportionate impacts due to their exclusion from formal social protection systems.

To address these challenges, there is a pressing need for social protection policies that are inclusive of informal workers, providing them with adequate security without compromising their ability to remain flexible in their employment. These policies must be designed to reflect the unique characteristics of informal work, such as its irregular income patterns and the diversity of activities that constitute the informal economy. Therefore, this chapter attempts to bridge the gap between the informal economy and social policy. It offers practical recommendations relevant to policymakers and researchers. The chapter contributes to the broader discourse on transformative social policy by examining the intersection between the informal economy and social policy, proposing a hybrid approach that strikes a balance between the sector's flexibility and worker protection while addressing the structural dependencies that have historically undermined comprehensive social policy development.

The objectives of the chapter are threefold: (1) to review the literature and case studies on the role of the informal economy in providing employment and economic stability, (2) to develop a conceptual framework toward bridging the informal economy and social policy that addresses the limitations of traditional social policy approaches, and (3) to offer actionable recommendations to policymakers on how to integrate these workers into social protection systems without disrupting their livelihoods while contributing to broader sustainable development objectives in Sub-Saharan Africa.

The chapter is structured as follows: the current introduction is followed by the methodology, literature review, conceptual framework, key findings, discussion, policy recommendations, and, finally, the conclusion and suggestions for future research.

13.2 Research Methodology

The chapter employs an integrative literature review, following the principles established in comparable research frameworks (Snyder, 2019). Articles were retrieved from reputable academic databases, including Scopus (for interdisciplinary and social science articles), Web of Science (high-quality peer-reviewed outputs), Google Scholar (policy briefs and gray literature), JSTOR (historical and policy-oriented studies), and institutional databases from the ILO and World Bank (region-specific reports and case studies). Databases were chosen for their reputation for providing peer-reviewed academic literature, a method supported by Xiao and Watson (2019). The search terms included "informal economy" AND "social policy," "Africa" AND "informal employment," "social protection" AND "Sub-Saharan Africa," and "case studies" AND "policy integration."

The book chapter selection process involved a systematic screening to ensure relevance and quality. In Stage 1, duplicates across databases were removed. Stage 2 excluded articles that did not focus on informal economies or social policy. Stage 3 consisted of a quality appraisal retaining peer-reviewed articles, institutional reports, and high-quality case studies. The inclusion criteria were as follows: (1) papers that addressed directly the nexus between informal economies and social policies, centering on perspectives from Africa or offering lessons that can be applied elsewhere, and (2) preference for papers published in high-impact journals or by reputable institutions like the ILO and World Bank. The exclusion criteria were set against papers not specifically for informal employment or social protections and had less rigorous methodology and applicability. The initial search yielded 148 scholarly articles and reports. After the relevance and quality checks, 77 articles met the inclusion criteria.

Data analysis was done thematically to bring out structural impediments, contributions from the informal economy, challenges against social protection, and innovative solutions. The use of thematic coding for data analysis corresponds with approaches described by Nowell et al. (2017), ensuring a systematic and transparent

Table 13.1 Overview of the most significant articles

Author(s)	Year	Title	Findings	Relevance	Citations
Chen, M. A.	2023	The Informal Economy in Comparative Perspective	Defines the informal economy as flexible but precarious, highlighting its role in income generation during crises	Offers theoretical insights into the informal economy's complexity and its interaction with social policies	400+
Canelas, C. & Niño--Zarazúa, M.	2022	Social Protection and the Informal Economy: What Do We Know?	Emphasizes the role of tailored, noncontributory social protection models for informal workers	Explores innovative strategies for integrating informal workers into social protection systems	300+
Adesina, J.	2021	Universalism, Social Policy, and Africa	Emphasizes the role of solidarity and universal social protections in addressing informality and inequality	Offers a framework for addressing systemic barriers and fostering social inclusion	800+
ILO	2022	Informal Employment Trends Worldwide	Provides statistical analysis of global and regional informal economy trends, emphasizing Africa's unique challenges and opportunities	Supplies essential statistical context for the prevalence of informal economies globally and in Africa	500+
Patel et al.	2023	Cash-Plus Programs and Social Protection	Examines the evolution of cash transfer programs and their relevance for informal workers in Africa	Explores innovative social protection models that address the unique needs of informal workers	200

Source: Authors' own compilation

identification of key themes and patterns in literature. Comparative analysis of case studies from Africa and other regions, such as Ghana, South Africa, Kenya, and Zambia, highlighted best practices. Critical evaluation ensured relevance and policy applicability. Ethical considerations included proper citation, adherence to intellectual property rights, and transparent reporting of findings. Limitations included reliance on secondary data and a geographic focus on Sub-Saharan Africa, with less coverage of North Africa.

Summary of Selected Articles

Table 13.1 provides an overview of the most significant articles included in the chapter, detailing their key findings, relevance to the intersection of informal economies and social policy, and citation metrics.

13.3 Literature Review

13.3.1 *Definition, Characteristics, and Types of Informal Economic Activities*

Current definitions of the informal economy lack a comprehensive understanding of the complexity of informal work. While frameworks provided by the ILO (2022), Polanyi (1957), Medina and Schneider (2018) have some established bases, they inadequately capture the diverse traits that define informal work within modernized economies. The ILO's (2022) classification of informal employment as work outside formal labor regulations—such as casual labor, street vending, and small-scale agriculture—presents a static view of inherently dynamic and contextual work. This framework assumes a sharp demarcation between formal and informal sectors, missing nuances within the informal economy where legal activities sometimes border on illegal ones, and workers' participation in informal labor has varying motivations. Polanyi (1957) sees informal work as voluntary and social, neglecting the exploitation reality for many workers. This view ignores how fragile informal employment is and what economic forces drive workers into these sectors. In addition, Schneider et al. (2025) and Williams (2023) assume necessity drives informal work, overlooking that it can be strategic due to flexibility and market opportunities.

These framework inadequacies have significant policy implications. A better understanding of informal work is essential. This chapter develops a dynamic framework that distinguishes types of informality through necessity and choice attributes: vulnerability, worker autonomy, and socioeconomic context. In developing countries, informal workers operate in weak institutional settings, where structural gaps allow informal survival. In developed countries, informal workers may seek economic benefits, such as tax avoidance or flexible hours. This difference is significant for creating nuanced social protection models since motivation and vulnerability differ. Current views of informal work inadequately guide policies. Policymakers must design effective social protection systems that address the diversity of informal work and its associated risks and opportunities across regions. Policymakers need contextualized definitions of the informal economy to design effective interventions. Without this analytical approach, efforts to bring informal workers into formal systems will likely fail.

Legal nonrecognition is well acknowledged in the literature (Meagher, 2022; Canelas & Niño-Zarazúa, 2022), leaving informal workers without labor rights, social security, and health care, creating cyclical risk. Lacking formal status means informal workers miss important protections and formal credit systems, worsening economic insecurity. Income instability, linked to seasonal demand and regional variations, structurally weakens informal labor markets (Meagher, 2022). This chapter argues that informal worker vulnerability to economic shocks stems from underdeveloped infrastructure, political instability, and inadequate financial systems—factors creating high vulnerability to external shocks. The chapter demonstrates how income fluctuations connect to larger socioeconomic conditions and perpetuate poverty cycles. Though informal work provides flexibility and identity, it denies workers access to long-term benefits, such as pensions and healthcare (Canelas & Niño-Zarazúa, 2022). This chapter highlights trade-offs between economic flexibility and social protection, noting that community-based schemes may alleviate but not solve structural inadequacies. The chapter advances discussion by integrating legal exclusion, income precariousness, and absent social protections in a holistic framework that underscores the structural origins of informal workers' precarity. It argues for policy reforms integrating informal workers into formal social protection systems while addressing immediate and structural needs. Without such interventions, informal workers will continue to be marginalized in national economies.

Informal work types in Sub-Saharan Africa reveal sectoral structural complexity. Self-employment in urban areas, including street vending, small-scale agriculture, casual labor, and small businesses, is examined through literature review and case studies in Kenya, Zambia, South Africa, and Ghana. Lack of formalization hinders market access, credit facilities, and legal protection, perpetuating poverty and limiting economic opportunities (ILO, 2022). This excludes over 70% of informal street vendors from formal credit systems and legal protection, limiting business expansion. Small-scale agricultural workers experience low yields and precarious incomes, exacerbated in Zambia by inadequate infrastructure and financing. Improved market access and rural infrastructure could increase agricultural productivity by up to 35% (Noyoo & Boon, 2018), achieving income stability. Casual laborers, including domestic and construction workers, lack contracts or job security due to inadequate labor protection. Limited market access and unfair competition constrain artisanal sector growth. Absent formal protection and credit access hinders economic mobility and exacerbates socioeconomic instability. Combined approaches are proposed to formalize processes through simple assistance and gradual integration of informal workers into official systems. Policies should facilitate small loan provision, legal safety, and skills training for informal workers, providing stepwise guidance while remaining flexible. This approach could reduce poverty by 25% within 5 years, based on Kenyan data. The chapter enhances informal sector knowledge and internal dynamics understanding, facilitating barrier identification for economic mobility among informal workers in Sub-Saharan Africa. The policy-oriented hybrid formalization model provides actionable policy recommendations to effectively integrate informal workers into social protection systems, enabling sustainable economic mobility.

13.3.2 Differences Between Formal and Informal Economies

The formal economy and informal economy are two distinct sectors that operate under contrasting conditions, which affect workers' legal recognition, economic security, and access to social protection. The formal economy functions within a well-defined regulatory framework that includes taxation, labor laws, and social protection policies. Formal workers benefit from job security through legal contracts that outline wage payments, working hours, and job conditions. They are entitled to social protections such as health insurance, pensions, and unemployment benefits, which provide a measure of financial stability and economic predictability (ILO, 2022). These protections help mitigate income volatility, ensure basic living standards, and reduce the potential for exploitative practices. In contrast, the informal economy consists of unregistered businesses and self-employed individuals who operate outside of these formal legal frameworks. Informal workers lack access to labor protections and social benefits. Consequently, they face income instability, as their earnings are influenced by fluctuating market conditions, seasonality, and demand shifts (Meagher, 2022). This absence of legal recognition not only undermines workers' economic security but also exacerbates their vulnerability to economic shocks and poor working conditions. Informal workers' lack of access to social protections intensifies their exposure to risks, as they are not covered by basic provisions such as healthcare or retirement funds (Noyoo & Boon, 2018). This exclusion from formal systems contributes to a cycle of poverty and inequality, as informal workers have limited recourse to government programs or legal mechanisms when their rights are violated. The absence of labor law enforcement and the inability to access benefits make informal workers highly vulnerable to exploitation, thus reinforcing the structural divide between the two economies.

To better understand the distinctions between formal and informal economies, the following table provides a concise summary of their key characteristics. Table 13.2 highlights differences in regulation, employment conditions, economic contributions, and their respective societal roles.

13.3.3 Barriers to Formalization and Integration into Social Protection Systems

The informal economy in Sub-Saharan Africa is a vital source of employment, with approximately 70% of the workforce engaged in informal work (ILO, 2021). While it cushions the economy during economic shocks, it also faces challenges to

Table 13.2 Key differences between formal and informal economies

Aspect	Formal economy	Informal economy
Regulation	Heavily regulated, legally compliant	Unregulated, operates outside legal frameworks
Employment	Contractual, benefits included	Casual, lacks benefits and job security
Economic role	Significant GDP contribution, well-documented	Substantial but underrepresented in statistics
Social impact	Structured, excludes vulnerable populations	Inclusive, vital for marginalized groups
Dependency	Relies on formal contracts and labor laws	Provides cost-effective labor for the formal sector

Source: Authors' own compilation

formalization. The major obstacles are the exorbitant costs of registration that small businesses and low-income workers cannot afford (Dube & Moyo, 2021). Most informal workers are engaged in precarious economic activities where earnings are irregular; thus, the costs of formalization seem more feasible than the actual benefits (Cohen et al., 2019). Additionally, there is low trust in government institutions; therefore, workers perceive formalization as a burden due to fear of higher taxes and exploitation (Williams, 2018). Furthermore, informal workers lack social protection, making them highly vulnerable to economic shocks. Proposed systems of micro-contributions could accommodate them by allowing flexible small payments to social protection systems based on earnings. Mobile platforms may facilitate these contributions, making it easier for informal workers to access the social services they need without incurring high costs.

Gender inequality exacerbates the difficulty of formalization. For example, women in domestic work and agriculture have less access to resources despite facing greater vulnerabilities (Mosoetsa, 2021). Policies addressing gender inequality, for example, provide subsidized registration and easier access to financial services to help improve the condition of women and thereby reduce the differences between men and women regarding formalization rates (Williams, 2023). Political resistance also slows the process, as broad policies do not adequately address the informal economy. Policies from different sectors that address the specific challenges in various industries—such as agriculture or street vending—would be crucial in providing practical support (Cohen et al., 2019). Governments can achieve this by reducing the regulatory burden and building trust through targeted support, such as micro-contributions and gender-sensitive initiatives, which will increase participation in the formal economy and thereby enhance economic security and stability (Table 13.3).

13.3.4 Innovative Approaches to Social Protection for Informal Workers

African countries have demonstrated creativity in developing models to incorporate informal workers into national social protection systems; however, scalability remains limited. One example is Kenya's NHIF, which is currently expanding coverage to informal workers through grassroots-level cooperatives, allowing them to make income-based, sliding-scale premium payments (Canelas & Niño-Zarazúa, 2022). This has significantly improved service delivery in urban areas where cooperatives are more established. Zambia has also made progress by linking informal dairy farmers to the social protection system, allowing them to access pensions without formalizing their employment (Miti et al., 2023). While this model works well in rural areas, it struggles to scale in urban settings where informal employment is very diverse. This highlights the limitations of sector-specific approaches and underscores the need for more adaptable models.

Table 13.3 Barriers to formalization and integration into the social protection system

Barrier	Description	Suggested solution	References
Regulatory constraints	Informal workers perceive high registration fees and complex compliance	Simplified registration processes, subsidized fees, or waivers for small businesses. Mobile-based platforms enabling flexible micro--contributions to social protection systems	Dube and Moyo (2021), Cohen et al., (2019)
Mistrust of government institutions	Informal workers fear exploitation, excessive taxation, and inefficient or corrupt public institutions. There is a lack of clarity on the benefits of formalization compared to the perceived risks	Transparent outreach programs to educate informal workers on the long-term benefits of formalization, such as access to social services, credit, and wider markets. Streamlining bureaucratic processes to reduce corruption and build trust	Williams (2018)
Gender inequality	Women in informal sectors face discriminatory practices, restricting their access to financial resources and business formalization training. The gendered division of labor limits their ability to formalize their businesses	Gender-sensitive policies offering subsidized business registration costs, financial services, and skills training for women to reduce the gender gap in formalization rates	Mosoetsa (2021)
Political resistance	Policies often treat the informal economy as homogeneous, failing to recognize the diversity of the sector. This leads to a lack of tailored interventions that address specific barriers faced by different informal workers	Tailored policies that reflect the diverse needs of informal workers, recognizing the complexity of the informal sector and providing sector--specific interventions	Cohen et al., (2019)

Source: Authors' own compilation

Free systems, along with reasons for slow formalization, could successfully increase social protection among the informal sectors. For instance, mobile systems like M-Pesa in Kenya demonstrate how technology can connect informal workers to essential services (Meagher, 2022). A mixed model combining mobile systems with social protection services, such as pensions and job loss insurance, could enhance coverage for informal workers through agreements between mobile companies, banks, and governments. Flexible policy frameworks are essential for accommodating the authentic and diverse nature of informal work. The rigidity of traditional models often misses the flexibility essential to the livelihood of workers. Policies oriented toward partial formalization, i.e., encouraging or allowing mobile platforms for contributions, could increase coverage among informal workers. Hybrid social protection models that combine noncontributory systems with technology-driven solutions are more likely to succeed in scaling up support for informal workers. Finally, the future of social protection for informal workers in Africa hinges on creating flexible systems that integrate aspects of both formal and informal. The obstacles to complete formalization must be removed, as hybrid models will improve access to essential social services for millions, thereby making them more inclusive and sustainable.

13.3.5 Informal Sector and Employment Creation, Poverty Alleviation, and Economic Stability

13.3.5.1 Employment Creation

The informal sector employs over 60% of the world's workforce (ILO, 2018); therefore, it is particularly essential in Sub-Saharan Africa. However, obstacles to formalization—regulatory obstacles, lack of education, and financial exclusion—block its path toward integration into formal social protection systems (Chen, 2012; Vanek et al., 2014). An in-depth analysis of mobile insurance schemes, including M-TIBA in Kenya, reveals that while these platforms reduce transaction costs and enable payments in small increments, their outreach is limited, particularly in rural areas (Mas & Radcliffe, 2018). More specifically, sector-specific initiatives, such as cooperatives, have achieved success at the local level but have not been able to surmount larger national challenges. A merged model that integrates mobile platforms with community-based cooperatives could improve both coverage and scalability. Monetary incentives for informal enterprises could encourage gradual formalization. Ghana and Nigeria demonstrate that easy registration, accompanied by monetary inducements, significantly increases participation in social protection schemes (Miti et al., 2023). This bolsters the idea that incremental formalization approaches can be practical. An understanding of the process of incorporating informal workers into social protection systems has been provided. Through the advocacy of hybrid models that combine technology, community initiatives, and targeted financial incentives, we suggest an approach to extension that is flexible in terms of providing coverage without complete formalization—the first step toward inclusive social protection in Sub-Saharan Africa.

13.3.5.2 Poverty Alleviation

The informal economy is critical for poverty alleviation, providing income to low-skilled and marginalized workers, particularly in rural areas. However, despite its importance, workforce formalization remains difficult in Africa due to regulatory barriers, limited access to education, and financial exclusion. In Zambia, for example, 53.8% of agricultural employment is self-employed (Miti et al., 2023), highlighting the sector's role in self-reliance. Yet, informal workers face low productivity, food insecurity, and a lack of social protection, undermining the sector's potential for poverty reduction. Policy solutions should focus on reducing formalization barriers, introducing targeted social protection schemes, and improving productivity through training and access to markets, ensuring that informal workers can move toward more secure and sustainable livelihoods.

13.3.5.3 Economic Stability

The informal economy plays a crucial role in stabilizing economies, especially during crises, by providing employment and maintaining consumer demand when the formal economy contracts. However, the lack of integration into national economic frameworks and regulatory oversight poses risks to broader economic stability (Canelas & Niño-Zarazúa, 2022). One of the key challenges is workforce formalization, which is hindered by regulatory barriers, limited access to social protections, and low financial inclusion in many African countries (Miti et al., 2021). To overcome these barriers, policies must simplify the business registration process, provide formalization incentives, and expand access to social protection systems for informal workers. Innovative approaches, such as mobile-based social insurance and informal worker cooperatives, could also provide pathways for more inclusive protections. These measures would enable the informal sector to leverage its strengths while addressing its vulnerabilities, ultimately fostering a more stable and inclusive economy.

13.4 Social Protection Systems and Transformative Policies in Sub-Saharan Africa

The social protection systems of Sub-Saharan African countries have had limited success in integrating informal workers, mainly due to weak administration, limited resources, and the absence of formalization. This has been noted by Noyoo and Boon (2018). Although post-apartheid South African systems have increased coverage, regional inequality persists (Motala & Treat, 2014). In Zimbabwe, the system of social protection has failed to address issues of informality and rising unemployment (Nhede, 2018). Malawi and Zambia are examples of hybrid models that combine community-based and formal systems to achieve greater inclusivity. It is little surprise, then, that such models demand prodigious amounts of adaptation to local contexts if they are to be scalable and sustainable. A context-sensitive framework is therefore proposed, one that utilizes technology, especially mobile platforms, for flexible contributions toward meeting administrative challenges and expanding coverage in resource-limited areas. When models are designed with technology integration, it can make the systems more accessible and Sustainable for Social Protection, providing informal workers with a reliable system of health, pensions, and unemployment benefits. This illustrates a straightforward approach to integrating informal workers into national systems; regional variations are encouraged to establish a long-term economic safety net for the vulnerable population in Sub-Saharan Africa.

Transformative policies in Sub-Saharan Africa aim to address systematic inequalities while fostering long-term resilience. These policies often empower marginalized populations by extending social protection to informal workers, women, and youth—groups typically excluded from conventional systems. In South Africa, introducing a basic income grant is viewed as a transformative policy toward income security, particularly for those who are informally employed without decent work opportunities and social protection (Motala et al., 2024). Similarly, Botswana has initiated targeted interventions in cash transfers to support vulnerable groups, particularly orphans and the elderly in rural areas (Noyoo & Boon, 2018). The social cash transfer program in Malawi, known as Mtukula Pakhomo, has proved very useful in ultra-poor household support led by women, thereby showing transformative policy potential at the household level for building resilience (Nhede & Marumahoko, 2023). Though there are exceptions to this, much of the research highlights the limited sustainability of most transformative policies in ensuring the mainstreaming of the informal sector into national frameworks, thereby constantly leaving informal workers vulnerable to economic shocks. Existing studies (Nhede & Marumahoko, 2023; Gumede, 2019) often focus on specific programs or countries, with limited regional comparative analysis to identify best practices. Another critical gap is the lack of data on the intersectionality of vulnerabilities, such as age, gender, and geography interact to exacerbate exclusion from social protection systems.

Social protection systems in Sub-Saharan Africa struggle to cover informal workers, who make up a large portion of the workforce but are excluded from national schemes. The apparent reasons for such exclusion are inconsistent employment patterns and a lack of proper documentation (Noyoo & Boon, 2018). This leaves most informal workers out of pensions and health care, further increasing their vulnerability. To worsen the disparities, gender inequality also places women, who are mainly in caregiving roles, away from formal employment and adequate protection (Keune, 2024). The obstacles to formalization are many and varied. Administrative difficulties, political will, and the sheer magnitude of the informal sector are a brake on contribution collection and, thus, social protection system sustainability. Even though technologies like blockchain and mobile payments ensure more efficient distribution of benefits, they do not address these structural issues at their root. Technology alone cannot untangle such deeply rooted problems. Policy reforms promoting informal workers' formalization can help bridge these gaps. Strategies should be informed by local context and needs, emphasizing vulnerable groups such as women, rural workers, etc. Only through comprehensive and context-sensitive policies will social protection systems become truly inclusive and sustainable.

13.5 Exclusion of Informal Workers from Existing Social Policies

13.5.1 Structural Challenges in Social Policy Design

Informal workers comprise over 60% of the world's labor force (ILO, 2018; Motala et al., 2024). Most of them are left out of formal social protection, which is usually designed for and assumes a person to have a formal employment contract. They are left out due to the country's systemic barriers that complicate how to channel contributions from informal workers who often fall outside the tax net and whose income shows significant variability. In South Africa, increasing social grants does not stop informal workers from being vulnerable to income shocks in periods of economic distress. Thus, it indicates a gross inefficiency of the social protection systems. Apart from the fact that informal workers are not only left out of formal employment, this challenge is also escalated by existing system architectures that have glaring inadequacies and do not consider the variability and informality characterizing their livelihoods. In countries like Nigeria, informal workers rely on traditional safety nets. There is a sharp contrast between community-based support systems and formal systems, which cannot reach these workers due to their exclusion from the official economic structure (Noyoo & Boon, 2018). For policymakers to effectively address this issue, they must first understand how to navigate the inherent tension between preserving the flexibility of informal employment and ensuring access to essential social protection. This calls for policy innovation that encourages formalization—such as a simple tax registration process for small businesses. It also needs to develop and expand inclusive safety nets responsive to the realities of informal work without undermining its intrinsic flexibility.

13.5.2 Economic and Social Impacts of Exclusion

Indeed, excluding informal workers has profound economic and social consequences since most are at greater risk of falling into poverty during a downturn or crisis. This was explicitly indicative of the COVID-19 pandemic, where informal workers did not enjoy unemployment benefits or stimulus packages in any part of the world, leading to a massive income loss (Hart et al., 2024). In Pakistan, this exclusion from social safety nets left millions vulnerable, with NGOs scrambling to fill gaps that should have been addressed by state policies (Khan, 2024). Furthermore, it worsens inequality, as most informal workers are women. They are in occupations like household work or vending, which have been disproportionately affected by the pandemic, and bear additional barriers to healthcare and maternity benefits due to the lack of coverage by specific policies. For example, informal female workers in Ghana reported that their inability to contribute to pension schemes left them with no safety net in old age (Noyoo & Boon, 2018). Addressing these disparities is critical to achieving broader social equity and reducing economic vulnerabilities.

13.5.3 *Pathways to Inclusion*

To bridge these gaps, governments must revisit the structure of social protection systems. One approach is the adoption of worldwide basic income schemes, which do not rely on employment status and thus provide a safety net for all citizens, including informal workers. South Africa has debated such a measure, with proponents arguing it would alleviate poverty and reduce dependency on inconsistent incomes (Hart et al., 2024). Another strategy is leveraging digital technologies—to facilitate micro-contributions to social protection schemes from informal workers, making it easier to integrate them into national systems (Khan, 2024). Policymakers must also address the unique needs of informal workers by tailoring programs to their circumstances. For example, flexible pension plans and subsidized health insurance schemes could provide essential security without imposing undue financial burdens. Countries like India, through the Rashtriya Swasthya Bima Yojana program, have shown that targeted interventions can successfully extend social protection to informal workers without dismantling traditional systems (ILO, 2022).

The knowledge gap lies in effectively integrating informal workers into social protection systems without compromising their flexibility. More than 60% of the world's workers are informal (Motala et al., 2024); however, they benefit from only 17% of social protection coverage (ILO, 2022). Most existing research reviews barriers such as fluctuating incomes and exclusion from tax systems but provides little regarding scalable and adaptable mechanisms such as digital micro-contributing tools. Gendered impacts of exclusion, particularly of women in domestic or vending roles, have been under-explored. Addressing these gaps would ensure the design of more equitable and pliant social protection systems.

13.6 Key Challenges and Obstacles to Integration in Social Policy Frameworks

13.6.1 *Fragmentation of Social Policy*

The fragmentation of social protection systems in Sub-Saharan Africa makes it hard to integrate informal workers. Overlapping systems, poorly coordinated among many countries, have inefficiencies and coverage gaps. In South Africa, certain PPPs have resulted in duplication and reduced the system's overall effectiveness

(Haurovi, 2021). Since the formal system does not cover informal workers, they face a challenge with insufficient data sharing among agencies, which hampers effective identification of where there are gaps in coverage. The administrative inefficiencies over which no targeted policy for the informal sector complements each other to perpetuate exclusion. For that, coherent, unified strategies that mandate easy data sharing and reduce overlaps are required to solve those challenges.

13.6.2 Financial Constraints and Sustainability Issues

A key obstacle in extending social protection is the financial sustainability of social policies under limited public resources. For example, although South Africa is growing its economy, it would not be able to support the social grant system. More than 18 million recipients in that country have raised concerns about the economy's slow growth and the long-term viability of this system (Haurovi, 2021). External financing inflates volatility toward positive changes in external economic conditions among aid-dependent countries. Informality challenges financing because informal dominant employment generates very low tax revenue (Kasera et al., 2024). Expanding the base also requires careful consideration, as an additional burden on low-income payers may not be politically expedient. Simpler registration for paying taxes by informal workers and direct benefits may be an innovation needed.

13.6.3 Resistance to Formalization and Policy Implementation Gaps

The fear of increased taxes and lost benefits is the holdback in bringing informal workers into formal systems. Barca and Alfers (2021) state that this hesitancy to go formal is made worse by a lack of trust in government programs because enforcement is uneven and governance is weak. In many cases, governments do not reach out to informally employed workers when creating policies, thus creating policies that are far from the workers' realities. This creates disengagement and inefficiencies that block the path into formal systems. Added to this is a financial constraint because most African countries do not have enough internal resources for social protection; hence, they depend on external aid, exposing them to global economic changes. This would be an example of South Africa's overextended social grant system, which shows how these commitments leave little room for informal worker coverage. To overcome these obstacles, policies ought to motivate the formalization of the workers without burdening them through gradual tax contributions or benefits that suit them. Secondly, governments must improve enforcement, better governance, and multi-stakeholder involvement toward inclusive policies related to economic realities concerning informal labor.

13.7 The Intersections of Informal Economics and Social Policy

13.7.1 The Importance of Including Informal Workers in Social Policy Frameworks

Integrating informal workers into social protection systems is essential for equitable economic growth and social justice. In Sub-Saharan Africa, where informal labor accounts for as much as 90% of nonagricultural employment (Canelas & Niño-Zarazúa, 2022), the exclusion of this workforce from formal social policies makes them very vulnerable to economic shocks, health crises, and old-age poverty. The paramount problem arises from the irregular income of informal workers and the absence of employer contribution social insurance programs. This prevents them from being unable to take part in contributory social insurance like pensions or unemployment benefits. While attempts have been made to develop noncontributory social protection exemplified by universal health coverage in Thailand, such initiatives are often underfunded and inadequate in addressing the needs of informal workers (Canelas & Niño-Zarazúa, 2022). Mistrust of government systems is the primary factor impeding the formalization of informal workers, more so because they fear increased taxation or regulation. This explains why policy design requires more inclusiveness and participation. Governments need to work hand in hand with informal workers to understand their challenges and develop solutions that will not further compound their already difficult circumstances. Such successful models—like India's Self-Employed Women's Association (SEWA)—prove that cooperative-based approaches can help bridge the gap between informal workers' needs and formal protection systems (Kabeer, 1996). Financial limitations also constrain several African countries from having sufficient resources for inclusive policies; however, innovative solutions like earmarked taxation or public-private partnerships can create sustainable sources. Kenya has demonstrated this through a partnership between its National Hospital Insurance Fund and local cooperatives, wherein health coverage has been extended to informal workers (Canelas & Niño-Zarazúa, 2022). These examples show that despite the challenges, there are viable strategies for integrating informal workers into social protection systems, which would make them more productive and help build more social cohesion and stability around economies.

13.8 Economic and Social Benefits of Integrating Informal Workers Without Enforcing Rigid Formalization

13.8.1 Economic Benefits

The mainstreaming of informal workers in the national economies can provide traction in growth while preserving sectoral flexibility. In Sub-Saharan Africa, it has been reported that more than 30% of GDP is from the informal sector, which provides stability during economic downturns (Canelas & Niño-Zarazúa, 2022). However, formalization can sacrifice this adaptability for a less important economic role. Countries like Rwanda have increased revenues through a flexible tax system that does not burden informal workers (Kong & Wei, 2023). This shows that innovative fiscal policies can promote inclusion without disruption, and resistance to formalization arises from fear of increased taxes and reduced flexibility. The key aspect is reconciling informal sector adaptability with formal systems. Informal markets survived during the COVID-19 pandemic; disrupted formal supply chains were unsuccessful (May & Mentz-Coetzee, 2021). This calls for policies that give tax incentives and simple registration procedures under which private-public cooperation ensures economic integration while preserving the sector's autonomy.

13.8.2 Social Benefits

Including informal workers in social protection systems improves economic stability and social equity. Such workers often do not have provisions for healthcare and pensions, but including them would significantly improve a country's quality of life and productivity, as Ghana's National Health Insurance Scheme exemplifies (Canelas & Niño-Zarazúa, 2022). This, therefore, reduces economic inequalities; it also enhances social cohesion, which leads to a better society at large. Informal workers rely on community-based organizations' support during crises; however, this cannot cover all risks. Government intervention by integrating such networks into national frameworks would improve their efficiency while maintaining worker autonomy. For instance, in Kenya, accessing credit and financial stability has been improved through voluntary savings groups, showing that flexible real-life solutions increase financial inclusivity, implying that informal arrangements integrated with formal protection may yield more sustainable social policies.

13.8.3 Avoiding Rigid Formalization

Overregulation often exterminates businesses and jobs. The high cost and the complicated registration process scare away many small entrepreneurs. According to a study on informal firms in South Africa, businesses consider the formalization process imposing much administrative cost and reflecting no access to formal credit systems, threatening their survival (Maduku & Zerihun, 2022). An approach that can be flexible, say streamlined processes, or provides for gradual formalization incentives, would address these concerns and encourage incremental integration. Flexible integration policies also recognize that informal work is diverse and ranges from small-scale street vending to home-based businesses. A uniform formalization policy would not fit these varied circumstances. By acknowledging the heterogeneity of informal work, appropriate interventions can be designed to target each specific form of work, offering the necessary support it needs while at the same time respecting the autonomy of workers.

13.9 Economic Consequences of Failing to Integrate Informal Workers into Social Protections

13.9.1 Impact on Economic Growth

Excluding informal workers from social protections significantly hampers economic growth. Informal workers contribute a substantial share to GDP, particularly in developing nations, yet their lack of inclusion in formal social safety nets creates a vulnerability, destabilizing economic output during crises (Davidescu et al., 2024). For instance, during the COVID-19 pandemic, the inability to extend financial support to informal workers exacerbated economic downturns, as they lacked savings and state support (Loewenson & Mukumba, 2023). Without access to protections such as unemployment insurance or healthcare, informal workers are less resilient and often forced to withdraw from economic activities when faced with health or financial crises, further reducing national productivity. Moreover, countries lose potential tax revenues by failing to integrate informal workers. Gumede (2019) reveals that in South Africa, the informal sector constitutes approximately 18% of GDP but remains untaxed mainly due to its informal nature. These lost revenues could finance public services, creating a vicious cycle where the lack of formal inclusion perpetuates underfunded social programs.

13.9.2 Increased Social Inequality

The absence of social protection for informal workers increases income inequality and thus prolongs poverty. Largely, informal workers are mostly low-wage workers, and disproportionately lack access to essential services such as healthcare, education, and pensions, creating a persistent wealth gap (Gupta et al., 2024). The lack of inclusion widens social stratification and restricts upward mobility in regions where informal workers are excluded from health and retirement benefits (Loewenson & Mukumba, 2023). In countries like Kenya, where such tries have taken place for informal workers' savings and credit schemes, they are fragile due to the absence of comprehensive inclusion policies, and the workers become more vulnerable and tied to exploitative labor conditions (UNEN Policy Brief, 2024). The vast overrepresentation of women in the informal sector means that they are particularly exposed to patterns of extreme precariousness associated with the absence of maternity benefits and unenforced protection against unfairly paid work (Davidescu et al., 2024).

13.9.3 Strain on Public Systems

Integrating informal workers' burdens puts additional strain on public systems, particularly healthcare. When informal workers are not covered under employer-provided health insurance, or when there are no state-sponsored programs, care becomes more dependent on overburdened public health systems or delayed—resulting in higher costs for emergency services (Loewenson & Mukumba, 2023). Such pressures are even heavier in crises, as witnessed in East and Southern Africa, where underfunded health systems have little public financing due to inadequately taxed informal sectors (Loewenson & Mukumba, 2023). Governments have additional costs related to social unrest. The exclusion of informal workers will lead to more dissatisfaction and unrest, leading to political instability and economic progress (Gumede, 2019). For example, the absence of support for the informal laborer in India resulted in significant protests during the pandemic, underlining the perils of neglecting this important workforce segment.

Informal workers remain a significant constituent in economic growth, yet their exclusion from the social protection system has long-term effects. The short-term impacts on output and inequality of informal workers' exclusion from social protection are well documented. However, more research should be done to better understand the benefits of integrating informal workers into social protection schemes. Current policies hardly adopt flexible, inclusive approaches that would consider the needs of informal labor. Governments should initiate noncontributory social protection schemes for informal workers, wherein mobile health and pension systems could be included alongside tax incentives for voluntary registration. National frameworks should also include informal savings and mutual aid networks to enhance financial stability at this level and reduce pressure on public services.

13.10 Conceptual Framework for Bridging Informal Economy and Social Policy

The conceptual framework addresses the integration of informal economies into social policy frameworks in Sub-Saharan Africa, where informality constitutes the backbone of livelihoods. Informal workers, often excluded from formal social protection systems, face vulnerabilities that perpetuate poverty and inequality. The framework aims to bridge the gaps by fostering inclusivity, flexibility, and sustainability in social protection policies (Miti et al., 2023). The framework encompasses eight interrelated components, each addressing distinct challenges and opportunities within the informal economy.

13.10.1 Informal Economy

The informal economy refers to economic activities and employment not covered by regulated legal provisions. This is where much employment occurs, especially in Sub-Saharan Africa, with around 80% of nonagricultural employment being informal and contributing about 36% of GDP (ILO, 2022). Despite the economic importance of the informal sector, workers are usually left out of proper formal social protection coverage, thus vulnerable to economic shocks (Canelas & Niño-Zarazúa, 2022; Wango, 2024).

13.10.2 Social Policy Intersections

Social policy intersects with the informal economy by addressing inequalities and vulnerabilities. In Sub-Saharan Africa, social protection systems often overlook informal workers due to their irregular income patterns and unregulated employment conditions. Social policies aim to extend protections such as healthcare and pensions to this underserved group while fostering equity and inclusivity (Miti et al., 2023).

13.10.3 Economic Contributions

The informal economy's contributions to GDP and employment are significant. Activities like street vending, artisanal work, and small-scale agriculture sustain livelihoods and economic stability in regions with limited formal employment opportunities. Recognizing these contributions is essential for policy frameworks for inclusive growth and productivity enhancements (Jackson, 2012; Wango, 2024).

13.10.4 Social Protection Gaps

These are acute in the informal economy, as less than 18% of the population of Africa receives full social protection benefits (ILO, 2022). Informal-sector workers do not have these facilities, for instance, health insurance, pensions, and unemployment benefits; therefore, poverty and inequality follow them throughout their lives. These gaps can only be filled by noncontributory schemes and policies sensitive to the informal nature of employment, as recommended by some scholars (Miti et al., 2023; Canelas & Niño-Zarazúa, 2022).

13.10.5 Transformative Social Policy

Transformative social policy emphasizes long-term resilience and equity by addressing systemic barriers to social inclusion. It merges the inclusion of informal workers within social protection while maintaining the flexibility of informal economies. This approach uses measures based on promoting the development of inclusiveness, technology, and community-based results that are sensitive to gender (Nhede & Marumahoko, 2023; Wango, 2024).

13.10.6 Components of Policy

Key elements of transformative social policy include (i) gender sensitivity, targeting disparities women face in informal work, such as unequal pay and lack of benefits (Nhede & Marumahoko, 2023); (ii) technology integration, utilizing digital platforms like Kenya's M-Pesa to enhance access to financial services and social protections (Jackson, 2012); and (iii) economic inclusion, structuring support systems, such as microloans and grants, to recognize and formalize the contributions of informal workers (Miti et al., 2023).

13.10.7 Outcomes

Implementing this framework is expected to result in (i) economic growth, which enhances productivity and integration of informal workers into formal systems; (ii) social equity, which includes reduction of poverty and inequality through inclusive policies; and (iii) efficiency, which includes streamlined and adaptive systems that balance flexibility with formal protections (Noyoo & Boon, 2018).

13.10.8 Challenges and Mitigation Strategies

Key challenges in integrating informal economies into social policy include resistance to formalization, resource limitations, and fragmented systems. Overcoming these obstacles is critical for extending protections to informal workers while maintaining sector adaptability (Miti et al., 2023). Mitigation strategies focus on participatory policy design, innovative financing, and system integration (Diagram 13.1).

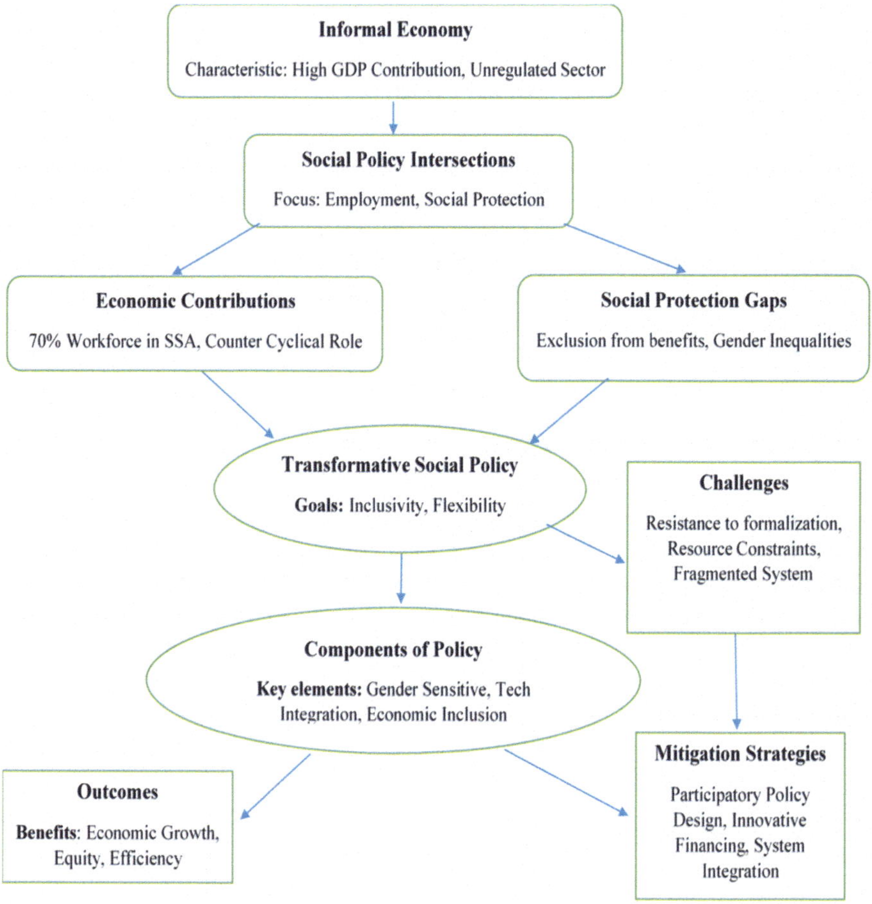

Diagram 13.1 Conceptual framework of informal economies and transformative social policy

13.11 Findings

This section presents key findings from the literature on the informal economy. The findings highlight key issues, including the role of the informal economy, the exclusion of informal workers from formal social protection systems, and gender disparities. Additionally, the literature underscores the importance of adapting social policies to the specific needs of informal workers, addressing resistance to formalization, and ensuring inclusive economic growth.

13.11.1 The Role of the Informal Economy

Findings show that the informal economy is vital for the economic stability of developing countries, particularly in Sub-Saharan Africa. However, since informal workers do not have access to formal social protection, they become more vulnerable; hence, income instability and economic shocks also intensify. In the absence of contributory pension and health insurance schemes, income remains volatile; hence, exposure to shocks increases, and long-term security remains negated. The provision of formal protection is important because its absence weakens resilience at an individual level and creates pressure on macroeconomic stability. Informal workers were noted in this chapter to be more financially vulnerable than their formally employed counterparts. This calls for policy reforms that will take them as part of the formal social protection systems. What significantly contributes to knowledge is that Zambia has successfully incorporated informal dairy farmers into its social security reforms (Miti et al., 2023). This proves that inclusion increases economic security. Extensions are made in other informal sectors wherein adequate social protection reduces vulnerability and improves economic stability. It ultimately makes a case for the inclusion of informal workers in formal protection systems as an imperative way of building both individual and societal resilience, hence supporting the need for policies that are inclusive and responsive to the specific needs of informal workers.

13.11.2 Employment Creation, Economic Stability, and Poverty Reduction

This chapter sheds new light on the dual aspect of the informal economy in creating jobs and supporting economic resilience, particularly in developing regions such as Sub-Saharan Africa. Findings prove that informal employment not only reduces unemployment levels but also plays a counter-cyclical role, particularly during economic crises. For example, during the COVID-19 pandemic, informal markets served as critical buffers, mitigating economic and social stresses for many workers (Loewenson & Mukumba, 2023). The findings have shown that informal employment is often precarious but not incompatible with generating employment; it usually perpetuates low incomes and poor working conditions, thereby undermining its effectiveness in reducing poverty. The findings have also shown that informal work can be vital in sustaining livelihoods in agriculture, domestic labor, and small-scale retail, but mainly at the expense of the long-term economic stability of the workers (Miti et al., 2023). This chapter makes a significant contribution by designing a framework to address the challenges through flexible social protection measures for informal workers. In summary, the findings expand the literature on theory by providing empirical evidence on how the informal economy contributes to employment creation and economic stability—while suggesting new policies to mitigate its inherent risks. It demonstrates a significant need for universal social protection systems that serve informal workers, ensuring their long-term welfare and strengthening the broader economic systems of developing countries.

13.11.3 Adapting Social Policies to the Needs of Informal Workers

This chapter uncovers significant misalignments in the social protection policies of informal workers in Sub-Saharan Africa. The existing frameworks hardly ever consider the peculiar characteristics of informal work—variable incomes and absence of formal employer-employee relationships—thus producing ineffective policies for a large portion of the workforce. Even though conditional cash transfers, community health insurance, and other successful practices from South Asia and Latin America have been beamed in with an idea to replicate them, they genuinely require drastic tailoring to Africa's socioeconomic conditions. The National Hospital Insurance Fund is known for local innovation that improves healthcare access to informal workers without complete formalization (Kong & Wei, 2023). South Africa's programs indicate that policy design should involve informal workers to increase trust and participation (Maduku & Zerihun, 2022). However, many African countries lag in this—inadequate administrative capacity, insufficient funds, and poor coordination. The findings demonstrate that the dynamics of Sub-Saharan Africa's informal economy must intervene; otherwise, formal social protection will not provide adequate coverage, for it has been left on the outside looking in. Therefore, it indicates a strong need for policies that reflect the characteristics of informal work and insert these peculiar workers into wider social protection systems as an instrument of economic stability.

13.11.4 Economic Costs of Exclusion

The findings show that excluding informal workers from any social protection scheme directly affects a nation's productivity, revenue, and social equity. Informal workers have a significant contribution to GDP; however, they are vulnerable during economic shocks. Such vulnerability reduces the workers' resilience and increases instability in the informal sector. For instance, there was no unemployment insurance during the COVID-19 pandemic, and income loss intensified (Loewenson & Mukumba, 2023). This exclusion, particularly of women and rural laborers, causes public health development income disparities. These findings call for the urgent need for policies that take informal workers as partakers in their social protection systems and recognize their economic and social roles. Integrating this workforce into social protection systems will secure individual security and promote collective economic stability, as an informal workforce often faces insecurity. The exclusion of this workforce will cause development equity and social progress to be held back; therefore, intervention targeted at resilience and sustainable growth of the region is necessary.

13.11.5 Social Policy Gaps for Informal Workers

It is evident from the findings that informal workers are largely left out of the social protection systems because these systems operate as a prerequisite of formal employment; hence, they do not consider irregular income and lack of documentation as typical attributes of informal work. Only 17% of informal workers globally have social protection (ILO, 2022). While successful initiatives in some African countries have proven to be feasible in integrating the informal workforce into social protection, they have been strictly limited due to barriers such as inadequate funding and administrative inefficiencies (Nhede & Marumahoko, 2023). This underscores how policies relevant to the realities and practicalities of informal work are urgently needed. The government should tailor its social protection schemes to address these specific needs and ensure benefit accessibility and system sustainability.

13.11.6 Economic and Social Benefits of Integration

The findings have shown that incorporating informal workers into social protection systems brings about much gain, as demonstrated by programs in countries like Brazil and Kenya. For instance, Brazil's Bolsa Família program has successfully reduced poverty and improved children's health and educational attainment in families engaged in the informal sector (Canelas & Niño-Zarazúa, 2022). In turn, cooperative-based health insurance initiatives in Kenya prove through their customized approaches that they extend to enhance workforce productivity and well-being. The findings also highlight how very costly and complicated regulations in formalizing workers tend to intimidate the informal workers into joining. Such impediments would remove the importance of the sector's flexibility—how it bounces back and grows (Maduku & Zerihun, 2022). The finding underlines the need for policies that balance inclusiveness and economic flexibility, whereby informal workers can easily access social protection without forfeiting or losing the economic flexibility that characterizes their contributions.

13.11.7 Gender Disparities and Fragmentation

The findings further highlight grave gender differences in the informal economy; women are precariously employed—more domestic workers and street vendors, and more often, these workers do not have access to the basic social protection they urgently need—healthcare, maternity benefits, and pensions (Noyoo & Boon, 2018). This inadequacy is made doubly so by the jagged social policy architecture in Sub-Saharan Africa, marked by redundant initiatives and poor coordination that breed inefficiency. The resources are frequently misused; more than what is needed for the better-served populations, they must be taken for the most vulnerable informal workers (Haurovi, 2021). Such systemic inefficiencies explain why inequality and economic insecurity continue to haunt women and marginalized groups. To do this successfully, there is a need for targeted interventions that will take cognizance of the particular vulnerabilities of women in informal employment. This could come in the form of social protection policies, among others, that cover maternity leave benefits and accessible healthcare directly related to women workers while improving the efficiency of the social protection systems in the region.

13.11.8 Resistance to Formalization

The findings show that workforce formalization becomes a big challenge in bringing informal workers under social protection coverage. They raise concerns about higher taxes on them, and once they lose their flexibility, along with general mistrust in government systems, all these factors work against formalization (Rogan & Skinner, 2019). The chapter also pinpoints successful cases, like the Self-Employed Women's Association (SEWA) in India, where cooperative structures have successfully abated these worries by providing micro-pensions and health insurance yet retaining the flexibility attributes of informal work. The chapter underlines the need for building trust and having policy-making processes that are open, inviting, and responsive to the realities of informal work. Policies that underline the autonomy of workers while gradually building informal workers into formal social protection systems will be able to address their resistance against formalization and, by extension, encourage them to participate more in social protection programs.

13.12 Discussions and Policy Recommendations

This section discusses key findings from the literature on the informal economy. The section also identifies recurring themes, such as gender disparities and the economic vulnerability of informal workers. It highlights critical gaps in existing research, particularly around the long-term impacts of exclusion and resistance to formalization. Drawing on these insights, this section further proposes policy recommendations to improve social protection for informal workers through flexible, inclusive, and gender-sensitive approaches.

13.12.1 Significant Outcomes from the Literature

The existing literature has fairly articulated the crucial role of informal employment in the world's economies, especially in developing countries. However, it leaves a significant lacuna on how vulnerable informal workers are to the existing social protection frameworks. This chapter contributes to that discussion by reviewing the linkage between informal labor and social protection systems, focusing on the challenges faced by workers in Sub-Saharan Africa. While informality accounts for about 60% of employment worldwide and a significant proportion of regional GDP—36% in Sub-Saharan Africa (IMF, 2020)—the beneficiaries of these two indicators remain an overwhelming majority excluded from any form of social protection mechanism as well as healthcare, retirement benefits, and unemployment support (ILO, 2022). Such exclusion continues to multiply new cycles of insecurity and poverty with high dynamism in economies at the initial stages of development, with fewer formal jobs available.

This chapter, therefore, looks at the gaps in social protection policies, which do not consider the precarious income dynamic and the atypical employer-employee relationships in informal work. The chapter also critically evaluates gender inequality in informal employment, mostly glaring exclusions of women engaged in low-paid sectors (e.g., domestic work) from existing protection schemes. In addition, through an assessment of community-based advances such as Kenya's National Hospital Insurance Fund (NHIF) and conditional cash transfers that can be furthered to develop social safety nets for informal workers within that region, scalable context-specific solutions are proposed (Canelas & Niño-Zarazúa, 2022). New insights from this analysis help reorient policy designs toward integrating informal workers into national social protection systems for economic security and poverty reduction among vulnerable groups.

13.12.2 Recurring Themes, Trends, and Patterns

This chapter traces the very significant role of informal employment in the global economy, particularly in developing countries, and thus reveals the vulnerabilities that informality exposes workers to. For long, informal work has been viewed as a key safety net during economic crises for those excluded from the formal labor market (ILO, 2022); however, the risks associated with informality—particularly precariousness—have not been adequately addressed in policy discourse. This chapter attempts to measure the level of exclusion of informal workers from social security protection schemes and finds that an overwhelming majority do not have access to basic social protection, pensions, and health insurance—inadequate to withstand economic shocks and chronic poverty.

The chapter amplifies that body of literature by examining the play of gender and informal work. It shows that women, already left out in the formal job setting, face even more exclusion from the few social protection systems for informal workers. Such an absence does not just stop the cycle of gender inequality but also worsens economic insecurity for women, particularly in low-income sectors where informal employment dominates (Noyoo & Boon, 2018). Drawing on data from qualitative case studies and quantitative national surveys, this chapter generates concrete insights into possible reforms in social protection systems to better include informal workers—particularly women—in a way that addresses their specific economic vulnerabilities. Findings indicate that inclusive policy interventions, such as expanding health insurance coverage and retirement benefits to informal workers, would enormously reduce levels of poverty and inequality, thereby enhancing the resilient economies desired in emerging regions.

13.12.3 Gaps and Controversies

This chapter fills in the missing links in the literature concerning integrating informal workers into social protection systems. Though informal work remains critical in maintaining economic stability during a crisis, the workers are largely excluded from social protection, adversely affecting long-term development, public health, and income inequality. Using original data from case studies, the chapter quantifies how informal sector workers have faced economic and social vulnerabilities during and after COVID-19. The findings show that these workers were disproportionately affected by the pandemic; inadequate social protection increased their vulnerability to income loss and long-term poverty. The fear of informal workers taking on formal social protection arises from their fears of being taxed, losing their flexibility, and a general regime of mistrust toward government systems (Rogan & Skinner, 2019). After analyzing participatory approaches and community solutions, a framework is proposed to integrate informal workers into social protection. This presents the case of the Self-Employed Women's Association (SEWA) in India, which illustrates scalable community-based approaches despite resource-oriented and political challenges. The chapter posits that reforms be inclusive; that is, provision for health insurance and pensions should cover informal workers for any meaningful reduction in poverty levels and inequality in developing economies. The chapter ends with policy recommendations for governments to restructure social protection systems around informal workers to build more resilient economies.

13.12.4 Relevance to Research Article

The gaps identified in the literature, particularly around the absence of models that can be scaled to integrate informal workers into social protection systems and scanty exploration of resistance to formalization, set the objectives of this chapter. In an attempt to propose a hybrid approach to social protection for informal workers while balancing flexibility with security, this chapter sought to fill these gaps. Thus, in this vein, the conceptual framework was proposed to make real and implementable suggestions to policymakers on how they can design policies that include the special characteristics of informal work in their countries while ensuring that the workers are not left out of vital social protections.

This chapter responds to the existing literature gaps on integrating informal workers into social protection systems, highlighting scalable models and the resistance of informal workers toward formalization. The chapter presents a hybrid social protection approach from case studies, surveys, and participatory model data that reflects the trade-off between flexibility and security aspects of informal work. Findings from this research indicate that inappropriate informal workers from conventional systems are exposed and vulnerable to greater impacts during times of economic crises, as in the recent COVID-19 pandemic. The chapter examines how informal workers avoid becoming formal, mainly because they worry about taxes, lose flexibility, and do not trust the government. It closely examines words and numbers of data to find obstacles to joining in and offers solutions led by communities. The Self-Employed Women's Association (SEWA) in India shows the ability of community models to scale up despite resource constraints and political challenges. This research initiates a conceptual framework for including informal workers in social protection systems. It further elaborates policy recommendations on how governments should design inclusive systems catering to the informal economy's peculiarities. The chapter argues that providing and extending health insurance and pension coverage to informal workers in developing economies is an imperative measure for poverty reduction, inequality, and economic instability amelioration. In this respect, it provides actionable insights for policymakers to foster resilience and equity.

13.12.5 Policy Recommendations

This chapter discusses the difficulties and prospects of including informal workers in social protection systems, pointing out the shortcomings of current models that orient to formal employment. Through case studies, the chapter discovers that old ways are inadequate for the new complexities of informal work. It appraises localized noncontributory social insurance schemes—Kenya's National Hospital Insurance Fund and Brazil's Bolsa Família—based on their significant improvements in access to healthcare and reduced poverty levels among informal workers; such programs could be a model for other countries striving to broaden coverage.

A significant finding is the increased precarity of women in informal employment, particularly in paid domestic work and street vending. The chapter quantifies gender imbalances and the inadequacies of present social protection systems. It proposes targeted interventions that could reduce gender inequality—for example, expanding maternity benefits and healthcare provisions for women in high-risk sectors—as illustrated by a pilot program in Brazil that enhanced healthcare access for informal female workers.

The chapter also looks into why workers in the informal sector do not want to become formalized. Informal workers fear paying taxes, losing work flexibility, and do not trust the government. It examines the obstacles of informal workers in cities and rural areas. The chapter proposes policies for an easier shift toward formalization, like gradual integration strategies, which keep workers flexible while increasing social protection benefits. The Self-Employed Women's Association (SEWA) in India shows a community-based model that offers health insurance and micropensions while maintaining autonomy. This suggests there could be scalability in other regions.

Thus, the chapter shows the economic and social effects of leaving out informal workers from social protection coverage, particularly in times of crisis, such as the recent COVID-19 pandemic. Data already collected on the effects of this pandemic proves that such a form of exclusion increases not only these workers' precariousness regarding income loss and health risk but also long-term poverty; hence, there is an urgent need for comprehensive research on their vulnerabilities. Therefore, the chapter argues that more investment be made by governments and international organizations in data collection that can inform future policy interventions.

13.13 Conclusion

The informal economy is vital for employment and economic stability, particularly in developing regions that contribute significantly to GDP and serve as a safety net during crises. However, workers in the informal economy are often excluded from social protection systems and, therefore, remain unprotected, resulting in poverty, inequality, and economic insecurity. In essence, excluding informal workers puts additional strain on the economy and society, worsening systematic inefficiencies and inequalities. Key findings highlighted that inflexible social regulations do not accommodate the peculiarities of informal work, including fluctuating earnings and no written or formal contracts. Resistance to formalization, caused by mistrust and fear of taxation, makes integration efforts more difficult. Such challenges must be addressed by designing appropriate and sensitive social-protective policies for informal workers. To address these challenges, policymakers must adopt flexible, inclusive social protection frameworks tailored to informal workers' realities. Participatory approaches, gender-sensitive interventions, and leveraging digital technologies can foster trust and increase accessibility. Further research should target the comprehension of the socioeconomic effects of exclusion and the adjustment of strategies to be integrated whilst keeping the sector flexible. Implementation by governments can leverage equalizing growth, reducing poverty, and building social trust for sustainable development for all.

References

Adesina, J. (Ed.). (2021). *Social policy in the African context*. CODESRIA.

Barca, V., & Alfers, L. (2021). *Including informal workers within social protection systems: A summary of options*. Social Protection Approaches to COVID-19 Expert Advice Service (SPACE). DAI Global UK Ltd. https://socialprotection.org/sites/default/files/publications_files/SPACE_Including%20informal%20workers%20within%20social%20protection%20systems_A%20summary%20of%20options_0.pdf

Canelas, C., & Niño-Zarazúa, M. (2022). *Social protection and the informal economy: What do we know?* United Nations ESCAP, Social Development Division. http://www.unescap.org/kp

Chen, S. (2012). Creating sustainable international social ventures. *Thunderbird International Business Review, 54*(1), 131–142. https://doi.org/10.1002/tie.21445

Chen, M. A. (2023). The informal economy in comparative perspective: Theory, policy and reality. *The Indian Journal of Labour Economics, 66*(2), 395–420.

Cohen, Z. D., Ashar, Y. K., & DeRubeis, R. J. (2019). Moving beyond "One Size Fits All." In S. Dimidjian (Ed.), *Evidence-based practice in action: Bridging clinical science and intervention* (pp. 111–132). The Guilford Press

Davidescu, A. A., Manta, E. M., Geambasu, C. M., & Birlan, I. (2024). Informality and inclusion: Assessing the effects of the shadow economy and informal labour in Europe. *Management & Marketing, 19*(2), 307–335.

Dube, F., & Moya, C.G. (2021). The Right to Electricity in South Africa PER / PELJ 2021(24) - http://dx.doi.org/10.17159/1727-3781/2022/v25i0a11839

Gumede, V. (2019). Social policy in post-apartheid South Africa. *Journal of Public Administration, 54*(4), 499–520.

Gupta, R., Hajabdollahi-Ouderji, Z., Zhibin Y.U., et al. (2024). Machine learning for sustainable organic waste treatment: a critical review. *npj Materials Sustainability, 2*(5), 1–18. https://doi.org/10.1038/s44296-024-00009-9

Hart, T., Motala, S., & Ngandu, S. (2024). *Social security review Vol. 2: Social security in the time of COVID-19*. Department of Social Development.

Haurovi, M. (2021). Social protection and a developmental state in South Africa. *Journal of Public Administration and Development Alternatives, 6*(1), 13–27.

ILO. (2018). *World Employment and Social Outlook: Trends 2018*. International Labour Office – Geneva: ILO, 2018

ILO. (2022). *Transition from the informal to the formal economy-* Theory of change. ILO. https://www.ilo.org/resource/brief/transition-informal-formal-economy-theory-change. (Accessed 14/11/2025).

International Labour Organization (ILO). (2021). *Women and men in the informal economy: A statistical picture*. ILO.

IMF. (2020). *International Monetary Fund Annual Report 2020*. IMF. https://doi.org/10.5089/9781513538211.011

Jackson, T. (2012). Cross-cultural management and the informal economy in Sub-Saharan Africa: Implications for organization, employment, and skills development. *International Journal of Human Resource Management, 23*(14), 301–319.

Kabeer, N. (1996). Agency, Well-being and Discrimination: Reflections on the Gender Dimensions of Poverty. *IDS Bulletin, 27*(1), 11–21.

Kasera, O. A., Odhiambo, M. O., & Oloo, B. C. (2024). Africa in global public policy: Theoretical perspectives and the role of international law in shaping public policy in Africa. *International Journal of Research and Innovation in Social Science, 8*(6), 910.

Keune, M. (2024). *The state of European social rights and European social citizenship in the EU*. EuSocialCit Flagship Report 1.

Khan, F. S. (2024). *Beyond safety nets: Provision of comprehensive social security in Pakistan. Conference Proceedings*.

Kong, X., & Wei, J. (2023). A review of informal employment research. *Open Journal of Applied Sciences, 13*(12), 2263–2272.

Loewenson, R., & Mukumba, C. (2023). Recovering lost tax to meet the health financing gap for universal public sector health systems in East and Southern Africa. *BMJ Global Health, 8*, e011820. https://doi.org/10.1136/bmjgh-2023-011820

Maduku, H., & Zerihun, M. F. (2022). Qualitative analysis of transiting from informal to formal sector: The case of manufacturing and service businesses in Gauteng province of South Africa. *Cogent Economics & Finance, 10*(1), 2095765. https://doi.org/10.1080/23322039.2022.2095765

Mas, I., & Radcliffe, D. (2018). Mobile money and mobile banking: The policy challenges. *Handbook of Digital Finance and Financial Inclusion*, 411–437.

May, J., & Mentz-Coetzee, M. (2021). Re-imagining resilient food systems in the post-COVID-19 era in Africa. *Sustainability, 13*(19), 10752.

Meagher, K. (2022). Crisis narratives and the African paradox: African informal economies, COVID-19 and the decolonization of social policy. *Development and Change, 53*(6), 1200–1229.

Medina, L., & Schneider, F. (2018). Shadow economies around the world: What did we learn over the last 20 years*? IMF Working Papers, 18*(17), 1.

Miti, J. J., Perkiö, M., Metteri, A., & Atkins, S. (2021). The informal sector and social protection. In E. Schüring & M. Loewe (Eds.), *Handbook on social protection systems* (pp. 389–402).

Miti, J. J., Perkiö, M., Metteri, A., & Atkins, S. (2023). Implementing a public policy to extend social security to informal economy workers in Zambia. *Forum for Development Studies, 50*(3), 425–443.

Mosoetsa, S. (2021). 10 Years in BRICS: South Africa Holds its own and looks into the Future, in Post Covid-19 Recovery: BRICS Academic Review. National Institute for the Humanities and Social Sciences (NIHSS) 2021. Available at https://www.nihss.ac.za/sites/default/files/BRICS/BAR_5_e-version.pdf (Accessed: October 14, 2024).

Motala, E., & Treat, J. (2014). Misrepresenting Unemployment: Evidence and Implications. *Post-School Education Journal, 1*(1), 16–8.

Motala, S., Hart, T., & Ngandu, S. (2024). Social security in the time of COVID-19. *Social Security Review, 2*.

Nhede, N. T. (2018). Improving public sector supply chain management in South Africa. *Journal of Public Administration, 53*(3), 543–556.

Nhede, N. T., & Marumahoko, S. (2023). Redesigning social security systems in some selected sub-Saharan African countries. *African Journal of Governance and Development, 12*(1), 80–97.
Nowell, L. S., Norris, J. M., White, D. E., & Moules, N. J. (2017). Thematic analysis: Striving to meet the trustworthiness criteria. *International Journal of Qualitative Methods, 16*(1), 1–13.
Noyoo, N., & Boon, E. (Eds.). (2018). *Indigenous social security systems in Southern and West Africa*. Sun Press. https://doi.org/10.18820/9781928357919
Polanyi, K. (1957). Reciprocity and redistribution in informal networks. *The Great Transformation.*
Rogan, M., & Skinner, C. (2019). *The informal economy: Is policy based on correct assumptions?* Econ3x3. https://www.econ3x3.org/sites/default/files/articles/Rogan%20%26%20Skinner%202019%20Informal%20sector%20policy%20-%20FINAL%20PDF.pdf
Schneider, J., Kern, M., & Lorenz, T. (2025). *Unifying work values: establishing a circular framework based on basic human values*. Frontiers in Psychology, 16,1526799. https://doi.org/10.3389/fpsyg.2025.1526799
Snyder, H. (2019). Literature review as a research methodology: An overview and guidelines. *Journal of Business Research, 104*, 333–339.
United Nations Economist Network. (2024, March). *Taxing for tomorrow: Aligning fiscal policies with the Sustainable Development Goals*. United Nations. https://www.un.org/sites/un2.un.org/files/2024/06/unen_policy_brief_march_2024_may_14_revised.pdf
Vanek, J., Chen, M. A., Carré, F., Heintz, J., & Hussmanns, R. (2014). *Statistics on the Informal Economy: Definitions, Regional Estimates and Challenges*. WIEGO Working Paper (Statistics No. 2). WIEGO, Cambridge, MA.
Vorng, P. (2024). Public administration and the informal economy. *Global Political Review*, 1–19.
Wango, N. C. (2024). The influence of informal economy on voluntary contributions to informal social organisations. *International Journal of Global Sustainability, 8*(1), 1–15.
Williams, F. (2018). Care: Intersections of scales, inequalities and crises. *Current Sociology, 66*(4), 547–561.
Williams, C. (2023). *The informal economy*. Cambridge University Press. https://doi.org/10.1017/9781911116325
Xiao, Y., & Watson, M. (2019). Guidance on conducting a systematic literature review. *Journal of Planning Education and Research, 39*(1), 93–112.

Chapter 14
Loan Acquisitions and Repayments Among Rural Women Participating in Informal Economics of Village Community Banks (VICOBAs) in Busega District, Tanzania

Deborah Andrew Ngusa

14.1 Introduction

Village Community Banks (VICOBA) originated from the idea of community-based microfinance aimed at empowering rural and low-income communities by providing access to savings and credit services (Gajijime, 2023). As Africa is confronted with persistent poverty, inequality, and the limitations of traditional social policy approaches (see Ndhlovu & Mokone; Chipenda & Tom in this volume), VICOBAs represent innovative grassroots responses that have emerged to fill critical gaps in formal financial systems. These informal economic structures have become significant in addressing development challenges where conventional social protection mechanisms have proven inadequate. This concept is rooted in the broader movement of informal savings and credit groups that have been part of African, Asian, and Latin American communities for centuries (Shand & Colenbrander, 2018; Polloni-Silva et al., 2021). In Africa, Niger was the first country to launch VICOBAs in 1991 (Magali & Barhe, 2022). VICOBA was known as "Mata Masu Dubara" (MMD), which translates to "Women on the Move," aimed to empower women economically (Ibid).

VICOBAs operate as part of the informal economies. As African countries are confronted by poverty, inequality, and the lingering effects of global economic disruptions simultaneously, these informal economies are often excluded from national statistics and policy designs, leading to underrepresentation in social welfare schemes. Informal economies include commercial activities that are not regulated by formal government laws or institutions. However, their significance in addressing development challenges has become increasingly apparent, especially in the context of achieving sustainable development goals related to poverty reduction,

D. A. Ngusa (✉)
Institute of Development Studies, University of Dodoma, Dodoma, Tanzania
e-mail: deborah.ngusa@udom.ac.tz

E. Ndhlovu, C. Chipenda (eds.), *Redefining the Social Policy Discourse*,
https://doi.org/10.1007/978-3-032-24067-5_14

gender equality, and economic growth. VICOBAs are especially prominent in East African countries like Tanzania, Kenya, Uganda, and Rwanda. For instance, in Uganda, VICOBAs are well established to promote financial inclusion, particularly in rural areas (Musinguzi, 2016). Rwanda has adopted the VICOBA model as part of its efforts to boost financial inclusion and poverty reduction (Isaac & Remmy, 2017).

Furthermore, VICOBAs as informal economic sectors have emerged as essential tools for financial inclusion and poverty alleviation in rural Tanzania (Shau, 2022). In Tanzania's development trajectory, where traditional social policy approaches have shown limitations in addressing systemic challenges (see Kessy in this volume), VICOBAs have come to represent grassroots innovations that demonstrate the potential for adaptive governance and locally-defined solutions. These banks are community-based and are operated by the members themselves through informal setups. Despite their informal nature, VICOBAs play a crucial role in providing financial services to communities. Within these VICOBAs, women constitute a significant proportion of members, highlighting their importance in community development and poverty reduction efforts (Kilamlya & Utonga, 2024). VICOBAs represent a grassroots approach to financial services, providing savings, credits, and other services to populations where access to formal banking is limited (Magali & Barhe, 2022; Mgongolwa et al., 2023).

The examination of VICOBAs becomes relevant within contemporary Africa's social policy discourse, where the continent's performance on achieving SDGs presents a mixed picture, with persistent challenges in poverty reduction and gender equality highlighting the need for more comprehensive approaches. VICOBAs are commonly perceived as a woman-centered initiative as they reduce vulnerability and promote investment in household welfare, particularly through income-generation activities (Maghina et al., 2023; Rwekaza, 2024). Savings serve as the foundation for providing loans to members, enabling them to access credit for various purposes, including starting or expanding businesses, investing in agricultural activities, meeting household needs, and addressing emergencies (Omona, 2021). Ultimately, the economic empowerment of rural women through loan acquisitions not only improves their own lives but also has a ripple effect, leading to community development, poverty reduction, and increased resilience to economic shocks (Kapoor, 2019; Pakkanna et al., 2020). VICOBAs provide a valuable opportunity for economic growth, particularly when participants invest loans in productive activities.

VICOBAs operate on the principles of mutual trust and collective responsibility, allowing women to contribute a portion of their income to the group's savings pool. Access to loans within VICOBAs is followed by informal processes for saving and borrowing (Saefullah et al., 2022; Gomera & Oreku, 2020). The loan repayment stage poses a challenge for borrowers, particularly those with uncertain incomes. Loan repayments for rural women have been a prominent focus in development discourse, particularly within the context of microfinance initiatives like VICOBAs (Pamuk et al., 2022). However, the ways in which these informal economic activities operate have not managed to improve the economic status of most rural women participating in these undertakings. This persistent challenge reflects broader issues in Africa's social policy landscape, where the absence of integrated frameworks has resulted in fragmented approaches that fail to address systemic problems. This is due to the fact that accessing loans without adopting social policy frameworks leads to failure to repay. The need for comprehensive social policy frameworks thus becomes critical, especially given Africa's experience with addressing interconnected development challenges. Social policy frameworks in loan acquisition and repayment provide principles, rules, and guidelines that enable participants to plan productive investments and manage to repay.

This chapter focuses on understanding the complexities of loan repayment as crucial for ensuring the success and sustainability of VICOBAs run by rural women. Understanding these complexities is essential for developing transformative social policy responses that can address multiple development challenges while building foundations for sustainable and inclusive transformation. Failure to repay loans affects group accountability and solidarity that force them to create informal mechanisms among themselves (Ojong & Simba, 2019; Sangwan & Nayak, 2021). Specifically, this chapter discusses the processes of loans acquisitions and repayments within the context of VICOBAs members, by targeting rural women residing in Busega District of Tanzania. Busega District represents a microcosm of Tanzania's broader development challenges, where VICOBAs have become essential for informal economic empowerment, particularly for women involved in small-scale agriculture and informal sector activities (Rugeiyamu et al., 2021; Mchembe et al., 2024). The district's context reflects the continent's characteristic challenges of low coverage of social protection systems and persistent food security challenges, making the study of VICOBAs relevant for understanding alternative approaches to social policy implementation. Busega District is highlighted because, like other areas in Tanzania, traditions and customs continue to place the primary responsibility of household care on women. Once economic resources are scarce, women face greater challenges in saving and repaying loans when they participate in these informal economic sectors, such as VICOBAs (Tesha et al., 2023). According to Munoz Boudet et al. (2018), a higher percentage of women (18.7%) in rural areas live in poverty compared to men (17.9%). As such, their involvement in VICOBAs is crucial to generate income and reduce poverty for themselves and their households.

This examination of VICOBAs contributes to contemporary discourse on social policy transformation in Africa, where innovative approaches addressing multiple development challenges through integrated interventions are increasingly necessary. According to Kim et al. (2019), the common dilemma encountered by VICOBA members—rural women who face considerable difficulties in fulfilling their repayment obligations post loan acquisition—primarily originates from their income status (Kim et al., 2019). It is worth noting that while there are positive outcomes associated with VICOBAs, resulting from loan utilization, these may not uniformly reflect the diverse utilization patterns exhibited by borrowers (Komba & Benson, 2018; Uisso et al., 2021; Kotoroi, 2022). The significance of examining loan acquisitions and repayments among rural women lies in understanding how these processes present structural challenges that extend beyond individual financial capacity

(Ngoyo et al., 2023). While existing scholarship emphasizes the benefits associated with VICOBAs, the complexities of loan acquisition and repayment challenges require systematic investigation to inform policy development.

This chapter addresses two research questions: First, what processes are applied for loan acquisition by rural women participating in VICOBAs in Busega District, Tanzania? Second, what challenges characterize loan repayments among rural women participating in VICOBAs? These questions are answered by examining reasons for joining VICOBAs, loan acquisition processes, eligible borrowing amounts, loan utilization, and repayment challenges among rural women participating in VICOBAs. Exploring these aspects provides a comprehensive understanding of the operations of VICOBAs for rural women. The study's findings contribute to broader discussions on redefining social policy discourse in contemporary Africa by providing the baseline for creating social policy frameworks to improve the operations of these informal economic sectors and to improve rural women's loan acquisitions and repayments.

14.2 Theoretical Foundation on Loan Acquisition and Repayment

Given the need for comprehensive social policy frameworks to address Africa's interconnected development challenges, especially in informal economic sectors, this study adopted Microfinance and Poverty Reduction Theory to examine the process of loan acquisitions and repayments. The concept of microfinance that governs loan acquisitions and repayments is often credited to Muhammad Yunus, an economist from Bangladesh (Ullah & Haq, 2018). He pioneered the idea of providing small loans to impoverished individuals without requiring collateral, to help them start or grow small businesses (Aisyah, 2024). While the Microfinance and Poverty Alleviation Theory doesn't have a single founder, it is largely associated with the work of Muhammad Yunus, who is often credited with popularizing the concept of microfinance as a tool for poverty alleviation (Oikawa Cordeiro, 2020). This idea was formalized in the founding of the Grameen Bank in 1983 (Mutoko & Mutoko, 2020).

The Microfinance and Poverty Alleviation Theory represents a conceptual framework that advocates for providing financial services, such as small loans (microloans), savings, and insurance, to low-income individuals, particularly those excluded from traditional banking. According to Ali (2024), the Microfinance and Poverty Alleviation Theory posits that providing financial services, particularly small loans, to underserved populations, especially women, can foster economic empowerment. By enabling investment in small-scale businesses, these services help improve income generation and enhance economic stability. The theory further suggests that access to microloans and related financial support allows individuals to build sustainable entrepreneurial ventures, ultimately increasing their income, reducing dependency, and lifting themselves out of poverty (Begum et al., 2019; Kapoor, 2019; Dhawan et al., 2022).

The debate surrounding Microfinance and Poverty Alleviation Theory often focuses on the roles of savings and borrowing with their implications for poverty reduction. Supporters argue that microfinance enables low-income individuals to access credit, which can empower them to invest in income-generating activities, improve their livelihoods, and escape from poverty (Nyamboga et al., 2019; Khursheed et al., 2021; Maganga, 2021; Babalola et al., 2023). Borrowing can provide the necessary capital to start or expand small businesses. However, critics contend that while access to credit can be empowering, it can also lead to over-indebtedness, particularly if borrowers take on more debt than they can manage to repay (Green, 2020; Green & Bylander, 2021). This can exacerbate financial vulnerability rather than alleviate it.

Many advocates emphasize the importance of savings in microfinance. They argue that microfinance institutions (MFIs) should encourage savings alongside lending to promote financial stability and resilience among clients (Alshebami & Rengarajan, 2020; Morshed et al., 2020; Tehulu, 2023). Savings can help individuals build assets and provide a buffer against shocks. In contrast, some critics argue that MFIs often prioritize lending over savings, leading to a lack of focus on building clients' savings habits (Khan et al., 2021). They suggest that without adequate savings, clients may still struggle financially even if they have access to credit (Laureti, 2018).

Advocates argue that microfinance can be a sustainable model for poverty alleviation if designed correctly (Arandia & Hepp, 2021). They suggest that microfinance can generate profits while promoting social good, especially when clients are able to save and manage their borrowing effectively (Ahmed et al., 2018; Geremewe, 2019). However, detractors caution that the commercialization of microfinance can lead to high interest rates and a focus on profit over social impact (Alper et al., 2020; Tang et al., 2020; Caballero-Montes, 2023). This can undermine the original goals of poverty alleviation and client welfare.

Many microfinance programs target women, arguing that empowering women through access to credit and savings can have a multiplier effect on household well-being and community development (Illangakoon et al., 2021; Daher et al., 2022). However, some researchers argue that while women may gain access to credit, they often remain responsible for household expenses and may not experience significant empowerment or decision-making power (Al-Shami et al., 2018; Murshid, 2018; Cornish et al., 2021). The burden of repayments can also lead to stress and conflict within households.

Supporters of microfinance point to numerous case studies and positive outcomes that demonstrate the potential for microfinance to reduce poverty and improve financial literacy (Kabonga et al., 2021; Bosu, 2024). However, empirical studies have produced mixed results, with some research indicating that microfinance may improve livelihoods or not significantly reduce poverty for all clients. Critics advocate for more rigorous evaluations and caution against overselling the effectiveness of microfinance as a panacea for poverty reduction (Chowdhury et al.,

2019). Despite these critiques, this study adopts the Microfinance and Poverty Alleviation Theory to underscore the idea that access to financial services can empower rural poor women, enhance their livelihoods, and reduce poverty, but its effectiveness depends on the broader informal economics and social context. This theoretical framework is relevant in the context of Africa's social policy challenges, where fragmented approaches have limited the effectiveness of development interventions. As such, this study bridges the gap by examining the characteristics of rural women's loan acquisitions and repayment challenges in VICOBAs as an informal economic sector.

14.3 Methodology

14.3.1 Study Area

The research was conducted within Busega District, located in the Simiyu Region of Tanzania. Its administrative hub is located in the town of Nyashimo. Busega District's strategic location is noteworthy, being bordered to the north by Lake Victoria and Bunda District, to the east by Bariadi District, and to the south by Magu District. Positioned at a latitude of 2° 10′ south and a longitude of 33° 340′ east. The geographical placement of the district renders it a pivotal nexus linking various neighboring districts and the expansive Lake Victoria (see Fig. 14.1).

Busega District, placed among the placid rural landscapes of Tanzania, embodies a community deeply rooted in agrarian traditions, where the majority of its residents are actively involved in small-scale agricultural activities. These endeavors, ranging from subsistence farming to smallholder crop cultivation, form the backbone of the local economy, sustaining livelihoods and fostering community resilience. A study was conducted in Busega District due to its distinct traditional customs and practices, which significantly influence societal roles and, in particular, limit women's ability to own and manage large enterprises. The rationale for selecting Busega District is the presence of cultural norms that present challenges to women's economic empowerment and contribute to disparities in entrepreneurial opportunities between men and women in the region (Madoshi, 2022). The specific villages visited include Yitwimila A and Yitwimila B in Kiloleli Ward; Bulima and Bukabile villages in Nyashimo Ward, and Kabita and Nyamikoma villages in Kabita Ward.

This study employed a cross-sectional research design to examine women's loan acquisitions and repayment challenges within VICOBAs operating in Busega District, Tanzania. The cross-sectional approach was selected for its efficiency in capturing data at a specific point in time, allowing for the investigation of loan acquisition processes and repayment challenges through single visits to respondents in their residences. This design facilitated direct interactions and enhanced the quality of collected data by enabling deeper insights into respondents' experiences and perspectives.

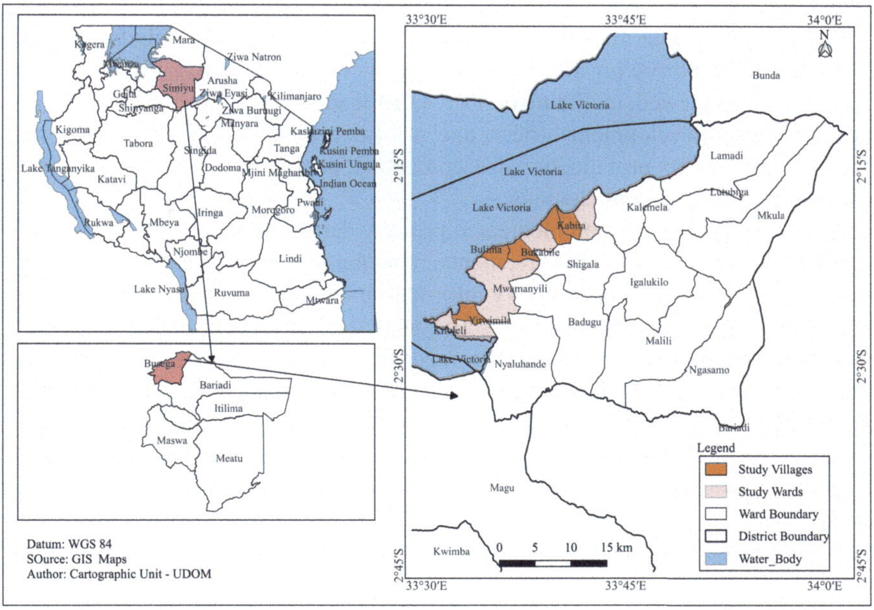

Fig. 14.1 Busega District map showing the selected villages for the study

The sampling frame comprised women actively participating as VICOBA members in Busega District, specifically targeting individuals aged 18 years and above who possessed the capacity to analyze pertinent issues concerning loan acquisitions and repayment challenges. Given the informal nature of VICOBAs and the absence of exact membership records, the sample size was calculated using the formula for estimating population proportions: $n = Z^2 \times P \times (1 - P)/E^2$, where Z represents the Z-score for 95% confidence level (1.96), P is the estimated population proportion (15.2%), and E is the desired margin of error (0.05). This calculation yielded a sample size of 198 respondents.

Purposive sampling was employed to select participants, with rural women identified through the district community development officer, who maintained lists of some registered groups, while other unregistered groups were identified through local networks. This sampling method ensured the inclusion of relevant participants who met the specific criteria of being active VICOBA members, while maintaining diversity in perspectives and experiences.

Data collection utilized multiple methods to ensure comprehensive coverage of the research questions. The primary approach involved structured and semi-structured surveys administered to the 198 participants, conducted during meetings facilitated by the community development officer at the district office. In-depth interviews were conducted with a subset of participants to obtain rich qualitative data on their loan acquisition and repayment experiences, exploring deeper perspectives on motivations, challenges, and strategies within VICOBAs. Two focus group discussions were conducted with six participants each, totaling 12 participants, to foster dialogue on loan-related topics and explore shared norms, beliefs, and attitudes concerning borrowing and repayment practices. Additionally, seven key informants were interviewed, including two VICOBA leaders, two informal loan officers, one community development officer, one ward executive officer, and one village executive officer, to provide broader contextual perspectives on women's loan acquisition and repayment challenges.

Documentary analysis complemented primary data collection through examination of VICOBA records, loan agreements, repayment schedules, and other relevant documents, providing supplementary information on loan terms, repayment performance, and organizational practices. The analytical approach employed content analysis for qualitative data to identify recurring themes, patterns, and narratives, while quantitative data underwent descriptive statistical analysis to examine numerical patterns and relationships. This dual-method approach ensured a comprehensive analysis that combined rich contextual understanding with quantitative insights, providing a thorough examination of women's experiences within VICOBAs in Busega District.

14.4 Findings

The study findings shed light on the significant role of VICOBAs in providing financial services to women in Busega District. It identifies the reasons for joining VICOBAs, the loan acquisition processes, eligible amounts for borrowing, utilization of loans, and challenges of repayments among rural women participating in VICOBAs.

14.4.1 Reasons for Joining VICOBAs

The findings indicate that a significant majority of rural women, accounting for 83.84%, joined VICOBAs primarily to secure loans, driven by their need for finance. However, the decision to join VICOBAs is influenced by a variety of factors, showcasing the complexity of individual motivations among members. Specifically, 11.62% of participants highlighted the importance of accumulating savings, underscoring their recognition of financial security and stability. Inclusion of 2.53% of respondents expressed their interest in VICOBA as a platform for developing entrepreneurship skills, acknowledging the role of business expertise in achieving economic empowerment. Moreover, 1.51% emphasized the value of social support and networking opportunities offered by VICOBAs, emphasizing the significance of community engagement and collaboration. Lastly, 0.50% cited miscellaneous reasons for joining VICOBAs, reflecting the diverse range of individual motivations that contribute to the organization's growth and vitality (Fig. 14.2).

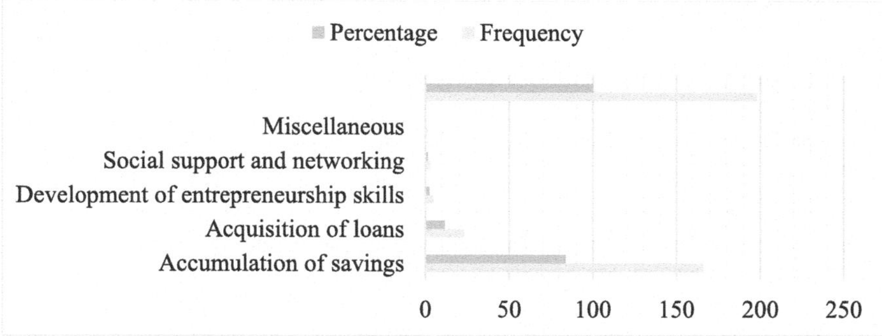

Fig. 14.2 Reasons for joining VICOBAs

Rural women who are members of VICOBAs have emphasized their commitment to attending meetings on a monthly basis, a practice that was substantiated by the testimonies from G. as she stated.

> I am dedicated to attend VICOBA meetings once per month. Our group is called *Wanawake Nuru.* The engagement in meetings creates the commitment to foster financial empowerment for the members. This regular attendance highlights the significance of gathering as a platform to gain knowledge, skill-building, and networking among members (Participant G., Bulima, 16/10/2024).

Another woman also testified to the importance of these meetings, as she explained.

> The consistent participation in VICOBAs' monthly meetings fosters a robust sense of ownership and engagement within the movement. We acknowledge that our collaborative efforts to establish a platform for savings and borrowing represent a viable strategy to alleviate income poverty for both ourselves and our households (Participant B., Bukabile, 17/10/2024).

14.4.2 Loan Acquisition Processes in VICOBAs

The majority, comprising 84.84% of women VICOBA members, reported that the processes of applying for and receiving loans from their groups were straightforward. However, a minority, constituting only 16.16% of respondents, explained that they encountered some difficulties in securing loans (Table 14.1). Additional information from focus group discussions revealed procedural difficulties involved in acquiring loans from VICOBAs. Participants mentioned the initial stage, which required members requesting loans to meet the criteria established by the VICOBA group. Eligibility often included regular attendance at meetings, adherence to savings requirements, and active participation in group activities. The eligibility requirements were further explained by S., who stated that

Table 14.1 Loan acquisition processes in VICOBAs

Status	Frequency	Percentage
Loan application and reception processes are straightforward and accessible	166	84.84
Loan application and reception processes are difficult to manage	32	16.16
Total	198	100.00

> Once eligibility is confirmed, we submit loan applications detailing the purpose and the proposed repayment plan. These applications are then reviewed by our VICOBA loan committee members, who assess the feasibility of the requests and their alignments with the group's objectives (Participant S., Kabita, 18/10/2024).

The approval was also required in order to ensure the sustainability of the given loans. K. explained that

> Upon approval, the loan amount is disbursed to the member, often accompanied by guidance on proper utilization and repayment expectations. Throughout the loan term, regular monitoring and communication is required to ensure that members receive necessary support and guidance, while also upholding their commitment to timely repayment. Despite the guidance offered by our leaders to borrowers, there is still a problem with loans repayments because most of the group members have low incomes. (Participant K., Nyamikoma Village, 22/10/2024).

The study discovered that the structured process not only facilitates access to financial resources but also promotes responsibilities to borrowers.

14.4.3 Eligible Amount for Borrowing

The community development officer explained that eligibility for the amount of borrowing differed among the VICOBAs. Some VICOBAs required borrowers to take twice the amount that they had previously saved at an interest rate of 10%. They mentioned that the process of accumulating their savings was difficult, especially for those who didn't have reliable income-generating activities, as it has been evident in H. words.

> While VICOBAs undoubtedly offer valuable support to their members, particularly in terms of financial inclusion and community empowerment, it's evident that one of the most formidable hurdles lies in the initial stages of investment. This critical phase presents various challenges, including but not limited to securing sufficient capital to launch income-generation activities, navigating market dynamics, and developing viable business plans (Participant H., Kabita Village, 18/10/2024).

The Ward Executive Officer expressed that accessing the needed amount of funds to initiate entrepreneurial endeavors can be discouraging, especially for individuals with low savings. J. stated that

> Despite having the opportunity to save and take loans, it is not easy to secure enough funds to start production activities. All the groups lend based on the amount of savings each

member has contributed. As a result, I end up borrowing a small amount of money, which is insufficient to make meaningful investments in petty trading. (Participant J., Yitwimila A, 23/10/2024).

The Village Executive Officer highlighted the need for addressing the barriers as VICOBAs seems to be the tangible strategy for poverty reduction to these rural women, as he detailed.

The Village Executive Officer also highlighted the starting point of savings that members contribute in order to access loans. He stated

From my experience, I observe that women are increasingly struggling with the debts from VICOBAs. Their ability to secure sufficient funds to start economic activities is limited because they lack the capacity to save large amounts of money (Village Executive Officer, Busega District, 24/10/2024).

14.4.4 Utilization of VICOBA Loans by Rural Women

The findings revealed that 92.93% of rural women in the study area allocated loans towards short-term expenditures, encompassing essential needs such as food, clothing, rent, children's tuition fees, health services, and utility bills such as electricity and water. In contrast, 7.07% of respondents utilized loans for long-term expenditures, including savings, miscellaneous expenses, and investments in tangible assets such as electronic appliances, furniture, and personal devices like mobile phones, as well as real estate ventures like plots of land and materials for house construction (Table 14.2). This indicates that the utilization of loans differed among these rural women based on the household needs. The diversity in utilization was detected during the FGDs as expressed through the discussions. They explained

The utilization of loans from VICOBAs varies significantly among members. A substantial portion of borrowers prioritize meeting their households' basic needs, leaving limited resources to invest in business ventures. This financial obligation, often driven by immediate household demands, presents a considerable challenge for many individuals seeking to use the loans as capital for business start-ups. (Focus Group Discussion, Bulima, 25/10/2024).

They recognized the ones who use the borrowed money as capital for running petty trades, as they explained.

However, the smallest number of beneficiaries are often those who strategically allocate the funds to expand their business capital. By reinvesting in their enterprises, these individuals can stimulate growth, enhance productivity, and ultimately generate sustainable income

Table 14.2 Utilization of VICOBA loans by rural women

Loan expenditure pattern	Frequency	Percentage
Short-term	184	92.93
Long-term	14	7.07
Total	198	100.00

> sources. This approach not only fosters personal financial advancement but also contributes to the economic development of the broader community (Focus Group Discussion, Bulima, 25/10/2024).

14.4.5 Challenges of Loan Repayments among Rural Women

The findings revealed that loan repayment experiences varied significantly among these rural women. One of the VICOBAs leaders expressed that those who utilized the borrowed funds to address household short-term needs encountered challenges in the repayment processes. This observation was underlined by a woman who elaborated on her struggles to repay the loan, which she had primarily utilized to cover her child's educational expenses, as M. explained.

> Securing a loan to cover my son's educational fees was essential for me, especially considering that I do not engage in trading activities and rely on assistance from my brothers. However, repaying the loan proved to be challenging. Despite my earnest efforts, I often found myself with limited funds available for repayment, as the majority of my resources were allocated towards meeting ongoing educational expenses for our son. This situation posed a significant challenge in fulfilling my loan repayment obligations (Participant M., Bulima Village, 25/10/2024).

The findings shed light on the challenges of loan repayments among rural women participating in VICOBAs. Social factors, representing 64.15% of respondents' perspectives, indicate that household support and social pressure significantly impact loan repayment initiatives. Economic factors, identified by 31.82% of participants, underscore the importance of income levels and economic shocks in determining loan repayment capabilities. Variations in income stability and unexpected financial scarcities may pose challenges to timely repayments.

Support mechanisms for loan repayments, mentioned by 2.02% of respondents, signify the presence of organized assistance programs or initiatives within VICOBAs aimed at facilitating loan repayment processes. Regarding support mechanisms, there are also some women who complained about how their husbands hinder them from investing in petty trades and generate income for gaining profits and managing to repay loans. This was explained by participant MJ as she explained that

> When I take a loan, I always have business goals, such as increasing capital for my petty trade of selling cloths. However, when I mention that I have taken a loan, my husband insists that I give him money to contribute toward paying water bills, electricity, and school fees for the children, even though he has a salary (Participant MJ., Nyamikoma Village, 22/10/2024).

The study discovered that, for such couples, there is always a challenge in decision-making on how their income should be distributed in their households. As a result, the man in a position of power decides unilaterally, without the woman's consent, on how to use the money borrowed from the VICOBA.

Group solidarity, cited by 1.01% of participants, emphasizes the cohesive bonds and mutual responsibility fostered within VICOBA communities. Collective

Table 14.3 Challenges of loan repayments among rural women

Factor	Frequency	Percentage
Social factors (e.g., household support, social pressure)	127	64.15
Economic factors (e.g., income level, economic shocks)	63	31.82
Support mechanisms for loan repayment	4	2.02
Group solidarity	2	1.01
Extension services provided by VICOBAs	2	1.01
	198	100.00

commitment to loan repayment obligations, coupled with shared accountability, fosters a supportive environment that encourages adherence to repayment schedules and promotes financial stability among members. Extension services provided by VICOBAs, mentioned by 1.01% of respondents, indicate the value placed on educational resources and guidance offered by the organization in enhancing members' financial literacy and empowerment (Table 14.3). This implies that inadequate support mechanisms, with weak group solidarity and insufficient extension services provided to rural women VICOBA members, have posed significant challenges for loan repayments. These barriers have hindered their ability to manage loans effectively, resulting in financial strain and, ultimately, the failure of many to sustain active participation.

14.5 Discussion

Loans are crucial for those who are planning to start or expand small businesses and meet their financial needs. Beyond just financial gain, joining VICOBAs also offers women opportunities to build financial literacy, increase their economic independence, and enhance their decision-making power within households and communities. The supportive structure of VICOBAs fosters trust and collaboration among members, further enabling them to pool resources and collectively overcome economic barriers. This alignment with the principles of mutual trust and collective responsibilities demonstrates how VICOBAs create adaptive governance structures that respond to locally defined solutions, especially in contexts where traditional social policy approaches have shown limitations. This is similar to Madaha's (2018) observation that VICOBAs create a networking structure for women seeking opportunities to improve their income. Therefore, while the immediate goal for women joining VICOBAs is often to save and obtain loans, the long-term benefits include personal empowerment, improved livelihoods, and enhanced community resilience.

The data suggests that for most women in VICOBAs, the processes of applying for and receiving loans are efficient and accessible. This indicates that these community-based lending groups are functioning effectively in providing financial services to their members. This effectiveness reflects the grassroots innovation potential of VICOBAs as informal economic structures that address critical gaps in formal financial systems, especially in rural areas where access to conventional banking remains limited. This was also discovered by Magufuli et al. (2024) as they reported that by offering small loans, microfinance institutions empower people to start or expand small businesses, invest in education, or improve their households' living standards. On the other side, some individuals remain skeptical and hesitate to suggest that savings and loans conducted by low-income populations can effectively reduce poverty (Page & Pande, 2018). This remains a debated area within the informal economy, particularly regarding savings and loan systems that operate without the oversight of financial experts and formal institutions. However, within contemporary Africa's social policy discourse, where the continent's performance on achieving SDGs presents persistent challenges in poverty reduction and gender equality, VICOBAs represent essential tools for addressing these interconnected development challenges. This study advocates for rural women's loan acquisition and repayments, as they are among the few viable options for accessing credit in developing countries. In the absence of more inclusive financial systems, these groups provide essential support to rural women by facilitating access to loans and promoting economic empowerment. This ease of access is likely to encourage greater participation and financial inclusion for rural women, enabling them to meet personal or business-related financial needs. This is supported by the Microfinance and Poverty Alleviation Theory that emphasizes providing financial services, particularly small loans. However, a minority of rural women face challenges in these processes, implying that there are barriers such as inadequate financial knowledge, administrative difficulties, or other structural issues within most VICOBAs. Addressing the concerns of this minority becomes important for developing transformative social policy responses that can address multiple development challenges while building foundations for sustainable and inclusive transformation, as has been argued by Tekwa (in this volume). This may also point to the need for targeted support, such as the establishment of social policy frameworks, financial literacy training, or simplified procedures, to assist those who find that the processes are challenging.

The findings indicate that borrowing eligibility within VICOBAs varies, with some groups requiring members to borrow double the amount they have saved, at an interest rate of 10%. This structure suggests that the more a member saves, the greater the borrowing capacity. This structure incentivizes higher savings among participants. However, the requirement to take on loans at a fixed multiplier may place a financial burden on some borrowers, particularly those who may not be ready to manage larger loans or repay them with interest. The 10% interest rate, though relatively low compared to some formal institutions, could still be a significant cost for those with limited income, potentially affecting their ability to repay loans on time. These conditions highlight the structural challenges that extend beyond individual financial capacity, especially in contexts where traditions and customs continue to place primary responsibility of household care on women, creating additional barriers to effective loan management. These conditions highlight the importance of altering loan amounts to individual circumstances and providing financial education to help members make informed borrowing decisions that align with their financial capacity and goals.

The funds obtained by rural women through loans from VICOBAs are primarily directed towards addressing immediate, short-term needs such as household expenses, school fees, or medical bills. This pattern reflects the broader reality that women in rural areas face greater challenges in saving and repaying loans when economic resources are scarce, given their primary responsibility for household care. While these loans provide critical financial relief in the short run, they often do not lead to a sustainable increase in income since a significant portion of the borrowed money is not channeled into income-generation activities or entrepreneurial ventures. As such, this focus on meeting immediate needs rather than making long-term investments limits the potential for economic growth and wealth accumulation. This limitation highlights the need for comprehensive social policy frameworks that provide principles, rules, and guidelines to enable participants to plan productive investments and manage repayments effectively. When loans are not invested in productive activities, such as expanding small businesses, improving agricultural outputs, or starting new entrepreneurial endeavors, it limits economic growth for individuals and households. The borrowers are less likely to experience significant financial returns. Consequently, this pattern creates a cycle where members continue to rely on short-term borrowing to cover recurring expenses, rather than building a stable, long-term income base.

The findings indicate that each woman explained the circumstances preventing them from repaying loans on time. They highlighted several interconnected challenges they faced, including unstable and seasonal income sources, unpredictable household expenses, and health emergencies. These challenges reflect the broader issues in Africa's social policy landscape, where the absence of integrated frameworks has resulted in fragmented approaches that fail to address systemic problems (see Chipenda & Tom in this volume). Many women relied on small-scale agricultural activities or informal businesses, both of which are susceptible to market fluctuations and low returns, limiting their ability to generate consistent income. Additionally, household responsibilities often took financial priority, with resources diverted to immediate needs such as healthcare or education, making loan repayment secondary. Cultural factors, such as limited financial decision-making power within households and a lack of access to financial literacy or business support, further complicate timely loan repayment. Understanding these dynamics is essential for developing transformative social policy responses that can address the systemic nature of these challenges while building foundations for sustainable economic empowerment. The repayment of loans by rural women participating in VICOBAs who lack income-generating activities presents a significant challenge.

14.6 Conclusion

The study concludes that VICOBAs, as informal economic activities, play a critical role in providing financial services for women, particularly in rural settings. VICOBAs have proven to be an effective tool for fostering financial inclusion. They foster accessibility to small-scale loans, promote savings, and encourage participation in entrepreneurship among women who often face barriers to formal financial institutions. The loans offered enable the majority of rural women to meet various needs in their households. The minority uses loans to expand their petty trades, while also fostering a sense of economic independence and community support. The system incentivizes savings, and in most cases, the loan application process is efficient, encouraging wider participation for economic empowerment. This is also the emphasis of Microfinance and Poverty Alleviation Theory, as the provision of financial services empowers participants economically and reduces poverty, especially among poor communities. The social policy implication is that strengthening VICOBAs by establishing frameworks with principles, rules, and guidelines is the tangible strategy of enhancing financial inclusion for marginalized rural women. However, certain barriers, such as inadequate financial knowledge or rigid borrowing conditions, hinder some women's ability to fully benefit from these loans. These challenges, including high interest rates or the pressure to borrow beyond their comfort zone, highlight the need for a more flexible and supportive structure. Additionally, the common use of loans to address short-term needs rather than long-term investments restricts the potential for sustainable income growth. Therefore, as some women strive to establish VICOBAs as a means of enhancing their economic standing, these grassroots initiatives represent a significant opportunity for socio-economic transformation, especially in rural communities. While the formation of VICOBAs is a positive step, their long-term success and sustainability are centered on the availability of adequate support from community development agents.

14.7 Recommendations

14.7.1 Practical and Policy Recommendations

The government should develop comprehensive social policy frameworks that outline clear principles, rules, and guidelines for loan acquisition and repayment to ensure the sustainability of VICOBAs, especially in contexts where the absence of integrated frameworks has resulted in fragmented approaches that fail to address systemic problems. A well-designed social policy will guide women in project planning and implementation, enabling them to generate income, save effectively, and responsibly access loans for sustainable development. The government should also formulate an informal economic framework that will integrate VICOBA operations by establishing systems for registration, monitoring, and evaluation of savings and loans acquisitions and repayments. This approach would ensure effective oversight and support for informal financial services within communities, enhancing their contribution to local economic growth and financial inclusion.

Community development officers should establish literacy programs to equip women with essential knowledge on budgeting, managing debt, and making informed financial decisions. Additionally, capacity-building initiatives should focus on strengthening leadership, business planning, financial management, and operational efficiency within these informal economic entities that have emerged to fill critical gaps in formal financial systems.

14.7.2 Recommendations for Future Research

Future research could explore male participants in VICOBAs to examine their loan acquisition and repayment behaviors, drawing comparisons with women's experiences, considering how gender influences access to microfinance within the context of Africa's social policy discourse and sustainable development goals. Such a study would provide findings on how gender influences access to microfinance, financial decision-making, and repayment strategies. This would ultimately contribute to a deeper understanding of gendered dynamics in microfinance participation and sustainability. In addition, since the study focused on women without considering marital status, it is also possible to examine how married participants perceive their husbands' roles in either supporting or hindering their involvement in VICOBA activities. This is particularly important in contexts where traditions and customs continue to place the primary responsibility of household care on women.

Ethics Approval The approval to conduct this study was granted by the University of Dodoma following a thorough review process to ensure compliance with ethical research standards. This approval reflects the university's commitment to upholding academic integrity and ethical considerations in research.

References

Ahmed, Z., Pervez, M., & Jayed, B. H. (2018). A comparative study on the role of microcredit and social business in sustainable poverty reduction. *International Journal of Development and Sustainability, 7*(5), 1749–1758.

Aisyah, S. (2024). Social and economic transformation of women through microfinance: Lessons from the M. Yunus Economic Movement. *Peradaban Journal of Economic and Business, 3*(1), 91–103. https://doi.org/10.59001/pjeb.v3i1.158

Ali, A. (2024). Women's empowerment and the contribution of micro-finance. *Pakistan Journal of Gender Studies, 24*(2), 1–17.

Alper, E., Clements, B., Hobdari, N., & Moya Porcel, R. (2020). Do interest rate controls work? Evidence from Kenya. *Review of Development Economics, 24*(3), 910–926.

Al-Shami, S. S. A., Razali, R. M., & Rashid, N. (2018). The effect of microcredit on women empowerment in welfare and decisions making in Malaysia. *Social Indicators Research, 137*, 1073–1090.

Alshebami, A. S., & Rengarajan, V. (2020). An imperative need for a new social order in microfinance community-Towards ushering in a resilient eco-system in the battle against corona pandemic. *International Journal of Research, 8*(08), 107–123. https://doi.org/10.29121/granthaalayah.v8.i8.2020.986

Arandia, O., & Hepp, S. (2021). Poverty alleviation through financial practices: The importance of microfinance. In *Handbook on ethics in finances* (pp. 397–421). https://doi.org/10.1007/978-3-030-00001-1_38-1

Babalola, F. I., Mhlongo, N. Z., Obinyeluaku, M. I., Oladayo, G. O., & Daraojimba, C. (2023). Microfinance and economic empowerment in Nigeria: A critical review of impact and sustainability. *Finance & Accounting Research Journal, 5*(12), 381–404. https://doi.org/10.51594/farj.v5i12.640

Begum, H., Alam, A. F., Mia, M. A., Bhuiyan, F., & Ghani, A. B. A. (2019). Development of Islamic microfinance: A sustainable poverty reduction approach. *Journal of Economic and Administrative Sciences, 35*(3), 143–157.

Bosu, C. (2024). Unveiling the impact of microfinance on women's empowerment: Insights from Cumilla District, Bangladesh. *International Journal of Management and Accounting, 6*(3), 55–70.

Caballero-Montes, T. (2023). Integrating market conditions into regulatory decisions on microfinance interest rates: Does competition matter? *Annals of Finance, 19*(2), 201–232. https://doi.org/10.1007/s10436-022-00417-8

Chowdhury, M. S., Ahmmed, F., & Hossain, M. I. (2019). Neoliberal governmentality, public microfinance and poverty in Bangladesh: Who are the actual beneficiaries? *International Journal of Rural Management, 15*(1), 23–48. https://doi.org/10.1177/0973005218817657

Cornish, H., Walls, H., Ndirangu, R., Ogbureke, N., Bah, O. M., Tom-Kargbo, J. F., et al. (2021). Women's economic empowerment and health related decision-making in rural Sierra Leone. *Culture, health & sexuality, 23*(1), 19–36. https://doi.org/10.1080/13691058.2019.1683229

Daher, M., Rosati, A., & Jaramillo, A. (2022). Saving as a path for female empowerment and entrepreneurship in rural Peru. *Progress in Development Studies, 22*(1), 32–55.

Dhawan, S. M., Wilson, K., & Zademach, H. M. (2022). Formal micro-credit for refugees: New evidence and thoughts on an elusive path to self-reliance. *Sustainability, 14*(17), 10469. https://doi.org/10.3390/su141710469

Gajijime, H. J. (2023). Challenges facing women's who are participating in Village Community Bank in Mbarali District. *Asian Research Journal of Arts & Social Sciences, 21*(3), 175–188. https://doi.org/10.9734/arjass/2023/v21i3482

Geremewe, Y. T. (2019). The role of microfinance institution for poverty reduction in Ethiopia. *Journal of Economics and Sustainable Development, 10*(5), 36–44. https://doi.org/10.7176/JESD

Gomera, W. C., & Oreku, G. S. (2020). Mobile devices in supporting membersâ€™ participation to Village Community Banks: The design thinking approach. *Zambia ICT Journal, 4*(1), 6–15.

Green, W. N. (2020). Regulating over-indebtedness: Local state power in Cambodia's microfinance market. *Development and Change, 51*(6), 1429–1453. https://doi.org/10.1111/dech.12620

Green, W. N., & Bylander, M. (2021). The exclusionary power of microfinance: Over-indebtedness and land dispossession in Cambodia. *Sociology of Development, 7*(2), 202–229. https://doi.org/10.1525/sod.2021.7.2.202

Illangakoon, A. G., Azam, S. M., Khatibi, A., & Sudasinghe, S. (2021). Women empowerment in microfinance: Poverty alleviation tool. *South Asian Journal of Social Studies and Economics, 12*(4), 159–172. https://doi.org/10.9734/sajsse/2021/v12i430324

Isaac, I. M., & Remmy, M. N. (2017). Impact of village saving and loan associations (VSLAs) in Rubengera sector–Rwanda. *Journal of Agricultural Economics and Rural Development, 2*(1), 011–021.

Kabonga, I., Zvokuomba, K., & Dube, E. (2021). Internal savings and lending schemes (ISALS) and the empowerment of women in rural Shamva, Zimbabwe. *Rural Society, 30*(2–3), 101–118. https://doi.org/10.1080/10371656.2021.2006991

Kapoor, S. (2019). Entrepreneurship for economic and social empowerment of women: A case study of a self help credit program in Nithari Village, Noida, India. *Australasian Accounting, Business and Finance Journal, 13*(2), 123–142.

Khan, A. A., Khan, S. U., Fahad, S., Ali, M. A., Khan, A., & Luo, J. (2021). Microfinance and poverty reduction: New evidence from Pakistan. *International Journal of Finance & Economics, 26*(3), 4723–4733. https://doi.org/10.1002/ijfe.2038

Khursheed, A., Khan, A. A., & Mustafa, F. (2021). Women's social empowerment and microfinance: A brief review of literature. *Journal of International Women's Studies, 22*(5), 249–265.

Kilamlya, J., & Utonga, D. (2024). The efficacy of VICOBA intervention in alleviating poverty in emerging economies: A study of the ORGUT SEDIT VICOBA lending scheme project in Morogoro, Tanzania. *American Journal of Applied Statistics and Economics, 3*(1), 92–98. https://doi.org/10.54536/ajase.v3i1.2616

Kim, M., Sudhir, K., Uetake, K., & Canales, R. (2019). When salespeople manage customer relationships: Multidimensional incentives and private information. *Journal of Marketing Research, 56*(5), 749–766.

Komba, C., & Benson, S. (2018). The success of Village Community Banks in improving household welfare in Kigamboni District, Dar es Salaam Region. *Journal of Policy and Leadership, 6*(1), 1–21.

Kotoroi, N. L. (2022). The contribution of VICOBA to the improvement of food accessibility among Maasai women pastoralists in Chalinze District Council. *Asian Research Journal of Arts & Social Sciences, 18*(4), 69–80.

Laureti, C. (2018). Why do poor people co-hold debt and liquid savings? *The Journal of Development Studies, 54*(2), 213–234. https://doi.org/10.1080/00220388.2017.1299137

Madaha, R. (2018). Challenges and opportunities of Village Community Networks within the neoliberal context: A case study of women's networks in Africa. *African identities, 16*(1), 50–66.

Madoshi, J. P. (2022). Addressing gender disparities through folklore: A cultural study of female child appellation among the Bāsukuma of Tanzania. *Journal of International Women's Studies, 24*(1), 20.

Magali, J., & Barhe, R. (2022). Clients' perception of VICOBA informal social security roles: The case of Nambis Division in Mbulu District, Tanzania. *East African Journal of Management and Business Studies, 2*(2), 1–14. https://doi.org/10.46606/eajmbs2022v02i02.0007

Maganga, A. (2021). Impact of microfinance village savings and loan associations on women's empowerment and resilience against vulnerability in Malawi. *International Journal of Rural Management, 17*(2), 190–212. https://doi.org/10.1177/097300522097255

Maghina, M. G., Ruheza, S., & Mwangole, F. P. (2023). Contribution of microfinance institutions to women socio-economic empowerment: A case of BRAC Branch Iringa Municipal Council, Tanzania. *Asian Journal of Economics, Business and Accounting, 23*(22), 232–247.

Magufuli, M., Rwezaula, A., & Byaro, M. (2024). Do small businesses benefit from microfinance loans in the era of free market capitalism? A critical analysis of PRIDE, Tanzania. *African Journal of Economic Review, 12*(3), 134–146.

Mchembe, S., Mtae, H., & Magobe, M. (2024). Influence of social structure on participation of women in public credit programs in Tanzania. *East African Journal of Education and Social Sciences (EAJESS), 4*(4), 112–121. https://doi.org/10.46606/eajess2023v04i04.0309

Mgongolwa, B. O., Wiketye, E., & Mpogole, H. (2023). Factors affecting fund management among Village Community Banks (VICOBAs); A case of registered VICOBAs in Iringa Municipality. *International Journal of Finance and Accounting, 2*(1), 62–77. https://doi.org/10.37284/ijfa.2.1.1489

Morshed, Z., Habib, M., & Jubb, C. (2020). Does regulation influence microfinance institutions to be more client-responsive? *International Journal of Financial Studies, 8*(4), 63. https://doi.org/10.3390/ijfs8040063

Munoz Boudet, A. M., Buitrago, P., Leroy De La Briere, B., Newhouse, D. L., Rubiano Matulevich, E. C., Scott, K., & Suarez-Becerra, P. (2018). Gender differences in poverty and household composition through the life-cycle: A global perspective. *World Bank Policy Research Working Paper*, (8360).

Murshid, N. S. (2018). Microfinance participation and women's decision-making power in the household in Bangladesh. *Journal of Social Service Research, 44*(3), 308–318.

Musinguzi, L. K. (2016). The role of social networks in savings groups: insights from village savings and loan associations in Luwero, Uganda. *Community Development Journal, 51*(4), 499–516. https://doi.org/10.1093/cdj/bsv050

Mutoko, W. R., & Mutoko, P. (2020). Towards implementing the Grameen Bank Financing Model among micro enterprises in Botswana: Is it a possibility. *European Scientific Journal, ESJ, 16*(4), 29.

Ngoyo, L. D., Lubuva, E. E., & Lupala, F. (2023). Factors influencing sustainable economic empowerment of small-scale women entrepreneurs using a local government women's development fund in Tanzania. *Local Administration Journal, 16*(2), 177–194.

Nyamboga, T. O., Odhuno, E., & Okibo Bichanga, W. (2019). Influence of crediting strategy on growth of small and micro enterprises in Kenya: A case of women enterprise fund. *Asian Journal of Economics, Business and Accounting, 13*(1), 1–17. https://doi.org/10.9734/ajeba/2019/v13i130160

Oikawa Cordeiro, B. (2020). Global mobility of microfinance policies. *Policy and society, 39*(1), 19–35. https://doi.org/10.1080/14494035.2019.1659472

Ojong, N., & Simba, A. (2019). Fostering micro-entrepreneurs' structural and relational social capital through microfinance. *Journal of Small Business & Entrepreneurship, 31*(1), 1–20.

Omona, J. (2021). Poverty eradication and wealth creation: The role and challenges of savings and credit cooperatives (SACCOs) in Uganda. *African Journal of Business Management, 15*(8), 198–210. https://doi.org/10.5897/AJBM2021.9234

Page, L., & Pande, R. (2018). Ending global poverty: Why money isn't enough. *Journal of Economic Perspectives, 32*(4), 173–200. https://doi.org/10.1257/jep.32.4.173

Pakkanna, M., Rasulong, I., & Akhmad, S. I. W. (2020). Microfinance institutions and women empowerment: evidence in the rural areas of Tangerang, Indonesia. *International Journal of Scientific and Technology Research, 9*(2), 3994–3999.

Pamuk, H., van Asseldonk, M., Ruben, R., Kweka, T., Wattel, C., & Hella, J. P. (2022). Social ties, access to loans, and loan repayments in savings and loan associations: Evidence from rural Tanzania. *Agricultural Finance Review, 82*(5), 777–796.

Polloni-Silva, E., Moralles, H. F., Rebelatto, D. A. D. N., & Hartmann, D. (2021). Are foreign companies a blessing or a curse for local development in Brazil? It depends on the home country and host region's institutions. *Growth and Change, 52*(2), 933–962. https://doi.org/10.1111/grow.12484

Rugeiyamu, R., Shayo, A., Kashonda, E., & Mohamed, B. (2021). Role of local government authorities (LGAs) in promoting local economic development and service delivery to local community in Tanzania. *Local Government Study, 14*(2), 103–122.

Rwekaza, G. D. (2024). An insight of Village Community Banks as the tool of forest resource conservation: A case of selected villages adjacent Pugu and Kazimzumbwi forest reserve. *Social Science and Humanities Journal (SSHJ), 8*(07), 4418–4425. https://doi.org/10.18535/sshj.v8i07.1215

Saefullah, K., Sadikin, I., de Bekker, J. C. M., & Chirangi, M. M. (2022). Mapping over-indebtedness of microfinance clients in Indonesia and Tanzania: The importance of analysing microfinance beyond its outreach and sustainability. *Budapest International Research and Critics Institute (BIRCI-Journal): Humanities and Social Sciences, 5*(1), 4998–5011.

Sangwan, S., & Nayak, N. C. (2021). Factors influencing the borrower loan size in microfinance group lending: A survey from Indian microfinance institutions. *Journal of Financial Economic Policy, 13*(2), 223–238. https://doi.org/10.1108/JFEP-01-2020-0002

Shand, W., & Colenbrander, S. (2018). Financing the inclusive city: The catalytic role of community savings. *Environment and Urbanization, 30*(1), 175–190. https://doi.org/10.1177/0956247817751340

Shau, I. (2022). The assessment of challenges facing the growth of informal lending groups in Tanzania. The case of VICOBA. *African Journal of Applied Research, 8*(1), 1–14. https://doi.org/10.26437/ajar.03.2022.1

Tang, J. J., Quayes, S., & Joseph, G. (2020). Microfinance institutions, financial intermediation and the role of deposits. *Accounting & Finance, 60*(2), 1635–1672.

Tehulu, T. A. (2023). What drives microfinance institution lending behavior? Empirical evidence from sub-Saharan Africa. *International Journal of Emerging Markets, 18*(8), 1745–1765.

Tesha, J., Fabian, A., Mkuwa, S., Misungwi, G., & Ngalesoni, F. (2023). The role of gender inequities in women's access to reproductive health services: A population-level study of Simiyu Region Tanzania. *BMC Public Health, 23*(1), 1111. https://doi.org/10.1186/s12889-023-15839-w

Uisso, A. J., Chirwa, P. W., Ackerman, P. A., & Mbwambo, L. (2021). Non-carbon benefits as incentives for participation in REDD+ and the role of village participatory land use plans in supporting this: Insights from Kilosa District, Tanzania. *Journal of Environmental Planning and Management, 64*(6), 1111–1132.

Ullah, A., & Haq, N. (2018). Micro financing management and its prospects: A case study analysis on Bangladesh perspective. *TRANS Asian Journal of Marketing & Management Research (TAJMMR), 7*(7), 17–25.

Chapter 15
Transforming Masculinity and Mental Health in Rural Southern Africa: A Comparative Analysis of Gender Transformation and Social Policy in Zimbabwe and South Africa

Winnet Hokoza and Clement Chipenda

15.1 Introduction and Background

The intersection of gender transformation, masculinity, and mental health represents a critical area of inquiry in contemporary Southern African contexts, where traditional gender norms continue to shape social behaviours and health outcomes (Mashele et al., 2024). This chapter utilises two complementary theoretical frameworks, namely Transformative Social Policy (TSP) and Gender Transformation Theory (GTT), to examine the dynamic nexus between gender norms, mental health, and social policy in rural Southern African contexts.

TSP offers a paradigmatic shift from conventional welfare approaches by conceptualising social policy as a multidimensional instrument for societal transformation rather than merely addressing immediate vulnerability (Mkandawire, 2007; Adesina, 2011). TSP understands social policy as 'collective interventions that directly affect social welfare, social institutions, and social relations' (Mkandawire, 2010), moving beyond residual approaches to embrace social policy as integral to development processes through production enhancement, social protection, redistribution, reproduction and social cohesion (Adesina, 2011). Complementing this, GTT posits that achieving gender equality requires systemic change in gender relations, transcending the traditional narrow focus on women's empowerment

W. Hokoza (✉)
Department of Community and Social Development, University of Zimbabwe,
Harare, Zimbabwe
e-mail: whokoza@sociol.uz.ac.zw

C. Chipenda
South African Research Chair in Social Policy, College of Graduate Studies, University of South Africa, Pretoria, South Africa
e-mail: chipec@unisa.ac.za

E. Ndhlovu, C. Chipenda (eds.), *Redefining the Social Policy Discourse*,
https://doi.org/10.1007/978-3-032-24067-5_15

(Backman-Levy & Greene, 2024). GTT emphasises the need for changes in power structures and social relations that govern gender interactions, recognising that true gender equality requires addressing structural constraints affecting all genders, including men's mental health experiences in rural contexts (Backman-Levy & Greene, 2024).

Within this theoretical framework, the problem context reveals that in Zimbabwe and rural South Africa, traditional gender norms have long dictated expected roles and behaviours for men and women (Chitando et al., 2024; Marandure et al., 2023; Mashele et al., 2024; Pearson & Makadzange, 2008; Skovdal et al., 2011). Masculinity is often associated with strength, stoicism, and the ability to endure hardships without showing vulnerability, creating parallel challenges for men's mental health across both contexts (Chitando et al., 2024). Men are expected to be providers and protectors, roles deeply ingrained in cultural expectations and reinforced by social institutions, including family, community and government policies (Ezeugwu & Ojedokun, 2020). These rigid gender norms contribute to psychological distress among men, who are often discouraged from expressing emotions or seeking help when facing personal challenges (Cooper, 2023). In both Zimbabwe and South Africa, this manifests in increased rates of mental health issues among men, including depression, anxiety and substance abuse (Chitando et al., 2024). Secondary evidence from rural South Africa reveals similar patterns, where traditional masculinity norms create comparable barriers to mental health help-seeking behaviours (Mogano et al., 2025). The societal stigma surrounding mental illness, coupled with expectations that men should be strong and self-reliant, creates significant barriers to accessing mental health services across both contexts (Bilsker et al., 2018).

The mental health crisis is compounded by economic and social pressures, including high unemployment rates, poverty and the lingering effects of political instability and social transformation (Cooper, 2023). These stressors disproportionately affect men, who are often expected to be primary breadwinners and are judged harshly when unable to fulfil this role (Uwimana et al., 2023). The result is a vicious cycle where economic hardship, coupled with limited focus on men's mental health in social policies, exacerbates stigma surrounding mental illness, preventing men from seeking help across both contexts.

Critical knowledge and research gaps emerge when examining the current mental health infrastructure in both Zimbabwe and rural South Africa, which is inadequate to meet growing population needs, particularly in rural areas (see Kidia et al., 2017). Many rural districts face significant challenges in health service delivery, with limited mental health facilities, a shortage of trained professionals and a general lack of awareness about mental health issues, especially among men (Pearson & Makadzange, 2008; Skovdal et al., 2011). While gender mainstreaming is common in social policy formulation across Southern Africa, it often fails to account for gender identity dynamics and specific mental health challenges faced by men, especially in rural settings where traditional norms remain most entrenched (Kidia et al., 2017). Zimbabwe's National Gender Policy (2013–2017) emphasises gender equality but largely focuses on women's empowerment, with limited reference to men's mental health or masculinity norms' impact on well-being (Mumbire & Chakabwata, 2024). Similarly, South Africa's gender policies, while more comprehensive, demonstrate gaps in addressing men's mental health needs within rural contexts.

The integrated TSP and GTT framework suggests that this oversight reflects broader challenges of external dependency in African social policy development. Specifically, internationally driven gender mainstreaming approaches often fail to account for local contexts and the multifunctional nature of TSP interventions that comprehensively address gender relations (Chipenda, 2025). Much focus on gender transformation initiatives has been on improving women's status, with less attention given to specific challenges faced by men, especially regarding mental health (Kidia et al., 2017). Without addressing how rigid gender norms harm men across Southern African contexts, true gender transformation cannot be achieved in the region. The failure to include men in these discussions perpetuates harmful stereotypes and limits policy effectiveness in promoting mental health and well-being for all genders (Chitando et al., 2024).

Given these gaps, this chapter explores the dynamic relationship between traditional masculinity norms, mental health, and social policy in Southern Africa, with primary empirical evidence from Chipinge, Zimbabwe, and secondary comparative evidence from rural South Africa. The study addresses these key questions: How do masculinity norms influence men's mental health across Southern African rural communities? What barriers exist in accessing mental health services for men in these contexts? How effective are current social policies in promoting gender transformation regarding men's mental health? The chapter aims to identify gaps in policies and propose strategies for developing more inclusive and equitable frameworks that better support men's mental health needs across the region. Through this comparative Southern African lens informed by integrated TSP and GTT principles, the study advocates for a transformative approach that moves beyond residual welfare approaches to embrace social policy as a productive factor that can drive economic growth, social cohesion and political stability while addressing structural inequalities and transforming gender relations (Mkandawire, 2007). This integrated theoretical approach recognises that social policy interventions must simultaneously address immediate mental health needs while transforming both the structural constraints that perpetuate gender inequality and the gender relations that impact all members of society, contributing to the broader project of decolonising social policy discourse through assertion of African epistemologies and context-specific solutions (Adesina, 2011). The chapter thus offers best practices for fostering a more supportive environment in Zimbabwe and similar rural Southern African contexts, while contributing to the development of indigenous approaches that reflect African values, institutions and development priorities in mental health and gender transformation.

15.2 Literature Review

This literature review examines the dynamic nexus between gender transformation, masculinity norms, and mental health in Southern African contexts. As indicated earlier, it utilises an integrated theoretical framework of TSP and GTT. This integrated approach recognises that achieving meaningful change in men's mental health outcomes requires both structural policy transformation and shifts in gender relations. The review critically analyses how traditional masculinity constructions create barriers to mental health help-seeking, explores emerging gender transformation initiatives and evaluates current social policy frameworks in promoting inclusive mental health approaches for men across the region.

15.2.1 Mental Health Crisis Among Men in Southern Africa

Mental health has emerged as one of the most pressing public health challenges globally, affecting nearly one billion people and representing a significant burden on healthcare systems worldwide (Yusupova, 2023). This crisis is significantly acute in sub-Saharan Africa, where mental health disorders among men have reached alarming proportions, reflecting what TSP theorists would identify as a multidimensional social policy challenge requiring comprehensive intervention beyond traditional welfare approaches (Mkandawire, 2007). The male suicide rate in this region stands at 18 per 100,000, which is significantly higher than the global average of 12.4 per 100,000 (WHO, 2022a). These statistics highlight a deeper crisis extending beyond individual suffering to encompass broader societal and economic implications. It illuminates the interconnected nature of gender relations, mental health outcomes and social policy effectiveness in Southern African contexts.

The situation in South Africa exemplifies this crisis's severity, where sociocultural factors create additional vulnerabilities for men within a dynamic multicultural context. The country has a suicide rate of 23.5 per 100,000 people, has the third--highest suicide rate in Africa, and Africa has higher rates of suicide than any other continent (Sinisi, 2024). In 2019, men accounted for approximately 79% of the 13,774 mental health-related deaths in the country, representing 10,861 deaths (Sinisi, 2024). This disproportionate burden reflects not merely statistical variation but a dynamic nexus of social, cultural and structural factors that systematically disadvantage men's mental health outcomes. The WHO's definition of mental health as a state of well-being in which the individual realises his or her abilities, can cope with the normal stresses of life, work productively and fruitfully, and contribute to his or her community points out that mental health encompasses far more than the absence of mental disorders (Zielke et al., 2023). It represents the foundation of human flourishing and social contribution.

The economic implications are equally significant, with the World Economic Forum estimating the global cost of mental health issues at approximately $2.5

trillion, with projections indicating this figure will rise to $6 trillion by 2030 (WEF, 2025). In Africa, where competing health priorities, severe underfunding and inadequate infrastructure compound these challenges, the situation becomes critical. The WHO reported that 116 million people in the African region were living with mental health conditions in 2022, which is higher than the global average (WHO, 2022b). This burden is significant in South Africa, where current research indicates that nearly one in three people will experience a mental health issue during their lifetime (Harriman et al., 2022).

Despite these troubling statistics, allocations to the national health budget for mental health services remain low. This resource scarcity intersects with multiple barriers that specifically affect men's access to care, including stigma, shortage of mental health professionals, cultural beliefs and limited integration of mental health services into primary care systems. From a TSP perspective, this situation exemplifies how residual welfare approaches fail to address the multifunctional nature of social policy interventions needed to comprehensively tackle mental health challenges while simultaneously transforming gender relations that perpetuate these disparities (Adesina, 2011).

15.2.2 Masculinity Norms as Structural Barriers

The relationship between masculinity and mental health in Southern African contexts represents a dynamic web of cultural expectations, social norms and structural constraints that systematically discourage men from seeking help for mental health challenges. Through a GTT lens, these patterns reflect how rigid gender norms create structural constraints that affect not only women but also men, requiring systemic changes in gender relations to achieve true gender equality (Backman-Levy & Greene, 2024). Traditional constructions of masculinity across Zimbabwe and rural South Africa have been identified as significant barriers to men's willingness to engage with mental health services (Seidler et al., 2016; Chitando et al., 2024; Mankowski & Smith, 2016).

These dynamics are deeply rooted in cultural, social and historical contexts that shape not only individual health behaviours but also broader patterns of mental health outcomes across the region (Agochukwu & Wittmann, 2019; Reihling, 2020). Masculine norms in Southern African cultures emphasise traits such as stoicism, self-reliance, emotional restraint and toughness (Aharon et al., 2024; Mashele et al., 2024; Mankowski & Smith, 2016). These cultural expectations create a contradiction; while men may experience mental health challenges at rates similar to or higher than women, they are simultaneously discouraged from acknowledging these difficulties or seeking appropriate support (Mlambo et al., 2014; Phoebe et al., 2024). This reluctance leads to chronic underutilisation of mental health services and increased vulnerability to severe mental health conditions (Vogel et al., 2011; Agochukwu & Wittmann, 2019). Hegemonic masculinity in Southern African contexts actively discourages vulnerability and emotional expression, leading men to internalise negative beliefs about help-seeking behaviours (Barragán, 2024; Cooper, 2023). In rural South Africa, this manifests through the normalisation of violence as a means of maintaining dominance in interpersonal relationships, creating a cross-cultural cornerstone of masculinity that contributes to elevated rates of intimate partner violence and associated mental health complications (Bateman, 2012).

The paradox of masculinity, where conventional male gender roles encourage behaviours that ultimately compromise men's well-being, has been consistently observed across Southern African rural contexts (Addis & Mahalik, 2003; Chitando et al., 2024; Reihling, 2020). This paradox is further complicated by socioeconomic factors and the enduring legacy of colonialism and apartheid, which have disrupted traditional support systems while reinforcing problematic aspects of masculine identity. From a TSP perspective, these historical disruptions demonstrate how social policies must address not only immediate mental health needs but also the broader structural inequalities and colonial legacies that continue to shape gender relations and mental health outcomes in the region (Mkandawire, 2007).

Social norms within Zimbabwe and rural South Africa that demand conformity to masculine ideals often result in self-stigma, where men perceive seeking help as an admission of weakness or failure (Sheikh et al., 2024; Bilsker et al., 2018; Agochukwu & Wittmann, 2019). Cultural expectations that men should prioritise strength, toughness and self-sufficiency create conflicts with the vulnerability inherent in addressing mental health issues. Consequently, men often delay seeking healthcare services and prefer to cope with stress and mental health challenges through potentially harmful mechanisms such as substance use, especially alcohol (Graeme et al., 2023; Ngwenya & Sumbane, 2023). This preference for self-medication over professional help reflects not individual weakness but systemic failures to create mental health services that are accessible, culturally appropriate and sensitive to masculine identity concerns (Thulani et al., 2023). Many men in these contexts also suffer from inadequate mental health literacy, leading to poor recognition of symptoms and diminished perceived need for professional intervention (Graeme et al., 2023; Pearson & Makadzange, 2008). The reluctance to seek help is intrinsically linked to both the stigma associated with mental health conditions and the perception that seeking help represents a threat to masculine identity.

Zimbabwe presents entrenched masculine ideals deeply rooted in patriarchal structures that emphasise physical strength, economic provision and social dominance (World Bank, 2024; Marandure et al., 2023). The valorisation of physical strength and dominance often leads to aggressive behaviours and gender-based violence, with statistics indicating that 38% of women experience intimate partner violence in Zimbabwe (Dzinamarira et al., 2024). This environment fosters mental health problems among men as they struggle to reconcile emotional needs with societal expectations, creating cycles of violence and poor mental health outcomes that affect entire communities (Bateman, 2012). Similarly, in rural South Africa, masculinity is often linked to dominance and violence, with societal norms that normalise intimate partner violence, creating dynamic interactions between economic pressures and racial and gender factors that perpetuate cycles of abuse (Bateman, 2012; Machisa et al., 2016). The expectation that men serve as primary breadwinners reinforces economic inequalities while limiting women's access to resources and decision-making roles, creating additional pressure on men to maintain economic productivity even while experiencing mental health challenges (Moyo, 2024).

15.2.3 Gender Transformation Pathways

In recent decades, there has been growing recognition across Southern African contexts of the need to move beyond rigid gender binaries toward more fluid and inclusive understandings of gender identity and expression (Morrell & Ouzgane, 2005; Backman-Levy & Greene, 2024). This shift aligns with GTT's emphasis on the need for systemic changes in gender relations that address structural constraints affecting all genders, including men's mental health experiences in rural contexts (Backman-Levy & Greene, 2024). While traditional masculine norms continue to significantly impact mental health outcomes in Zimbabwe and rural South Africa, there is emerging acknowledgement of the need to expand concepts of masculinity to include emotional complexity and support for help-seeking behaviours (Agochukwu & Wittmann, 2019).

This gender transformation agenda has gained traction across various African contexts, driven primarily by feminist activism, queer theory, and growing recognition of the diversity of human experiences (Nyanzi, 2011). GTT posits that achieving genuine gender equality requires systemic changes in gender relations. This goes beyond traditional narrow focuses on women's empowerment to address structural constraints affecting all genders, including men's mental health experiences in rural contexts (Backman-Levy & Greene, 2024). However, the process of gender transformation faces significant challenges in conservative social and cultural environments across Southern Africa. From a TSP perspective, achieving meaningful change requires addressing deeply ingrained cultural norms while simultaneously providing accessible, men-centred mental health services that can support evolving masculine identities through multifunctional policy interventions (see Mkandawire, 2007). Many cultures maintain rigid gender norms that dictate behaviour and expectations, making it difficult for individuals to express non-conforming aspects of gender identity (Damiri, 2023). In multicultural families across the region, traditional roles often clash with modern values, leading to internal conflicts and resistance to change (Damiri, 2023).

In Zimbabwe and similar Southern African contexts, traditional gender norms remain deeply entrenched, with efforts to challenge masculine hegemony often met with significant resistance and backlash (Dziwa et al. 2020). The intersections of race, class and locality further complicate the landscape of gender identity and expression, with rural communities often maintaining more rigid gender scripts than urban areas (Epprecht, 2013). Zimbabwe's Constitution and the National Gender Policy (2013–2017) have sought to promote gender equality and challenge traditional gender norms, yet implementation and enforcement remain limited due to persistent societal resistance to dismantling patriarchal structures (Mumbire & Chakabwata, 2024).

This situation exemplifies the critical knowledge gap. While gender mainstreaming is common in social policy formulation, it often fails to account for specific mental health challenges faced by men, especially in rural settings where traditional norms remain most entrenched. Similarly, while South Africa's gender policies are more comprehensive than those of neighbouring countries, they demonstrate significant gaps in addressing men's mental health needs within rural contexts, reflecting broader challenges of external dependency in African social policy development, where internationally driven approaches often fail to account for local contexts (Chipenda, 2025). The shift toward expanding masculinity concepts could substantially improve mental health outcomes by encouraging men to access necessary resources and support systems while reducing the stigma associated with help-seeking behaviours. However, this transformation requires what TSP theorists advocate for, moving beyond residual welfare approaches to embrace social policy as a productive factor that can simultaneously drive economic growth, social cohesion and political stability while addressing structural inequalities (Mkandawire, 2007). As GTT emphasises, without addressing how rigid gender norms harm men across Southern African contexts, true gender transformation cannot be achieved, as the failure to include men in these discussions perpetuates harmful stereotypes and limits policy effectiveness in promoting mental health and well-being for all genders (Chitando et al., 2024).

15.2.4 Social Policy Frameworks and Mental Health

The intersection of social policy and mental health in Southern African contexts reflects broader global trends toward recognising mental health as both a public health priority and a human rights issue. It also highlights the need for transformative approaches that address structural inequalities rather than merely providing residual welfare responses. Economic analyses suggest that investing in mental health yields substantial returns, with estimates indicating threefold returns for every dollar spent (Gaynor-Brook, 2023). This economic imperative aligns with TSP principles that view social policy as a productive factor for economic growth and social cohesion, conceptualising social policy as 'collective interventions that directly affect social welfare, social institutions, and social relations' (Mkandawire, 2010).

Zimbabwe's social policy framework has evolved significantly since independence, building comprehensive mental health legislation and policies. The Zimbabwe National Strategic Plan for Mental Health Services (2019–2023) serves as the cornerstone of mental health policy, building on previous frameworks including the Zimbabwe Mental Health Policy (2007), the Mental Health Act (1994) and the Zimbabwe Mental Health Plan of Action (1984) (MoHCC, 2019). This strategy outlines ambitious goals for mental health service delivery and integration into primary healthcare. Strategic objectives include improving the quality of care, increasing mental health awareness, empowering communities in managing mental health patients, promoting research and development and reviewing existing legislation (MoHCC, 2019).

Key components of Zimbabwe's approach include decentralisation to increase access to mental health services, especially in rural areas, establishment of mental health units in provincial hospitals, promotion of early identification and treatment of mental illness and provision of rehabilitative and palliative services (MoHCC, 2019). Despite this comprehensive focus, the strategy faces challenges in addressing men's mental health needs within traditional cultural frameworks. Mental health has consistently been recognised as a priority in Zimbabwe, with provisions of the Mental Health Act (1994) included within successive National Health Strategies, demonstrating government recognition of mental health as a public health priority (Chibanda et al., 2016). These policies operate alongside National Gender Policies of 2004, 2013–2017, and the recent 2023–2030 framework, which serve as the country's primary mechanisms for promoting gender equality and transformation (GoZ, 2023). However, as identified in the chapter's problem context, gender mainstreaming in social policy formulation often fails to account for gender identity dynamics and specific mental health challenges faced by men, especially in rural settings where traditional norms remain most entrenched and substance use as a coping mechanism is prevalent.

The Traditional Medical Practitioners Act (Chapter 27:14 of 1996) has established a comprehensive framework for integrating traditional healing into formal healthcare through the Traditional Medical Practitioners Council, operating under the Ministry of Health and Child Care (MoHCC, 2024). Despite this institutional framework, integration between traditional and modern mental health services remains fragmented, especially in rural areas where traditional healers often serve as the first point of contact for mental health concerns, especially for men whose cultural norms favour traditional healing approaches over formal mental health services. The Zimbabwe Mental Health Investment Case, launched in July 2021, represents growing recognition of mental health as a public health priority within Zimbabwe's Special Initiative for Mental Health Strategic Framework (WHO, 2021). This framework combines mental health situation analysis, institutional context analysis and economic analysis to determine investment priorities and projected returns over multiple time horizons, emphasising cross-sectoral engagement including social welfare, education, civil society organisations and the private sector (WHO, 2021). However, the investment framework falls short in addressing masculine-specific mental health challenges, especially in rural contexts where traditional norms most strongly influence help-seeking behaviours.

In contrast to Zimbabwe's fragmented approach, South Africa has developed a more comprehensive legal and policy framework for mental health and gender wellbeing. The South African Constitution (1996) provides significant legal protections through Section 9 (Equality Clause), which prohibits unfair discrimination, including on gender grounds (RSA, 1996). It also includes Section 27 (Healthcare Rights), which guarantees access to healthcare services, including mental health, and Section

10 (Human Dignity), which protects human dignity as fundamental to mental health care and stigma reduction (RSA, 1996). South Africa's Mental Health Care Act 17 of 2002, implemented in 2004, has been recognised as one of the most progressive pieces of mental health legislation globally, focusing on rights-based approaches that emphasise user rights and least restrictive treatment methods (RSA, 2002). The National Mental Health Policy Framework and Strategic Plan 2023–2030 represents South Africa's most recent comprehensive policy approach, with specific attention to gender issues in mental health, stigma reduction and community-based interventions (RSA, 2022). The policy framework recognises that poor mental health among men is linked to high rates of gender-based violence, highlighting the interconnected nature of gender issues and mental health outcomes.

Despite South Africa's more progressive legal framework, both countries face similar challenges in rural contexts where traditional masculine norms continue to influence mental health outcomes. The policy landscape across Southern African contexts requires significant strengthening to address both resource limitations and social dynamics surrounding mental health care, necessitating TSP approaches that move beyond residual welfare models to embrace social policy as integral to development processes while simultaneously transforming gender relations that impact mental health outcomes for all members of society (see Mkandawire, 2007; Adesina, 2011).

15.3 Theoretical Framework

This study employs an integrated theoretical framework combining GTT and TSP theory to analyse the relationship between masculinity norms, mental health and social policy. This integrated approach recognises that meaningful change in men's mental health outcomes requires both shifts in gender relations and comprehensive transformation of social policy approaches that move beyond traditional welfare models to address structural inequalities.

15.3.1 Gender Transformation Theory

GTT represents a paradigmatic shift from conventional gender approaches, providing a comprehensive framework for examining how deeply entrenched gender norms influence social structures, individual behaviours and institutional policies (Hundera & Mudde, 2024). GTT posits that achieving gender equality requires systemic change in gender relations, transcending the traditional narrow focus on women's empowerment (Backman-Levy & Greene, 2024). At its foundation, the theory emphasises the redistribution of power and resources among genders, directly challenging existing hierarchical structures and power imbalances (MacArthur et al., 2022). Unlike traditional gender-focused approaches that primarily target women's empowerment through education and legal reforms, GTT advocates for active engagement of all genders in promoting gender justice and accountability (OECD/World Health Organization, 2022). This inclusive approach recognises that sustainable gender equality requires collective participation and shared responsibility across the gender spectrum. The theory distinguishes itself by emphasising structural focus, inclusive engagement, and transformative change, addressing structural and social determinants of gender inequality rather than solely targeting individual behaviours (Fisher & Makleff, 2022).

The framework advocates for meaningful inclusion of diverse gender identities and experiences in the development process (Ferdous, 2024), ensuring that intervention strategies reflect the complexities of gender dynamics across various sociocultural contexts. Instead of merely accommodating existing gender norms, GTT seeks to reshape power dynamics and collective responsibilities among all genders. For this study, GTT provides a sophisticated lens for understanding how traditional masculinity shapes mental health outcomes and help-seeking behaviours in Zimbabwe, enabling in-depth exploration of men's interactions with mental health services, especially how these experiences are mediated by prevailing gender norms and cultural expectations.

15.3.2 *Transformative Social Policy Theory*

TSP represents a critical theoretical approach of significant relevance for social policy discourse in the African context and the global South. Developed through monumental research by the United Nations Research Institute for Social Development (UNRISD), TSP has emerged as a sophisticated scholarly intervention. It is designed to address the systemic weaknesses inherent in conventional social policy approaches predominantly applied in African and global southern developmental contexts (Mkandawire, 2004; Adésínà, 2015). TSP departs from earlier social policy approaches such as Social Dimensions of Adjustment, Social Risk Management and Transformative Social Protection, which it argues were primarily rooted in neoliberal philosophies (Tom & Banda, 2023). These approaches were characteristically narrow, emphasising targeted interventions for the 'deserving poor' through restrictive safety net mechanisms, critically failing to interrogate or address the structural causes of inequality, marginalisation and poverty (Adesina, 2020; Chipenda, 2023). Available evidence demonstrates their inability to generate meaningful developmental transformation and broader societal well-being.

The TSP approach offers a prophylactic perspective on social policy, placing emphasis on preventative interventions that address vulnerabilities proactively rather than reactively (Adesina, 2007; Chipenda, 2024, 2025). Drawing upon Fischer's (2018) critical conceptualisation, the framework exposes and challenges social policy landscapes in developmental contexts in the global south, which it argues are characterised as being 'segmented, stratified, and segregated', offering a divergent course from conventional social policy paradigms detached from structural transformation imperatives. TSP is anchored on five multiple functions: production, reproduction, social protection, redistribution and social cohesion or nation building (Mkandawire, 2004, 2007). Through these multiple and interrelated functions, TSP extends beyond mere income maintenance and social security, aiming to enhance the productive potential of society members while balancing the burden of reproduction with other social tasks. It seeks to address economic inequalities by redistributing resources, protecting individuals against life uncertainties and contributing to social cohesion and nation-building (Adésínà, 2015; Mkandawire, 2004). This multifunctional approach thus conceptualises social policy as 'collective interventions that directly affect social welfare, social institutions, and social relations' (Mkandawire, 2010).

15.3.3 Integration of GTT and TSP Frameworks

The integration of GTT and TSP theory provides an analytical framework that is suited to examining men's mental health challenges in rural Southern Africa. This integrated approach recognises that addressing men's mental health requires both transformation of gender relations and comprehensive policy reform that moves beyond residual welfare approaches to embrace social policy as integral to development processes. The complementary nature of these frameworks is evident in several key areas. First, both theories emphasise structural transformation over individual-level interventions, recognising that sustainable change requires addressing systemic inequalities and power imbalances. GTT's focus on transforming gender relations aligns with TSP's emphasis on addressing structural causes of inequality, marginalisation, and poverty. Second, both frameworks advocate for inclusive approaches that consider diverse experiences and identities, with GTT promoting engagement of all genders and TSP emphasising comprehensive social policy that addresses multiple societal needs simultaneously. Third, the integration addresses the multifunctional nature of social interventions needed to tackle men's mental health challenges. While GTT provides insights into how rigid gender norms create barriers to help-seeking, TSP offers a framework for developing comprehensive policy responses that can simultaneously address mental health needs, transform gender relations, and contribute to broader developmental objectives. This integrated approach is relevant for understanding how traditional masculinity norms intersect with mental health policy implementation in rural contexts, where cultural factors significantly influence service access and utilisation.

The theoretical integration also addresses the historical and contextual factors that shape men's mental health experiences in Southern Africa. GTT's recognition of how colonial legacies and structural inequalities influence gender relations complements TSP's emphasis on addressing the enduring impacts of historical disruptions on social policy effectiveness. This combined framework enables analysis of how policy interventions must simultaneously address immediate mental health needs while transforming both the structural constraints that perpetuate gender inequality and the gender relations that impact all members of society. In addition, the integration supports the development of context-sensitive interventions that reflect African values, institutions, and development priorities. TSP's emphasis on moving beyond externally imposed policy models aligns with GTT's recognition of the need for culturally appropriate approaches to gender transformation. This integrated framework thus provides a foundation for developing inclusive and equitable policies that recognise mental health as a crucial component of social equity while addressing the specific challenges faced by men in rural Zimbabwe contexts.

15.4 Methodology

The study employed a mixed-methods approach, combining primary qualitative data collection with secondary data analysis. It aimed to explore the intersections of mental health, masculinity, and social policy in rural Southern African contexts, using primary empirical evidence from Chipinge, Zimbabwe, and secondary comparative evidence from rural South Africa. This integrated approach aligns with the TSP and GTT theoretical frameworks by examining both immediate experiences and broader structural patterns across the region. The primary qualitative research design allowed the researchers to engage deeply with participants' experiences, emotions and perceptions, providing rich and contextual insights (Dodgson, 2017). For this study, this approach allows for an in-depth understanding of men's lived experiences and perceptions regarding mental health, service utilisation and gender norms. The study also utilised the qualitative research design in order to capture the contextual factors influencing men's mental health help-seeking behaviours and the impact of traditional masculinity on these behaviours.

The secondary data component involved systematic analysis of existing literature, policy documents, and research findings from rural South African contexts to provide comparative perspectives and validate emerging patterns from the primary data. This secondary analysis specifically examined gender policies, mental health frameworks and masculinity studies from rural South Africa. It aimed to identify similarities and differences with the Zimbabwean context, thereby strengthening the regional applicability of findings and supporting the development of context-specific solutions aligned with African epistemologies. The research design included semi-structured interviews and focus group discussions that provided flexibility in exploring participants' perspectives while ensuring that key topics were covered (Aveling et al., 2015). The study was conducted in Chipinge district, which is located in the south-eastern part of Zimbabwe, in Manicaland Province. The district is affected by broader socio-economic issues, including high unemployment rates and poverty, which, when combined with societal expectations, expose men to a cycle of distress. Mental health infrastructure and service provision are also significant challenges in the district, and this impacts overall access to mental health services and support. This context provides a backdrop for exploring the intersection of masculinity and mental health, as traditional beliefs and modern challenges interact within the lives of men in Chipinge.

Participants for the study were selected using purposive sampling. The sample consisted of 40 men aged 18 and above, residing in Chipinge with varying educational levels, socio-economic backgrounds and marital status. Informed by quota sampling and referral sampling techniques, the study areas included Chipinge town, Paidamoyo, Mount Selinda, Musani, Chibuwe and Checheche. Participants were recruited from word-of-mouth referrals and local community organisations with interests and current engagements on mental health services, such as Platform for Youth and Community Development Trust (PYCD), Christian Blind Mission Zimbabwe, Medecins Sans Frontieres Doctors Without Borders. The inclusion criteria focused on men who have either sought mental health services or have been affected by mental health issues. The sampling techniques used are significant as they allow for targeted participant selection, exploration of diverse perspectives and are essential to reach the hard-to-get population (Aveling et al., 2015).

A thematic content analysis approach was utilised for data analysis and interpretation, proceeding through systematic stages of preliminary coding, iterative theme development and cross-validation. The analytical framework involved initial open coding of collected texts, followed by deep analysis that cross-referenced themes across different data sources and an examination of emerging patterns and contradictions. The final synthesis linked findings to the TSP and GTT theoretical frameworks, identifying critical dynamics and structural vulnerabilities impacting men's mental health across rural Southern African contexts. The analysis integrated primary qualitative data with secondary comparative evidence to develop a comprehensive understanding of how traditional masculinity norms influence mental health experiences and policy effectiveness across the region.

Rigorous ethical protocols were implemented throughout the research process. These included informed consent procedures, participant confidentiality measures and careful protection of vulnerable populations. The methodology maintained a critical awareness of potential biases, including power dynamics in narratives, resource constraints and potential limitations in data representation, especially when integrating primary and secondary data sources from different national contexts.

15.5 Findings

15.5.1 Masculinity Perceptions and Mental Health Challenges

15.5.1.1 Traditional Masculinity and Emotional Suppression

Primary evidence from Chipinge district revealed that deeply ingrained traditional masculinity norms significantly shape men's identities and mental health experiences. Participants consistently described intense societal pressure to embody masculine traits of strength, endurance and autonomy, with vulnerability being perceived as essentially feminine and therefore unacceptable. A Chipinge participant explained his upbringing and its lasting impact.

> Growing up, I was taught that men do not cry and they do not show emotions. My father and uncles would always tell us boys that showing tears or weakness is what separates men from women. I felt like I had to handle everything on my own, every problem, every challenge that came my way. Asking for help seemed unmanly, like I was failing in my duties as a man. I realized it too late that being tough does not make me happy. It just made me lonely and weak, isolated from everyone around me, including my own family.

This emotional suppression was further reinforced by cultural expectations of self-sufficiency, with participants indicating that these shared cultural ideas dictate their self-presentation, social interactions, and responses to challenges. The compulsion to fulfil provider and protector roles was quite evident, with another Chipinge participant reflecting on the burden of masculine expectations.

> To be honest with you, expressing my real emotions is often seen as a sign of weakness in our community. We are taught from when we are young boys to be strong and to hide our struggles, and this is what separates real men from weak ones. It is really difficult to admit when we are hurting, for me that makes me feel like I am failing my family and my role as a man. People will say you are not man enough if you show any form of vulnerability.

These findings align with research conducted among Northern Sotho male youth in the Limpopo Province of South Africa. The research established that masculine culture is significantly centred on socialisation, gender roles and behavioural expectations (Mogano et al., 2025). The cultural construction of masculinity mirrors patterns observed across other South African contexts, where elderly men and fathers significantly influence their sons' masculine identity formation through direct grooming and indirect household influence (Mogano et al., 2025). Research from colleges across KwaZulu-Natal, Limpopo, Mpumalanga, Gauteng and Eastern Cape Provinces also confirmed that men's childhood socialisation significantly influenced their ability to express emotions when troubled (Sikweyiya et al., 2025). The restricted emotionality creates cycles of internalised stress and prevents help-seeking behaviours, as evidenced by cultural practices like traditional circumcision (*koma)* among Northern Sotho communities, where failure to participate results in peer ridicule and questioning of manhood (Mogano et al., 2025). This emotional suppression is reinforced by findings from the Waterberg District in Limpopo Province, where black men's adherence to masculine social norms such as emotional control and self-reliance makes it challenging for them to seek help when in need (Masemola et al., 2022).

15.5.1.2 Anxiety, Stress and Economic Pressures

Economic hardship emerged as a primary trigger for anxiety and stress among Chipinge participants, with unemployment and limited economic opportunities creating psychological distress. Participants described feeling overwhelmed by the pressure to maintain facades of strength while struggling with financial inadequacy. A Chipinge participant detailed his experience with anxiety and the realisation of his mental health struggles.

> I always feel I must keep it all together at home, at work, and in my mind. The pressure is enormous, especially when you are struggling to provide for your family. I barely meet all their needs, and sometimes I must choose between buying food and paying school fees. I did not even realise how stressed I was until I started having panic attacks in the middle of the night. I thought I was just being a man, you know, handling my responsibilities. I always thought I had to tough it out for myself and my family, of course, but my body was telling me otherwise.

The disconnect between societal expectations and economic realities created additional stress, with participants feeling misunderstood even by their partners. Another Chipinge participant expressed his frustration with family expectations.

> I try to explain to my partner how stressing it is to try and make ends meet in a crippling economic situation like this, but she always says all she wants is for me to own up and be a father to my kids. It feels like she does not see how hard I am trying. She does not understand that I wake up every morning worried about where the next meal will come from, how I will pay rent, how I will keep the lights on. The stress is eating me alive, but I cannot show it because that would make me look weak.

This finding is corroborated by research from South Africa, which identified employment as crucial for attaining "real man" status, with a traditional masculine ethos requiring men to be independent breadwinners (Mogano et al., 2025). Evidence from Northern Sotho communities supports this, showing that unemployment negatively affects mental health and leads to shame and inadequacy, with unemployed men lacking respect due to their inability to provide for families (Mogano et al., 2025). Maladaptive coping mechanisms were common among Chipinge participants, with many resorting to alcohol and substance use for temporary relief. This pattern is well-documented in South African research, where men turn to harmful substances when facing psychological stress and gender role conflicts (Mogano et al., 2025; Sikweyiya et al., 2025). Research from South African college campuses documented that men engage in negative coping strategies, including harmful alcohol use and substance use, when dealing with bottled-up emotions (Sikweyiya et al., 2025).

15.5.1.3 Depression and Grief Processing

Depression emerged as a significant mental health challenge among Chipinge participants, with many reporting persistent hopelessness, lack of motivation, and sadness. The impact of traditional masculine norms on depression experiences was evident, as participants felt compelled to suppress emotions during profound losses. A Chipinge participant's experience following his wife's death illustrates the devastating impact of masculine expectations on grief processing.

> When my wife passed away 3 years ago, I was devastated. I felt like my whole world had collapsed, like I had lost everything that mattered to me. I wanted to talk about my grief, to

> express how much I missed her, how empty the house felt without her laughter, how I struggled to sleep in our bed alone. But I felt I had to be strong for our children, especially my teenage son, who was also struggling. My friends would always say I had to hold it together for the children, that crying would confuse them and make them think I couldn't take care of them. So, I buried my feelings deep inside. Till today it feels like I am carrying a weight I cannot bear, and sometimes I wonder if I will ever feel normal again.

This emotional suppression had lasting psychological consequences, with participants describing the inability to process grief effectively. Another Chipinge participant reflected on the long-term impact of suppressing emotions.

> After my father's death, I was expected to step up as the man of the house immediately. I was only 19 years old, but everyone expected me to be strong, to comfort my mother and siblings, to handle all the funeral arrangements. I never got the chance to properly mourn my father. Even now, years later, I still feel that pain inside me. It is like a wound that never healed because I never allowed myself to feel it properly.

These findings align with research from the Waterberg District in South Africa, which confirms that black men demonstrate awareness of depression symptoms, including sleep disturbances, loneliness, excessive stress and concentration difficulties (Masemola et al., 2022). The emotional suppression is reinforced by cultural beliefs documented across South African contexts, where men are taught that 'only weak boys and women cry' and that crying results in loss of respect (Mogano et al., 2025). Research from Northern Sotho communities confirms that strict adherence to masculine norms can discourage men from seeking mental health help, contributing to higher rates of untreated mental health conditions such as depression and anxiety (Mogano et al., 2025).

15.5.1.4 Anger and Trauma Responses

Anger was identified as a significant manifestation of underlying mental health struggles among Chipinge participants, primarily triggered by economic hardships and societal expectation pressures. Economic instability, unemployment, and job loss created feelings of inadequacy that manifested as frustration and anger. A Chipinge participant's experience after losing his job illustrates this connection.

> When I lost my job at the tea estate, I felt like I had let my whole family down. I had worked there for 8 years, and suddenly I was told they were cutting staff. Every time I looked at my wife and children, I got angry at myself and the situation. I felt like I was failing them, like I was not man enough to keep my job and provide for them. I was taught that real men do not show weakness, they do not cry, they do not complain. So when I got overwhelmed with the situation, when the bills kept coming, and there was no money, I got angry instead of admitting that I was struggling. I felt that was the only emotion I was allowed to show, the only option I had.

Post-traumatic stress disorder (PTSD) emerged from various traumatic experiences among Chipinge participants, including job loss, failure to meet familial expectations, exposure to domestic violence and loss of loved ones. Domestic violence experiences were particularly traumatic, with one Chipinge participant describing ongoing abuse and its lasting psychological impact:

> Ever since I lost my job and relocated to stay here, my wife has exposed me to a cycle of abuse for the decision I had made. She tells me I am worthless countless times, that I am not a real man because I can't provide for the family. She belittles me in front of our children, in front of neighbours, and it strips me of my honour, my dignity as a man. The worst day that still makes me scared till today is when she saw me at the shops sitting next to a random woman, just talking casually, and she came and beat the lady up and cut my arm with a broken bottle. She said she thought I was having an affair, but I was just talking to someone. I have all that drama and humiliation to deal with today. The most embarrassing thing a man can receive is such a treatment in front of other men, in front of the whole community. I find it difficult to recover from what she did and still does to me. I wake up sometimes in cold sweats, reliving that day.

This finding is supported by research from South African contexts showing that men's struggle to reconcile personal identity with societal expectations leads to unhealthy coping strategies, including aggression, as they attempt to assert masculinity (Mogano et al., 2025). The connection between unemployment and anger is well-documented in Northern Sotho communities, where unemployed men face disrespect due to their inability to provide for families, significantly impacting mental well-being (Mogano et al., 2025). The impact of trauma on daily functioning was significant, with Chipinge participants withdrawing from social interactions and fearing symptom misunderstanding. This social withdrawal pattern is consistent with South African college research where men bottle up emotions and avoid help-seeking, even risking poor mental health outcomes (Sikweyiya et al., 2025).

15.5.2 Help-Seeking Behaviours and Barriers

15.5.2.1 Reluctance to Seek Professional Help

Chipinge participants exhibited strong reluctance to seek professional mental health services, with cultural norms surrounding masculinity promoting the idea that men should handle problems independently. This led to shame and weakness associations with help-seeking behaviours. A Chipinge participant's reflection on traditional sayings and cultural expectations highlights this reluctance.

> Our tradition says '*kufa kwemurume kubuda hura*' (a man is really dead when his intestines are laid bare). My grandfather used to tell me this whenever I faced difficult challenges, and it keeps me pushing even when I feel like giving up. There are times that I feel I am overstretching myself so bad and that I really need to stop, to ask for help, to admit that I cannot handle everything. But then I remember that I am a man and I need to keep pushing, keep fighting. I fear becoming '*mbwende*' or '*gwara*' (coward) in the eyes of my community, so I struggle with my problems alone, even when I know I need help.

Instead of professional services, Chipinge participants relied on informal support networks, confiding in close family members and friends. This preference stemmed from the belief that discussing mental health issues with trusted individuals feels safer and less stigmatising than seeking professional help. A Chipinge participant explained his approach to seeking support.

> When I am struggling, I prefer to talk to my older brother or my close friend who has been through similar experiences. They understand what it means to be a man in our community, they know the pressures we face. Going to a stranger, to a professional, feels like admitting defeat, like saying I cannot handle my own problems. With my brother, I can share my struggles without feeling like I am being weak or failing as a man.

This finding is strongly supported by research across multiple South African contexts, including studies from Limpopo, Mpumalanga, and Eastern Cape provinces, which consistently document men's help-seeking avoidance behaviour (Mogano et al., 2025; Masemola et al., 2022; Molokoane & Modipane, 2025; Sikweyiya et al., 2025). Research from Northern Sotho communities confirms that participants support masculine self-reliance principles over seeking external assistance, with help-seeking perceived as weakness (Mogano et al., 2025). Research from Ehlanzeni district in Mpumalanga province corroborates this preference, where young men found it easier to confide in families and friends due to accessibility and comfort (Molokoane & Modipane, 2025). Help-seeking behaviours among Chipinge participants were predominantly crisis-driven, with participants only seeking assistance when mental health issues escalated to critical points. This pattern is consistent with Mpumalanga province findings where men only consulted mental health professionals when referred by physicians, when family offered payment or when transferred to mental health institutions (Molokoane & Modipane, 2025).

15.5.2.2 Societal Stigma and Awareness Barriers

Societal stigma surrounding mental health emerged as a crucial barrier among Chipinge participants, rooted in cultural norms defining mental health issues as weakness and failure. Participants expressed shame about mental health struggles and feared harsh judgment from disclosure. A Chipinge participant articulated this fear.

> I know I should seek help; almost everyone close to me knows my struggles, but the fear of being judged if I decide to seek real help holds me back. In this community, admitting to mental health struggles feels like showing weakness, like admitting you cannot handle your responsibilities as a man. I cannot bear the thought of what people might say, how they might look at me differently, how they might question my ability to be a husband and father. I would rather suffer in silence than face that judgment.

Lack of awareness regarding mental health issues and available services constituted another significant barrier among Chipinge participants. Many reported limited knowledge about mental health symptoms and available resources, leading to misconceptions about mental health conditions. A participant revealed his lack of understanding.

> I did not really understand what anxiety or depression is until someone highlighted it to me in a jocular manner during a casual conversation. I just thought I was stressed out from work and family responsibilities. Mental health was just a buzzword in my head, something I heard on the radio or television. I also thought talking about mental health was something that only women did, that it was a feminine thing. I did not realise it could be okay for men to do that, that it was normal for men to struggle with these things too.

This knowledge gap led to misconceptions about mental health, with participants believing mental health issues were personal failures rather than legitimate health concerns. Another respondent reflected on his misconceptions.

> I used to think that if you were having mental health problems, it meant you were weak, that you could not handle life like other men. I thought it was a choice, that you could just decide to be strong and everything would be fine. I did not understand that it was a real health condition, like having diabetes or high blood pressure, something that needed proper treatment and care.

These findings are supported by research from multiple South African contexts. The stigma barrier is consistently documented across South African contexts, with research from Limpopo, Mpumalanga and multiple provinces confirming stigma as a primary help-seeking obstacle (Masemola et al., 2022; Molokoane & Modipane, 2025; Sikweyiya et al., 2025). Research from Ehlanzeni district identifies two stigma types: public stigma reflecting community stereotypes, and self-stigmatisation where individuals internalise negative stereotypes (Molokoane & Modipane, 2025). The awareness barrier is supported by Waterberg District research showing that black men had a limited understanding of pharmacological treatments and the benefits of treatment options (Masemola et al., 2022).

15.5.2.3 Infrastructural and Policy Implementation Challenges

Mental health infrastructural inadequacies emerged as pressing barriers among Chipinge participants, with existing healthcare infrastructure often overlooking mental health concerns. Participants highlighted that rural communities face significant accessibility and availability issues, with clinics poorly equipped to provide the required mental health services. A respondent described the accessibility challenges.

> Rural communities like ours usually face accessibility and availability issues when we talk of mental health services. The clinics are not well equipped and resourced to provide a set of required mental health services especially for men. Our clinic has basic medical supplies, but when it comes to mental health, there is nothing. Another problem is that in some areas across the district one must walk for about 9 km to reach services such as the clinics, mental health facilities, police stations among other services. By the time you walk that distance, especially when you are already struggling mentally, you might give up on seeking help.

The gendered nature of service provision was problematic, as Chipinge participants noted that most services target women and children, with male-focused support groups concentrated in urban areas. A respondent expressed his frustration with the lack of male-focused services.

Mental health services in our country always seem to be underfunded. Our clinics prioritise physical health, so mental health programs suffer. There are very few support groups or programs meant to address mental health issues for men. Most services are for women and children, for mothers and their babies. Most male focused mental health support groups are found in big cities like Harare and Bulawayo, making them inaccessible for all of us here in rural areas. This actually means few men have access to adequate mental health services that understand our specific needs and challenges.

Poor enforcement of mental health policies emerged as another significant barrier among Chipinge participants. While policies like the National Mental Health Strategic Plan of 2019–2023 exist, their practical implementation often falls short. A key informant from Chipinge highlighted the gap between policy and practice.

The saddening part is that we have policies that talk of mental health, but they do not translate into real change on the ground. It always feels like they are just words on paper. The policies look good when you read them, but when you go to the clinic or try to access services, nothing has changed. It is the same old story of neglect and lack of resources.

The absence of clear guidelines for accessing mental health services created additional confusion and frustration among Chipinge participants. A participant described his experience navigating the system.

There are no clear guidelines communicated to people, specifically men, on how to access mental health services. You are left with just yourself to navigate through all these discriminatory and ignorant spaces to find mental health services. Nobody tells you where to go, who to talk to, what documents you need, or how the process works. This is confusing and frustrating, and it makes you want to give up before you even start.

These findings are widely supported by research across South African contexts. The infrastructural barrier is consistently documented, with research highlighting inadequate mental health infrastructure as a significant care obstacle (Masemola et al., 2022). Research from South African college campuses confirms that men feel available services do not resonate with their needs, highlighting the gendered nature of service provision (Sikweyiya et al., 2025). The accessibility challenge is corroborated by South African research indicating substantial treatment gaps, with very few people receiving needed care due to a significant lack of access (Masemola et al., 2022).

15.6 Discussion

The intersection of masculinity and mental health in Chipinge district reveals dynamics shaped by traditional gender norms and socio-economic pressures, reflecting broader patterns across Southern African rural contexts. Utilising the integrated TSP and GTT framework, this chapter examines how systemic change in gender relations is necessary for addressing mental health challenges confronting men. The integrated theoretical approach's emphasis on structural determinants of gender norms provides a critical lens for understanding how traditional masculinity shapes mental health outcomes and help-seeking behaviours in rural Zimbabwe, with comparable patterns documented in rural South African contexts, The findings from Chipinge demonstrate that the integrated TSP-GTT framework's focus on structural determinants effectively explains how traditional masculinity norms constrain men's emotional expression while perpetuating mental health stigma. The TSP dimension's emphasis on social policy as a multidimensional instrument for transformation, combined with GTT's focus on redistributing power and resources among genders, directly corresponds with participants' consistent descriptions of pressure to embody traits of strength and stoicism, leading to emotional suppression that triggers psychological distress. These patterns mirror findings from rural South African contexts, where Northern Sotho male youth experience similar socialisation pressures that restrict emotional expression.

Participants' experiences revealed that societal expectations of being primary providers create significant anxiety and feelings of inadequacy, especially when exacerbated by high unemployment rates. This economic strain affects both individual mental well-being and family dynamics, patterns that are consistently documented across rural South African communities where unemployment negatively impacts men's mental health and leads to shame and inadequacy. The convergence of these findings across Southern African contexts confirms how institutional practices and social norms perpetuate gender disparities in mental health outcomes. The economic pressure-mental health nexus identified in Chipinge creates a cycle of hopelessness that aligns with structural determinants of gender inequality highlighted in the integrated theoretical framework.

The documented prevalence of depression, anxiety, and PTSD among Chipinge men provides clear evidence of the broader societal implications of these pressures across Southern African rural contexts. Participants' accounts demonstrated that adhering to rigid masculine norms consistently leads to maladaptive coping mechanisms, including alcohol and substance use, further complicating the mental health landscape. These patterns are corroborated by research from South African college campuses, where men engage in negative coping strategies, including harmful alcohol and substance use, when dealing with bottled-up emotions. The grief processing challenges identified among Chipinge participants, especially the inability to mourn effectively due to masculine expectations, align with broader patterns documented in rural South African contexts where men are taught that only weak boys and women cry. The Chipinge study's identification of these patterns is essential for developing targeted interventions that address both symptoms and root causes of mental health challenges across the Southern African region.

The reluctance of Chipinge men to seek professional mental health services emerged as a critical challenge that reflects broader Southern African patterns. Cultural stigma surrounding mental health, as evidenced by participants' experiences and corroborated by research from Limpopo, Mpumalanga and Eastern Cape provinces, discourages open discussions, with many expressing specific concerns about judgment and being labelled weak. This reluctance is consistently documented across South African contexts, where men's help-seeking avoidance behaviour stems from masculine self-reliance principles and perception of help-seeking as weakness. This pattern validates the need for systemic change as outlined in the integrated TSP-GTT framework, which emphasises that social policy interventions must simultaneously address immediate mental health needs while transforming structural constraints that perpetuate gender inequality. While participants relied on informal support networks for some safety, a pattern also documented in rural South African contexts, where men prefer confiding in families and friends, they often lack the expertise necessary for effective mental health care across both contexts. This highlights the importance of creating accessible, culturally sensitive mental health services that encourage open dialogue. The findings from both Chipinge and rural South African contexts suggest that promoting awareness and understanding can help change the narrative around mental health and masculinity, making it easier for men to seek help without fear of stigma across the Southern African region.

15.7 Recommendations for Policy and Practice

Given the Chipinge findings and their alignment with broader Southern African patterns, there is an urgent need for policies that promote gender transformation and mental health awareness across the region. The TSP framework's conceptualisation of social policy as a multidimensional instrument for societal transformation suggests that current social policies must evolve to recognise the connection between economic pressures and mental health, incorporating economic empowerment initiatives alongside mental health support services. This moves beyond residual welfare approaches to embrace social policy as integral to development processes, requiring comprehensive systemic change and development of inclusive, equitable policies that view mental health as integral to social equity. The Chipinge study, supported by South African evidence, demonstrates that effective policy development requires a participatory approach engaging male community members, traditional leaders, and healthcare practitioners. Multi-stakeholder dialogues on gender and mental health are essential, grounded in regular community needs assessments and intervention effectiveness evaluations across Southern African contexts.

From a TSP perspective, Zimbabwe and similar Southern African contexts have several policy gaps regarding mental health issues, with health being key in any country's social policy architecture. The TSP framework's emphasis on social policy as collective interventions that directly affect social welfare, social institutions and social relations suggests that addressing these gaps is imperative, with policy reforms focusing on several key areas. Based on the Chipinge study, corroborated by South African evidence, and guided by the integrated TSP-GTT framework, the following represent policy imperatives within the region's social policy architecture: First, integration of gender-transformative approaches into mental health policies must address masculine norms and their impact on help-seeking behaviours. Second, stronger coordination mechanisms between traditional and modern mental health services are needed, especially in rural areas where traditional norms remain most entrenched across Southern African contexts. Third, capacity building for healthcare providers should include training in gender-sensitive approaches to mental health care. Finally, monitoring and evaluation systems must be strengthened to track gender-transformative mental health intervention effectiveness across the region.

Success in policy implementation requires addressing structural barriers, including resource constraints and cultural resistance to changing gender norms. The integrated TSP-GTT framework suggests that these policy considerations must be understood within Southern Africa's broader development challenges and the global mental health landscape. Resource constraints, competing health priorities, and deeply entrenched cultural norms all influence policy implementation effectiveness across the region. Addressing the masculinity-mental health intersection can potentially contribute to positive outcomes when supported by appropriate policy frameworks and implementation strategies, considering both global best practices and local cultural contexts that reflect African values, institutions and development priorities.

Service delivery innovation is crucial for Southern African contexts. Countries in the region and relevant stakeholders should focus on establishing male-friendly mental health spaces and integrating mental health services within primary healthcare systems. Developing mobile mental health units, if resources permit, can significantly help address access challenges in rural areas. Training healthcare providers in gender-sensitive approaches and developing networks of community mental health workers and peer support programs is essential across the region. Educational initiatives through community workshops on masculinity and mental health, school-based programs promoting emotional literacy and public awareness campaigns challenging stigma should be prioritised.

From the TSP perspective, structural healthcare policy reforms in Southern African countries are essential to be responsive to contemporary dynamics, local and international priorities, and best practices. The TSP framework's emphasis on social policy as a productive factor that can drive economic growth, social cohesion, and political stability while addressing structural inequalities suggests that, at all levels, mental health must be prioritised as an important issue. The study confirms that establishing male support groups and mentorship programs is essential across the region. Economic empowerment initiatives linked to mental health support are important for addressing the interconnected nature of financial and psychological well-being documented among participants across Southern African contexts.

15.8 Conclusion

The Chipinge study, situated within broader Southern African contexts, highlights the dynamic relationship between masculinity, mental health and social policy across the region. The findings reveal that traditional gender norms shape men's perceptions of vulnerability and their reluctance to seek help for mental health issues across both Zimbabwe and South Africa. These rigid constructs of masculinity, which valorise emotional stoicism and self-reliance, not only exacerbate mental health challenges such as anxiety, depression and PTSD but also create significant barriers to accessing necessary support. The reluctance of men across Southern African rural contexts to engage with mental health services stems from pervasive societal stigma, which discourages open discussions about mental health and reinforces harmful stereotypes about masculinity. This situation demonstrates the urgent need for a paradigm shift in how mental health is addressed within gender transformation contexts across the region.

Current social policy inadequacies across Southern African contexts are evident, as they often prioritise women's empowerment while neglecting men's unique mental health needs. Policy frameworks across the region, while well-intentioned, fail to adequately address the masculinity-mental health intersection, perpetuating cycles of stigma and silence around men's mental health issues. This oversight is concerning, given the documented rates of mental health problems among men, which is compounded by economic pressures and societal expectations. The study advocates for a transformative approach integrating mental health considerations into gender policies, ensuring men are included in discussions about mental health and well-being across Southern African countries. The study offers actionable recommendations for policymakers and stakeholders to create more inclusive frameworks addressing men's mental health needs. The integrated TSP-GTT framework suggests that addressing these challenges holistically is essential for creating equitable societies where all individuals, regardless of gender, can access mental health support without fear of stigma or judgment across the region. This contributes to the broader project of developing indigenous approaches that reflect African values, institutions and development priorities in mental health and gender transformation.

References

Addis, M. E., & Mahalik, J. R. (2003). Men, masculinity, and the contexts of help seeking. *The American Psychologist, 58*(1), 5–14. https://doi.org/10.1037/0003-066x.58.1.5

Adesina, A. (2020). Performance and sustainability overview of alkaliactivated self-compacting concrete. Waste Dispos. Sustain. Energy 1 (3). https://doi.org/10.1007/s42768-020-00045-w

Adesina, J. O. (2007). Social policy in Sub-Saharan African context: In search of inclusive development. In J. Adesina (Ed.), *Social policy in Sub-Saharan African context*. Palgrave Macmillan.

Adesina, J. O. (2011). Beyond the social protection paradigm: social policy in Africa's development. *Canadian Journal of Development Studies/Revue canadienne d'études du développement, 32*(4), 454–470.

Adésínà, J. O. (2015). Return to a wider vision of social policy: Re-reading theory and history. *South African Review of Sociology, 46*(3), 99–119.

Agochukwu, N. Q., & Wittmann, D. (2019). Stress, depression, mental illness, and men's health. In *Effects of lifestyle on men's health* (pp. 207–221). https://doi.org/10.1016/B978-0-12-816665-9.00010-X

Aharon, G., Aisenberg-Shafran, D., & Levi-Belz, Y. (2024). Adherence to masculinity norms and depression symptoms among Israeli men: The moderating role of psychological flexibility. *American Journal of Men's Health, 18*(3), 15579883241253820. https://doi.org/10.1177/15579883241253820

Aveling, E.-L., Gillespie, A., & Cornish, F. (2015). A qualitative method for analysing multi-voicedness. *Qualitative Research, 15*(6), 670–687.

Backman-Levy, J. K., & Greene, M. E. (2024). Gender-transformative programmes: A framework for demonstrating evidence of social impact. *BMJ Global Health, 9*(4), 014203. https://doi.org/10.1136/bmjgh-2023-014203

Barragán, J. J. (2024). Beyond 'manning up': The impact of hegemonic masculinity & gender norms on mental health help-seeking among college men. (Master's thesis, San Jose State University).

Bateman, C. (2012). SA men-time for introspection/renewed action? *South African Medical Journal, 102*(7), 597–598. https://doi.org/10.7196/SAMJ.6041

Bilsker, D., Fogarty, A. S., & Wakefield, M. A. (2018). Critical issues in men's mental health. *The Canadian Journal of Psychiatry, 63*(9), 590–596.

Chibanda, D., Weiss, H. A., Verhey, R., Simms, V., Munjoma, R., Rusakaniko, S., Chingono, A., Munetsi, E., Bere, T., Manda, E., Abas, M., & Araya, R. (2016). Effect of a primary care-based psychological intervention on symptoms of common mental disorders in Zimbabwe: A randomized clinical trial. *JAMA, 316*(24), 2618–2626. https://doi.org/10.1001/jama.2016.19102

Chipenda, C. (2023). Peasant livelihoods in times of Covid-19: A classical agrarian and political economy perspective. *Journal of Asian and African Studies, 58*(8), 1545–1570.

Chipenda, C. (2024). A transformative social policy perspective on land and agrarian reform in Zimbabwe. *Africa Spectrum, 59*(1), 89–118.

Chipenda, C. (2025). Social infrastructure in rural Zimbabwe: An exploration of services and development dynamics. *Discover Global Society, 3*(1), 60.

Chitando, E., Mlambo, O. B., Mfecane, S., & Ratele, K. (2024). Introduction: African men and masculinities. In *The Palgrave handbook of African men and masculinities* (pp. 1–30). Springer.

Cooper, R. (2023). Crystal meth in Zimbabwe–unemployment, stress & mental health linkages. *The Dyke, 17*(1), 1–6.

Damiri, D. (2023). Gender transformation in multicultural families. *Harakat an-Nisa: Jurnal Studi Gender dan Anak, 8*(1), 33–42. https://doi.org/10.30631/81.33-42

Dodgson, J. E. (2017). About research: Qualitative methodologies. *Journal of Human Lactation, 33*(2), 355–358.

Dzinamarira, T., Iradukunda, P. G., Saramba, E., Gashema, P., Moyo, E., Mangezi, W., & Musuka, G. (2024). COVID-19 and mental health services in sub-Saharan Africa: A critical literature review. *Comprehensive Psychiatry, 131*, 152465.

Dziwa, D. D., Postma, L., & Combrink, L. (2020). Gender conversations in Zimbabwe: A precursor of male gaze in visual art practices. *TD: The Journal for Transdisciplinary Research in Southern Africa, 16*(1), 1–9.

Epprecht, M. (2013). *Sexuality and social justice in Africa: Rethinking homophobia and forging resistance*. Zed Books.

Ezeugwu, C. R., & Ojedokun, O. (2020). Masculine norms and mental health of African men: What can psychology do? *Heliyon, 6*(12).

Ferdous, J. (2024). Transformative approaches to gender, migration, and development. In *Gendered migrations: Navigating challenges and opportunities for development* (pp. 1–27). Springer Nature.

Fischer, A. M. (2018). *Poverty as ideology: Rescuing social justice from global development agendas*. Zed Books.

Fisher, J., & Makleff, S. (2022). Advances in gender-transformative approaches to health promotion. *Annual Review of Public Health, 43*, 1–17. https://doi.org/10.1146/annurev-publhealth-121019-053834

Gaynor-Brook, L. (2023). Towards mental health as a human right: The key role of lived experience. *PLoS Medicine, 20*(10), e1004307. https://doi.org/10.1371/journal.pmed.1004307

GoZ. (2023). *Zimbabwe national strategy to prevent and address gender based violence 2023–2030*. Ministry of Women Affairs, Community, Small and Medium Enterprises Development.

Graeme, D., Smith, M., & Hebdon, M. (2023). Mental health help-seeking behaviour in men. *Journal of Advanced Nursing, 80*(3), 851–853. https://doi.org/10.1111/jan.15869

Harriman, N. W., Williams, D. R., Morgan, J. W., Sewpaul, R., Manyaapelo, T., Sifunda, S., Mabaso, M., Mbewu, A. D., & Reddy, S. P. (2022). Racial disparities in psychological distress in post-apartheid South Africa: Results from the SANHANES-1 survey. *Social Psychiatry and Psychiatric Epidemiology, 57*(4), 843–857. https://doi.org/10.1007/s00127-021-02175-w

Hundera, M., & Mudde, H. L. M. (2024). A comprehensive analysis of gender dynamics in organizations: Integrating social, structural and individual dimensions. *European Journal of Education, 59*, e12722. https://doi.org/10.1111/ejed.12722

Kidia, K., Machando, D., Mangezi, W., Hendler, R., Crooks, M., Abas, M., Chibanda, D., Thornicroft, G., Semrau, M., & Jack, H. (2017). Mental health in Zimbabwe: A health systems analysis. *The Lancet Psychiatry, 4*(11), 876–886.

Machisa, M. T., Christofides, N., & Jewkes, R. (2016). Structural pathways between child abuse, poor mental health outcomes and male-perpetrated intimate partner violence (IPV). *PLoS One, 11*(3), e0150986. https://doi.org/10.1371/journal.pone.0150986

MacArthur, J., Carrard, N., Davila, F., Grant, M., Megaw, T., Willetts, J., & Winterford, K. (2022, November). Gender-transformative approaches in international development: A brief history and five uniting principles. In *Women's studies international forum* (Vol. 95, p. 102635). Pergamon.

Mankowski, E. S., & Smith, R. M. (2016). Men's mental health and masculinities. In *Encyclopedia of mental health* (2nd ed., pp. 66–74). https://doi.org/10.1016/B978-0-12-397045-9.00182-8

Marandure, B. N., Mhizha, S., Wilson, A., & Nhunzvi, C. (2023). Understanding the nature of substance use in Zimbabwe: State of the art and ways forward: A scoping review protocol. *PLoS One, 18*(3), e0272240.

Masemola, H. C., Moodley, S. V., & Shirinde, J. (2022). Perceptions and attitudes of black men in a rural district of South Africa towards depression and its treatment. *South African Family Practice, 64*(1), e1–e9. https://doi.org/10.4102/safp.v64i1.5557

Mashele, T. H., Letsoalo, D. L., & Oduaran, C. A. (2024). *Effects of masculine culture on the mental health of Northern Sotho male youth.*

Ministry of Health and Child Care (MoHCC). (2019). *Zimbabwe national strategic plan for mental health services (2019–2023).* MoHCC.

Ministry of Health and Child Care (MoHCC). (2024). *Traditional medicine.* https://www.mohcc.gov.zw/?p=5884

Mkandawire, T. (2004). Can Africa have developmental states. In *Making the international: Economic interdependence and political order* (pp. 291–329). Pluto Press.

Mkandawire, T. (2007). Transformative social policy and innovation in developing countries. *The European Journal of Development Research, 19*(1), 13–29.

Mkandawire, T. (2010). Aid, accountability, and democracy in Africa. *Social Research: An International Quarterly, 77*(4), 1149–1182.

Mlambo, T., Munambah, N., Nhunzvi, C., & Murambidzi, I. (2014). Mental health services in Zimbabwe—a case of Zimbabwe National Association of Mental Health. *World Federation of Occupational Therapists Bulletin, 70*(1), 18–21. https://doi.org/10.1179/otb.2014.70.1.006

Mogano, N. T. H., Letsoalo, D. L., & Oduaran, C. A. (2025). Effects of masculine culture on the mental health of Northern Sotho male youth. *BMC Psychology, 13*(1), 605. https://doi.org/10.1186/s40359-025-02934-3

Molokoane, P., & Modipane, M. (2025). Young men's perspectives on expressing problems affecting their mental health to family and friends in Ehlanzeni district, Mpumalanga province. *Journal of Men's Health, 21*(3), 119–126.

Morrell, R., & Ouzgane, L. (2005). African masculinities: An introduction. In L. Ouzgane & R. Morrell (Eds.), *African masculinities* (pp. 1–20). Palgrave Macmillan. https://doi.org/10.1057/9781403979605_1

Moyo, I. (2024). HIV care services toward the attainment of sustainable development goals in Africa. In *SDGs in Africa and the Middle East Region* (pp. 1–25). Springer.
Mumbire, R., & Chakabwata, W. (2024). The intersections of gender, leadership, and ethnicity in Africa: A case of civil society organizations in Zimbabwe and South Africa. In *Minority women in K-12 education leadership: Challenges, resilience, and support* (pp. 192–218). IGI Global Scientific Publishing.
Ngwenya, W. M., & Sumbane, O. G. (2023). The urgency of access to men-centered mental healthcare services to address men's sensitive issues in the communities of South Africa. In *Healthcare access—New threats, new approaches*. IntechOpen. https://doi.org/10.5772/intechopen.108493
Nyanzi, S. (2011). Unpacking the (Govern)mentality of African Sexuality. In: *African sexualities*: A Reader (edited by Sylvia Tamale), 477–501. Cape Town: Pambazuko Press.
OECD/World Health Organization. (2022). Mental health care. In *Health at a glance: Asia/Pacific 2022: Measuring progress towards universal health coverage*. OECD Publishing. https://doi.org/10.1787/51a53e13-en
Pearson, S., & Makadzange, P. (2008). Help-seeking behaviour for sexual health concerns: A qualitative study of men in Zimbabwe. *Culture, Health & Sexuality, 10*(4), 361–376.
Phoebe, M., Zibusiso, S., & Moyo, M. (2024). *The Zimbabwe National Gender Policy (NGP) 2013–2017 and women's economic empowerment (WEE): A study of small-scale gold miners in the Bubi and Gwanda rural districts, Zimbabwe*. https://doi.org/10.21504/10962/434638
Reihling, H. (2020). *Affective health and masculinities in South Africa: An ethnography of (in) vulnerability* (1st ed.). Routledge. https://doi.org/10.4324/9780429328565
Republic of South Africa. (1996). Constitution of the Republic of South Africa, No.108 of 1996. Government Gazette.
Republic of South Africa. (2002). Mental Health Care Act No. 17 of 2002. Government Gazette.
Seidler, Z. E., Dawes, A. J., Rice, S. M., Oliffe, J. L., & Dhillon, H. M. (2016). The role of masculinity in men's help-seeking for depression: A systematic review. *Clinical Psychology Review, 49*, 106–118. https://doi.org/10.1016/j.cpr.2016.09.002
Sheikh, A., Payne-Cook, C., Lisk, S., Carter, B., & Brown, J. S. L. (2024). Why do young men not seek help for affective mental health issues? A systematic review of perceived barriers and facilitators among adolescent boys and young men. *European Child & Adolescent Psychiatry*. https://doi.org/10.1007/s00787-024-02520-9
Sikweyiya, Y., Mahlangu, P., Jewkes, R., et al. (2025). 'We do not like talking about our problems': Socialization and idealized masculinity as drivers of help-seeking avoidance among college men in South Africa. *BMC Public Health, 25*, 1091. https://doi.org/10.1186/s12889-025-22252-y
Sinisi, V. (2024). *SA has the third highest suicide rate in Africa—there are steps we can take to tackle it*. Retrieved February 10, 2025, from https://www.spotlightnsp.co.za/2024/09/20/sa-has-the-third-highest-suicide-rate-in-africa-there-are-steps-we-can-take-to-tackle-it/
Skovdal, M., Campbell, C., Madanhire, C., Mupambireyi, Z., Nyamukapa, C., & Gregson, S. (2011). Masculinity as a barrier to men's use of HIV services in Zimbabwe. *Globalization and Health, 7*, 1–14.
Thulani, R. D., Dzinamarira, M., Mutevere, S., Nyoka, S., Moyo, E., Muzenda, L., Kakumura, F., & Dzinamarira, T. (2023). "Creating a generation of equality": A stakeholder's perspective on power dynamics and gender-based violence in Zimbabwe. *Global Journal of Health Science, 16*(1), 1–9. https://doi.org/10.5539/gjhs.v16n1p1
Tom, T., & Banda, R. (2023). Women farm workers in Zimbabwe: The social policy outcomes two decades after the transformative fast track land reform. *Feminist Africa, 4*(1), 75–102.
Vogel, D. L., Heimerdinger-Edwards, S. R., Hammer, J. H., & Hubbard, A. (2011). "Boys don't cry": Examination of the links between endorsement of masculine norms, self-stigma, and help-seeking attitudes for men from diverse backgrounds. *Journal of Counseling Psychology, 58*(3), 368–382. https://doi.org/10.1037/a0023688
Uwimana, G., Elhoumed, M., Gebremedhin, M. A., Nan, L., & Zeng, L. (2023). Determinants of timing, adequacy and quality of antenatal care in Rwanda: A cross-sectional study using demographic and health surveys data. *BMC Health Services Research, 23*(1), 217.

WHO. (2021). *Scaling up the national response for mental health in Zimbabwe*. https://www.who.int/news-room/events/detail/2021/07/29/default-calendar/scaling-up-the-national-response-for-mental-health-in-zimbabwe

WHO. (2022a). *Integrated African health observatory. Analytical Fact Sheet 2022*. Retrieved February 11, 2025, from https://files.aho.afro.who.int/afahobckpcontainer/production/files/iAHO-HIV_Regional_Factsheet_November2022.pdf

WHO. (2022b). *Mental Health Atlas 2020 country profile: South Africa*. Retrieved February 12, 2025, from https://www.who.int/publications/m/item/mental-health-atlas-2020-country-profile%2D%2Dsouth-africa

World Bank. (2024). *Zimbabwe gender assessment*. World Bank. https://openknowledge.worldbank.org/entities/publication/5eeeb624-e8df-466b-b5e1-a6fb5e84f994

World Economic Forum. (2025). *Mental health*. Retrieved February 11, 2025, from https://intelligence.weforum.org/topics/a1Gb0000000pTDbEAM

Yusupova, M., (2023). *Traditional masculinity is a vague, unhelpful term we should abandon—Here's why*. Retrieved February 11, 2025, from https://theconversation.com/traditional-masculinity-is-a-vague-unhelpful-term-we-should-abandon-heres-why-202472

Zielke, J., Batram-Zantvoort, S., Razum, O., & Miani, C. (2023). Operationalising masculinities in theories and practices of gender-transformative health interventions: A scoping review. *International Journal for Equity in Health, 22*, 139. https://doi.org/10.1186/s12939-023-01955-x

Chapter 16
From Policy to Practice: Interventions for Learners with Disabilities in South Africa

Raymond Chirowamhangu

16.1 Introduction

Social policy refers to the outcomes of local and national decisions to meet societal needs through tailor-made and coordinated responses to social problems and the allocation of resources (Sewpaul et al., 1999; Manyire & Asingwire, 1998). This is aligned with Adesina (2007), who conceptualises social policy as collective public efforts to protect the social well-being of a constituency within a given territory. The provisions of social policy include social security, health care and education, which are the key components of this discussion. The linkages between social policy and the economic-political ecology are central aspects of the policy discourse. The latter is of prominent significance considering the historic apartheid legacy of South Africa, marred by racial and class discrimination, exacerbated by poverty and extreme disparities. This pretext is echoed by Kanga (1992), who alludes that understanding the interacting forces within the territory helps to shape the social policy and the attainment of the expected outcomes. In this chapter, the expected outcomes pertain to the provision of education for learners with disabilities.

Education is a significant instrument of social policy. The right to inclusive education is universal, and its realisation is mandatory for all children (Strohwald, 2021). The basis of children's rights, specifically the rights of children with disabilities, is entrenched in the Convention on the Rights of Persons with Disabilities (CRPD), African Charter on the Rights and Welfare of the Child (ACRWC) and Convention on the Rights of the Child (CRC). Vitally, inclusive education is supported by regional and international policy frameworks. The United Nations Sustainable Development Goal (SDG) 4 promotes inclusive and equitable education

R. Chirowamhangu (✉)
Department of Sociology, University of Pretoria, Pretoria, South Africa

E. Ndhlovu, C. Chipenda (eds.), *Redefining the Social Policy Discourse*,
https://doi.org/10.1007/978-3-032-24067-5_16

for all, including learners with disabilities (United Nations, 2015a, 2015b). This is affirmed by the CRPD, General Comment 4, which states that inclusive education is a fundamental right (CRPD, 2016). However, despite these reinforcements, learners with disabilities still experience educational exclusion in Africa.

The pre-eminence of the inclusive education discussion from a policy perspective is meticulously drawn from a context-based perspective, which considers the historical injustices which have prompted transformation in the education system of South Africa. The apartheid education system was characterised by discriminatory social and economic structures which impacted access to education for learners with disabilities (Deghaye, 2021; Daniels, 2010). Notably, the World Conference on Special Needs Education adopted the Salamanca Statement, which ignited policy reforms towards the global achievement of inclusive education. Notwithstanding those efforts, with over 30 years since its adoption, the Salamanca Statement is critiqued as an 'amalgam of deals' due to lack of conceptual clarity on the nature and scope of inclusive education at the national level, the misconception that inclusive education is 'expensive' and poor representation of all stakeholders during the policy formulation process (Magnússon, 2019; Amor et al., 2018; Nilholm & Göransson, 2017).

The rights of children with disabilities are violated when they are under-resourced and excluded from the learning process. From a continental perspective, the African Committee of Experts on the Rights and Welfare of the Child reveals that, especially in the rural areas of Africa, there are disparities in school enrolment and attendance of children with special needs (ACERWC, 2022a, 2022b, 2022c). Burgeoning literature, synthesised by researchers in South Africa, has explored the lack of inclusive education, especially in rural areas, due to the scarcity of resources, disparities in school enrolment and attendance for children with special needs (Donohue & Bornman, 2014; Musenyente et al., 2022). Special needs schools provide appropriate support at dichotomy levels to offer maximum support for learners experiencing challenges (Khumalo & Hodgson, 2017; Mahlaule et al., 2024). The focus on special needs schools in this research is based on the paucity of progress in the implementation of inclusive education in the post-democracy era of South Africa.

This chapter explores the extent of policy implementation to address the challenges in the inclusive education sector of South Africa. The rest of the chapter is organised into four sections. The next section discusses the methodology approach adopted in this chapter, which encapsulates a qualitative research design. This is followed by a comprehensive literature review which conceptualises inclusive education, provides an overview of the White Paper 6 policy on inclusive education in South Africa and deliberates on Matland's top-down and Easton's models of policy implementation. This is followed by the presentation findings, which highlight significant areas of concern, including a lack of special needs schools, exclusionary enrolment, impact of the Coronavirus disease (COVID-19), transport and infrastructure challenges. The final section provides an in-depth discussion of the findings.

16.2 Methodology

The study adopted a qualitative research design anchored on an interpretive paradigm. The method crystallises social policy by highlighting prevailing conditions exposed to learners with disabilities and considers the prevalent cultural-historical aspects of the research population. Scholars support that the qualitative design is flexible, responsive and provides detailed descriptions of the issues being investigated (Creswell & Creswell, 2022; Shikwambana & Fourie, 2023). In addition, the prevailing circumstances at the time were a major contributing factor, as the study was conducted from May to July 2020, at the peak of the COVID-19 pandemic. As a precautionary measure, the Government of South Africa, through the Disaster Management Act, introduced lockdown restrictions to contain and mitigate the virus (Republic of South Africa, 2020). Thus, online media platforms and telephonic methods were more feasible options for collecting data.

The research population was made up of 50 participants, including educators from special needs schools, participants from local NGOs, community leaders and government representatives from the Eastern Cape province of South Africa. Various factors motivated the research area selection; firstly, the province has the highest disability prevalence for children; secondly, it has the second highest rate of child poverty at 79%; thirdly, over 60% of the populace reside in rural areas; lastly, scholarly literature has described the state of the province as an 'education crisis' engraved in poverty, failed social welfare reforms and poor infrastructure (Stats SA, 2020; Ngumbela, 2021). Therefore, the research goes beyond reflecting on inclusive education as a focal area, but also incorporates perspectives on a demography which is often overlooked in social policy implementation, which is the rural populace.

In terms of the practical aspects, the participants were selected using the non--probability method, through snowball sampling, whereby each interviewee recommended a potential candidate for a follow-up interview. For instance, the participants from NGOs had a coalition network used when lobbying for the rights of children with disabilities; therefore, it was easier to get recommendations on participants who could assist with the interviews. As reinforced by research, the sampling technique is efficient and ensures that participants are competent with relevant knowledge on the issues under discussion (Kirchherr & Charles, 2018; Kumar, 2019). Furthermore, data were collected from both primary and secondary sources. In the case of primary data, semi-structured key informant interviews were used. This supported the integration of open-ended questions to allow for an interactive and transparent exchange, whereby the participants responded based on their own perspectives and experiences (Busetto et al., 2020). As mentioned earlier, due to the COVID-19 restrictions, some interviews were conducted telephonically and others using online platforms. Desktop research was adopted to collect data from secondary sources, including policy documents, reports and official statistics databases.

The data was comprehensively analysed using Atlas.ti 8 research software. The computer-assisted qualitative data analysis software allows researchers to effectively code data and create interacting themes to interpret data (Adelowotan, 2021).

Researchers have supported the use of qualitative content analysis to identify and categorise data into themes (Bengtsson, 2016). This proved to be a viable option considering the large volume of interview transcripts, which needed to be categorised into specific themes. Following the prescriptions of other previous studies, the research developed a six-stage data analysis process which involved familiarisation, defining themes, searching for themes, reviewing themes, coding and compiling the research report (Moosa-Tayob & Risenga, 2022; Jackson et al., 2016).

16.3 Literature Review

16.3.1 Conceptualising Inclusive Education

Inclusive education is concerned with promoting access to education and full participation by removing learning barriers for all learners (UNESCO, 2020; Murungi, 2015). This learning framework is grounded on the principle of 'no learner is left behind' and necessitates that education should accommodate the needs and circumstances of all children. Similarly, the United Nations SDG 4 advocates for inclusive, equitable, quality education and promotes lifelong learning opportunities for all (United Nations, 2015a, 2015b). Thereby, enabling learners to access educational opportunities based on equity, flexibility and adaptability.

The policy provisions of inclusive education are established on fundamental legislative underpinnings. The Universal Declaration of Human Rights (UDHR) declares that 'everyone has the right to education' (United Nations, 1948). Notably, the treaty in its commencement as a legal instrument did not recognise disability as a form of discrimination; in this defence, Herr (1980) argued that this omission was a portrayal of the prevailing ignorance and unawareness concerning disability at that time. However, since then, great strides have been achieved through the collective adoption of the CRC and ACRWC. As highlighted in Articles 2(2) and 23(1) of the CRC and Article 13(1) of the ACRWC, all children with special needs have the right to education without discrimination on any grounds (CRC, 1989; ACRWC, 1990). The focus on inclusive education is bolstered in Article 24(2)(a) of the CRPD, which calls for state parties to ensure a non-discriminatory, inclusive education system (CRPD, 2006).

Although most countries seem to share the same ideology and commitment toward inclusive education, the implementation has not been equally successful (Dyson, 2001; Swart & Pettipher, 2005; Balescut & Eklindh, 2005). For example, the CRPD gives specific guidelines to promote inclusive education, such as the use of braille, augmentative and alternative modes of teaching, but this has not been the reality, especially in South African rural schools. Ransom (2008), in a study of inclusive education, states that policies sometimes refer to inclusive education or learners with special needs, without tentative measures to meet their needs. As noted by Balescut and Eklindh (2005), exclusion is a culmination of several factors, including limitations in policy, lack of strategic delivery, and the erroneous assumption that inclusive education is expensive to implement.

16.3.2 White Paper 6: Inclusive Education Policy in South Africa

The White Paper 6 was adopted in 2001 to develop an integrated system and promote education access for children with special needs (Muthukrishna & Schoeman, 2010). During the apartheid era, learners with disabilities faced social exclusion in racially segregated special schools. The White Paper 6 reiterates accountability on inclusive education, including constructing more special schools and reforming the quality of education to address the disability challenges. Moreover, it develops an intersectional approach which recognises other factors, such as historical inequalities, social constructs, discrimination and poverty. Due to inequitable apartheid resourcing, most special schools inherited by the post-apartheid government were located in formerly white residential areas. According to UNESCO (1994), the policy draws inspiration from the Salamanca Statement, which sets a precedent for equal access, participation and inclusive learning. Moreover, the policy framework outlines the specific process to identify, assess and enrol learners with special needs (Department of Education, 2008). The initial stage is critical because when disabilities are identified at an early stage, it allows for the necessary resources and support structures to be allocated efficiently. Furthermore, it presents applicable teaching approaches and curriculum changes which meet the needs of a diverse range of learners (Department of Education, 2011).

Despite the enabling policy described above, the implementation of inclusive education in South Africa is a stagnant and partial process. Thus, in order to achieve the objectives of the White Paper 6, a conceptual shift is required to provide support for children with learning barriers because it lacks explicit details in its attempt to address the diverse needs of all learners who experience barriers to learning. In response, Polat (2011) suggests that resources and improved infrastructure are necessary but insufficient inclusion, through attitudinal barriers from educators, learners and community members, remains ubiquitous.

16.3.3 Matland's Top-Down and Easton's Models of Policy Implementation

Inclusive education as social policy reinforces the need for diversity, social equity and lifelong learning. Burnett (2009) asserts that inclusive education transforms learning institutions to cater for all learners regardless of ability, social class, race, economic status, sexual orientation and gender. Thus, the policy is designed to enhance social inclusion. Hall and Midgley (2004) state that social policy constitutes planned action that impacts the livelihoods of people. In support of this view, scholarly literature highlights that these collective interventions foster access to resources and transformation of social institutions and relations (Mkandawire, 2001; Adesina, 2015). As such, the transformative social policy assists in overcoming inequalities and social exclusion.

As previously argued by the contributions of Adesina and Mkandawire, social policy involves the transformation of institutions and social relations, thus making social policy more political in orientation as compared to other policies. In the case of South Africa, research shows the significant policy modifications from the apartheid to the post-1994 democratic era (Gumede, 2008). Whereby, during apartheid, social policy was rooted in European models of governance, which were discriminatory, and the advent of democracy ushered in a social policy anchored on transformation to redress the inequalities of the past. This solidifies the link between social policy and politics, in which Dror (2006) highlights that the two aspects 'closely interact, often overlap, and in part cannot be separated even analytically'. Moreover, policy makers who are largely elected representatives are responsible for making decisions that affect society and impact on the distribution of resources. Equally, the impact of the political system on social policy cannot be overlooked, as this determines the priorities and approaches to implementing social policy. In addition, Lang et al. (2011) show that the successful implementation of policy depends on country-level implementation efforts, political will and adequate government structures.

Savio (2010) supports Easton's model of policy implementation, which provides a detailed analysis, categorised into various stages, including initiation, deliberation, implementation and assessment. Scholars argue that the segmented approach might not always be visible in practice and can obscure the process (Wihlborg, 2010; Hill, 2007). South Africa's inclusive education policy has been known as 'symbolic implementation', which has yielded little success in the rural areas (Adewumi et al., 2019). In this view, to dissect the exploration 'from policy to practice' in this chapter requires emphasis on the implementation phase, which reflects on how policies are executed to support learners with disabilities. It is of paramount importance to note that no fruitful analysis can be conducted if the policy is not effectively implemented. Easton's model, also known as the classic model, is primarily top-down in nature, with emphasis on efficiency and accountability. Therefore, the authenticity of the policy is not questionable, but rather a microscopic view of the actors and institutions responsible for implementation is adopted. The delegation of responsibilities to policy implementers is a fundamental aspect of the model, which assists in thwarting power conflicts between policy makers and implementers (Kota et al., 2021). Though it can be argued that this model is autocratic in nature, it is actually more interactive in practice, because policy makers are mandated to draft policies which 'speak the language' and embrace the community needs. This is often a process anchored on a continuous engagement process.

Literature on the White Paper 6 exposes a lack of clarity on the extent of inclusion due to school-level and cultural barriers (Makoelle & Burmistrova, 2020; Hudson et al., 2019). For instance, scholars show that with regard to cultural barriers, children with disabilities are discriminated against because disability is attributed to family sin, witchcraft and conflict with the ancestors (Ndlovu, 2023; Yaacob et al., 2021). This makes it difficult to implement a bottom-up approach, as a result of cultural diversity, social and religious views on disability. Another illustration is based on the economics of implementing policy, as demonstrated by Dehnhardt et al. (2022), who show that the policy yields cost-effective benefits. But as challenged in recent studies, this seems unrealistic in South Africa, considering the substantial increase in provincial funding (Walton & Engelbrecht, 2022). Thus, the top-down approach streamlines policy objectives and delegates duties to ensure children with disabilities have access to basic education.

Therefore, Matland's top-down approach is an instrumental implementation tool because it provides clarity on the scope and reach of the policy (Hudson et al., 2019). This means the government and policymakers articulate fundamental principles, guided by specific mandates, and establish a mechanism to account for achieving the policy. This calls for clear and effective legislative frameworks to support inclusive education in South Africa (Ramango & Naicker, 2022). In the process, this helps to eradicate negative attitudinal perspectives from legislators, educators and learners to foster inclusive participation of children with disabilities in the mainstream school system. In a similar case study on inclusive education in the Eastern Cape province, the researchers concluded that there was a lack of political will, policy coherence and accountability between the provincial and national Department of Education (Walton & Engelbrecht, 2022; Bipath et al., 2021). However, Shukia (2020) argues that the approach fails to recognise complex implementation structures; thus, giving policy makers exclusive control, even when they lack a proper understanding of the realities faced by persons with disabilities.

16.4 Results/Findings

16.4.1 Limited Number of Schools and Exclusionary Enrolment

Inclusive education cannot be implemented in isolation, whilst ignoring other fundamental rights prescribed in the Constitution, White Paper 6 and ratified treaties. This section of the chapter presents three main themes on barriers to learning, as highlighted by the research participants. Firstly, the issue of insufficient available schools to cater for the needs of children with disabilities was raised at the preliminary stage. These reflections were illustrated in the interview excerpt.

> For special schooling here, we only have 2 Government special schools in Gqeberha. Therefore, you can understand there's a massive waiting list, or sometimes, you know a child is on a waiting list. And obviously, there are specific criteria which you need to perform to be placed in a special school. Most children or you know, are on waiting lists in stages. Not enough special schools to accommodate all our learner's needs. The waiting list of up to 550 learners per annum. Over 2500 children are still at home because they are intellectually challenged. (P5, special needs teacher, Ndevana)

As quoted above, a limited number of special needs schools are available in the rural areas, with only a few available schools located in the urban city of Gqeberha, which is approximately 260 km. Significantly, an excessively long waiting list of 550 learners per annum meant that over 2500 children with special needs did not have access to education. This similar trend was noticed in other rural areas, such as Matatiele and Nqanqarhu.

Apart from the limited number of special schools, the other area of contention during the research was the criteria used to enrol learners with special needs. This was expressed by one of the participants.

> But the specialized school takes children from specific ages from the age of 14, which is a school of skills basically, and that child needs to be on a Grade 5 level academically to be able to do those things. But some children do not fall on that level or fall above the percentile or below. And because of such a system they end up placed at schools that don't have an emphasis on academics, which kind of just exacerbates their motivation. (P9, special needs teacher, Gqeberha)

The response shows the criteria used were based on the classification of disability and age of the child; consequently, children with disabilities above 14 years might struggle to get access to inclusive education, if there is an age limitation below 14 years. Also, the criteria used to distinguish disability were inconsistent. This was reported by another participant.

> And that is an enormous problem class because there are no remedial (sic). There is only one special school in East London, (name withheld) but it's full of kids who are significantly disabled. So, they're either significantly intellectually disabled or physically disabled. And these kids are not they are just "slow" if you want to use that term. And the sadness for me is that the local Public School with less than a half kids will not give the kids the right standard of examination so that they get more than 50%. But in fact, they're not achieving. They do it so that they can make themselves look better. And those kids will proceed, whereas our kids who will be performing at the correct level will not be accepted. (P16, special needs teacher, East London)

The frustrations expressed above show various inequalities in the criteria used for the acceptance of children with special needs. Thus, leading to unfortunate remarks, in which children are discriminated and labelled as 'slow' learners. Discrimination against children with intellectual and physical disabilities was not only aggravated by other learners, but also involved educators who did not assess learners based on the national standards; thereby, making them complicit in exclusion.

16.4.2 Support, Transport and Infrastructure Challenges

The need for expert psychosocial support helps to easily identify learning challenges and quickly prescribe relevant assistance. The psychosocial concept denotes the interaction of social and environmental circumstances influencing the learner (Kirst-Ashman, 2010). In such cases, the learner requires specialised assistance, which often is beyond the scope of an educator. Unfortunately, as reflected in the responses below, there were few trained personnel to assist learners with disabilities:

> Let me give you a scenario in the Eastern Cape, we have 61 social workers and 63 psychologists, and we do not have enough people to provide psychosocial support. Even with the Learner Support agents, we have about 300 of them, so can imagine, versus the number of schools that we have. (P32, government representative, Bisho)

> Many of our learners also present with comorbidities such as physical, psychological, and neurological disabilities. We have social workers, a physiotherapist, an occupational therapist, an audiologist, an educational psychologist, and a nursing sister. All these are paid for by the school because of a lack of funding from the government. There is a lack of experts in the department to offer support and guidance to children with special needs. (P45, NGO representative, Ndevana)

The narration summarises challenges experienced in providing expert assistance for children with special needs, including a lack of funding to support specialised professionals with quality equipment to operate at full capacity. In addition, the research findings revealed that these services were only available in some schools and a large percentage of rural schools were not supported by expert-trained staff. This assertion makes Matland's top-down approach relevant, as its premise is on the availability of funds to train, recruit and delegate expert human resources to special schools in rural areas. Thus, making the actualisation of social policy initiatives a reality for all children.

Concerning scholar transport, children with disabilities have to travel long distances to the nearest school. As a result, the child might not be able to walk to school every day, to school and there is an additional cost required to warrant safe transport. These sentiments were shared by one of the participants.

> Some of the children can easily walk. But some of the special schools are in Uitenhage, Sunshine, Moakley and Besa, so we have to go abroad. They have to travel 320 km from the nearest city. Or in some areas, the nearest school would be 60 km daily We have to drive them every day usually two loads to transport all the children. Then on rainy days the rest of the staff just have to pull in and get in a car and drive the kids. (P8, community leader, Nqanqarhu)

Thus, children with special needs are severely disadvantaged when they have to travel long distances to school and yet children without disabilities will easily walk or have access to cheaper public transport options when travelling to school. As illustrated, some children had to walk to special needs schools and will require assistance to make sure they reach the school safely, as some of them have to cross rivers and dilapidated bridges to get to school.

Table 16.1 Number of schools with pit latrine toilets

Province	Schools with pit latrines
Eastern Cape	1585
Free State	156
Gauteng	0
KwaZulu Natal	1365
Limpopo	507
Mpumalanga	127
North West	145
Northern Cape	0
Western Cape	0
Total	3885

Source: Amnesty International (2020); Compiled by the researcher

Similarly, the infrastructure challenges also included aspects related to water and sanitation in the special needs schools. Section 27(1)(b) of the Constitution and Water Services Act 108 of 1997 highlight that access to clean and safe water is a basic right (Republic of South Africa, 1996). Access to basic education takes into consideration the state of sanitation and hygiene, which includes the availability of toilets in schools, the provision of water and hygiene. However, some frustrations were recalled by the participants, who echoed that

> But we're still sitting with over 1000 schools with pit latrines. The second timeline was on school infrastructure, we still have hundreds of schools with improper infrastructure, with poor quality materials from zinc, asbestos, and all those things. (P28, NGO representative, Mthatha)

The Sanitation Appropriate for Education (SAFE) programme, aimed at eradicating all pit latrine toilets by 2023, has yielded scattershot and uneven progress, as the Eastern Cape province still has the highest number of schools with pit latrine toilets (Amnesty International, 2020). Moreover, the support service disparities between provinces in South Africa are a common connotation.

Table 16.1 shows that the Eastern Cape province has 1585 schools with pit latrines, which remains the highest in South Africa. The statistics demonstrate the extent of inequalities between predominantly rural and urban provinces. For example, urban provinces, such as Gauteng, Northern Cape and Western Cape, have no pit latrine toilets recorded, whilst dominantly rural provinces, including Limpopo, Mpumalanga, KwaZulu-Natal and Eastern Cape still have pit latrines, which are a health and safety risk for learners with disabilities. This representation reflects the appalling school sanitation problems post-1994 democratic dispensation, which mirrors the apartheid spatial geography, whereby predominantly black and marginalised communities had a lack of access to sanitation.

Cognisant of the South African rural context, a common denominator is the lack of social and economic viability needed to sustain technological improvement (Cristobal-Fransi et al., 2020; Hall, 2019). Section 28(2) of the Constitution recommends that the interests of the child are an uncompromisable prerequisite; therefore, the safety of a child at school is paramount. However, issues pertaining to the lack of infrastructure development were cited during the research.

> In terms of the Norms and Standards, by 29 November 2016, all schools must have access to some form of power supply, water supply and sanitation. Schools entirely made of inappropriate materials, such as mud, metal, asbestos or wood, must be replaced with new schools. In the Eastern Cape province, a lot of our work has been around creating accountability that's linked to the law of norms and standards. (Equal Education, 2016; p. 5)

> The policy says we should have an inclusive education, but on the ground, they cannot accommodate even a child with a wheelchair, let alone a child with an intellectual disability. (P23, NGO representative, Seymour)

In terms of Section 4(3)(a) of the Norms and Standards Act, 'all schools built entirely from mud as well as those built entirely from materials such as asbestos, metal and wood must be prioritised' (Republic of South Africa, 2013). Infrastructure also includes school structures to improve the mobility of children with disabilities, such as ramps, walkways, restrooms and extra space to accommodate the free movement of wheelchairs in schools. In this way, the right to basic education necessitates the provision of appropriate school infrastructure, which is safe and does not endanger the life and welfare of children with disabilities (Grimes et al., 2023; Du Plessis & Mestry, 2019).

16.4.3 The Impact of the COVID-19 Pandemic

The first case of COVID-19 in South Africa was recorded on the 5th of March in 2020 (Giandhari et al., 2021; Pillay-van Wyk et al., 2020). In response to the high infection and death rate, the Government of South Africa declared a state of emergency in accordance with Section 37 of the Constitution, and the State of Emergency Act 64 of 1997. The COVID-19 pandemic was cited as a major challenge by 72% of the responses, as illustrated by some participants in their responses below.

> Because of such challenges, we are left behind, and with physical learning, due to Covid, they're only allowed to attend classes once or twice depending on the lockdown restrictions. In some cases, students have gone months without access to learning material. (P6, special needs teacher, Chintsa East)

> Currently, we are working in a climate that uses the hybrid model which also physical and virtual learning. The challenge is that working with learners from previously disadvantaged backgrounds means their economic status does not necessarily allow them to participate in virtual learning because of various reasons, such as they might not have cell phones or laptops and limited access to internet. (P38, special needs teacher, Nkwenkwezi)

The limitations stated above prompted the need for innovative teaching methods through the improvisation of online learning. Special schools' responses towards the pandemic were incapacitated by limited resources, poor contingency planning and lack of innovative learning, especially in the case of rural schools.

Research shows that during the COVID-19 pandemic, many schools had limited operationalisation due to a lack of online learning, with only 7% of households with children aged between 5 years and 18 years having internet access at home, and 66.8% of the households accessed the internet via smartphones (Stats SA, 2022; Dube, 2020). This proves that the chasm between the 'haves and the have-nots', especially when considering the socio-economic inequalities in South Africa, is actually true. Thus, as a recommendation, the government of South Africa supported inclusive education through online learning initiatives.

16.5 Discussion

Overall, the research findings found that children with disabilities in the Eastern Cape province faced multiple barriers to education, which are largely operational at the education system level. These are summarised into four key categories: namely, capacity, mobility, resources and disaster response on children with disabilities. This view is supported by literature from previous studies (Sumbane et al., 2023; Mizunoya et al., 2018; Simo Fotso et al., 2018). The capacity-related challenges include a limited number of special needs schools and exclusionary enrolment criteria. Similarly, Zwane and Malale (2018) revealed that the lack of education facilities for children with disabilities contributes towards overcrowding in the few available schools.

The limited number of special needs schools and mobility challenges are intertwined, because children with disabilities have to travel long distances, going as far as 60 km to the nearest school. Previous studies have documented scholar transport challenges in the Eastern Cape province, such as backlogs in transport provision and the use of unroadworthy vehicles (Chirowamhangu, 2023; Kota et al., 2021). This has been extensively articulated in existing literature, which explores other pertinent aspects, including continuous advocacy to address negative attitudes by transport drivers, as well as aspects pertaining to the safety, security and cost of scholar transport for learners with disabilities (Duri & Luke, 2022; Calle et al., 2021; Kett et al., 2020).

The resource-based challenges were echoed by Sumbane et al. (2023) in a study focusing on children with autism spectrum disorder in Limpopo, which, similar to the Eastern Cape, is predominantly a rural province with extensive poverty levels. As presented in this chapter, this category of barriers includes lack of water, poor sanitation, dilapidated infrastructure, psychosocial and expert support. In their assessment, Equal Education (2016) and Odeku (2022) assert that primary schools in the Eastern Cape province are exposed to pit latrine toilets with poor hygiene, sanitation and infrastructure, which are issues of concern for learners with special needs. Human Rights Watch (2021) criticises the Government of South Africa for dilapidated physical school infrastructure and its failure to implement safety measures (Human Rights Watch, 2021; Mpu & Adu, 2021). This includes schools with long staircases and those without wheelchair ramps, which make access difficult for learners with mobility disabilities.

In reference to psychosocial and expert support, this encompasses assistance from comprehensively trained personnel with best inclusive practices (Amod, 2015; Engelbrecht et al., 2015). This perspective is highlighted in prevailing discourse on inclusive education (Forsberg & Schultz, 2023; Pillay et al., 2023). As a researcher, I concur with this view from a structural functionalist perspective on disability, which advocates for a multi-participatory approach which envisages the complementary role of family and community structures.

Then, in relation to disaster response on children with disabilities, specifically, with regard to COVID-19, the World Health Organization confirmed that learners with disabilities were impacted more significantly by the pandemic (WHO, 2020). In view of online learning, research shows resource mobilisation and online learning were not prioritised timeously to incorporate the frequent physical contact required to carry, lift and feed students with disabilities whilst learning (Kuper et al., 2020; Tremmel et al., 2020). Consequently, when responding to disasters, such as in the case of COVID-19, governments should allocate all necessary resources and prioritise communication with all relevant stakeholders (Beckmann & Reyneke, 2021; Magedi et al., 2023).

Therefore, Matland's top-down approach to policy implementation consolidates inclusive social policy formulation with practice. This is achieved by clearly defining the policy objectives. As demonstrated in this chapter, using the White Paper 6 as a praxis, there is no lack of policy articulation on children with special needs, but there are no clear-cut guidelines and timelines to address the challenges raised by the policy. For example, the White Paper 6 from its inception had a 20-year provision to reach 380 special schools, but there are no guidelines and timeframes to address the resource based and disaster response challenges. The COVID-19 pandemic was meant to be a wakeup call, that if no tailor-made policies are in place, the most vulnerable are affected. Donohue and Bornman (2014), recommends that reducing the ambiguity in policies promotes the inclusion of children with disabilities in education.

16.6 Conclusion

The chapter reflects on the challenges facing learners with disabilities, and shows the extent to which the White Paper 6 policy on inclusive education has been implemented. Moreover, evidence was provided through relevant literature to contextualise inclusive education in South Africa and juxtapose the White Paper 6 as an important component of social policy. The discussion was predicated on Matland's top down and Eastwood's model of policy implementation; thus, it supports the correlation between the role of policy makers and agents of policy implementation. The former provided clarity on the context and accountability measures on implementing social policy on inclusive education, whilst the latter supports the contribution of educators, community leaders and NGOs as pillars of implementing policy, especially in rural communities. The findings highlight a multiplicity of distinct challenges including the limited number of special needs schools, exclusionary enrolment criteria, lack of psychosocial and expert support, poor scholar transport, dilapidated infrastructure and the COVID-19 pandemic. Whilst some of these challenges on learners with disabilities are entrenched in historical inequalities, recent events with ongoing effects, such as the COVID-19 pandemic pose important avenues of reflection on the adaptation of social policy during periods of global crisis.

Ethical Approval Approval was granted by the Ethics Committee, University of Pretoria (HUM035/0820).

References

ACERWC. (2022a). *Concluding recommendations by the African Committee of Experts on the Rights and Welfare of the Child (ACERWC) on the Republic of South Africa Report on the status of implementation of the African Charter on the Rights and Welfare of the Child.* Retrieved August 16, 2024, from https://www.acerwc.Africa/sites/default/files/2022-06/South-Africa-concluding-Observations.pdf

ACERWC. (2022b). *List of countries with have signed, ratified/acceded to the African Charter on the Rights and Welfare of the Child.* Retrieved September 5, 2024, from https://www.achpr.org/ratificationtable?id=46

ACERWC. (2022c). *Report of the African Committee of Experts on the Rights and Welfare of the Child.* Retrieved October 11, 2024, from https://www.acerwc.africa/sites/default/files/2022-10/

ACRWC. (1990). *African Charter on the Rights and Welfare of the Child.* Retrieved August 16, 2024, from https://au.int/sites/default/files/treaties/36804-treaty-african_charter_on_rights_welfare_of_the_child

Adelowotan, M. (2021). Software, method, and analysis: Reflections on the use of Atlas.ti in a Doctoral research study. *Eurasian Journal of Economics and Finance, 9*(4), 188–203.

Adesina, J. O. (2007). In search of inclusive development: Introduction. In J. O. Adésínà (Ed.), *Social policy in sub-saharan African context in search of inclusive development.* Palgrave Macmillan Ltd..

Adesina, J. O. (2015). Return to a wider vision of social policy: Rereading theory and history. *South African Review of Sociology, 46*(3), 99–119.

Adewumi, T. M., Mosito, C., & Agosto, V. (2019). Experiences of teachers in implementing inclusion of learners with special education needs in selected Fort Beaufort District primary schools, South Africa. *Cogent Education, 6*(1), 14–19.

Amnesty International. (2020). *Broken and unequal: The state of education in South Africa.* Amnesty International Ltd.

Amod, Z. (2015). South African foundation phase teachers' perceptions of ADHD at private and public schools. *International Journal of Environmental Research and Public Health, 12*(3), 3042–3059.

Amor, A. M., Hagiwara, M. K., Shogren, A., Thompson, J. R., Verdugo, M. A., Burke, K. M., & Aguayo, V. (2018). International perspectives and trends in research on inclusive education: A systematic review. *International Journal of Inclusive Education, 23*(12), 1277–1295.

Balescut, J., & Eklindh, K. (2005). Literacy and persons with developmental disabilities: Why and how, paper commissioned for the *EFA Global Monitoring Report 2006, Literacy for Life*. Retrieved October 21, 2024, from https://unesdoc.unesco.org/ark:/48223/pf0000145940

Beckmann, I., & Reyneke, I. (2021). Covid-19 challenges to access to education for learners living with severe disabilities: An education law perspective. *Perspectives in Education, 39*(1), 122–137.

Bengtsson, M. (2016). How to plan and perform a qualitative study using content analysis. *NursingPlus Open, 2*(4), 8–14.

Bipath, K., Tebekana, J., & Venketsamy, R. (2021). Leadership in implementing inclusive education policy in early childhood education and care playrooms in South Africa. *Education Sciences, 11*(12), 815–820.

Burnett, N. (2009). Foreword. In *Policy guidelines on inclusion in education*. UNESCO.

Busetto, L., Wick, W., & Gumbinger, C. (2020). How to use and assess qualitative research methods. *Neurological Research and Practice, 2*(14), 1–10.

Calle, C. A., Campillay, C. M., Araya, G., Ojeda, A. I., Rivera, B. C., & Dubó, A. P. (2021). Access to public transportation for people with disabilities in Chile: A case study regarding the experience of drivers. *Disability & Society, 37*(6), 1038–1053.

Chirowamhangu, R. (2023). *Orphans, vulnerable children and access to basic education with reference to the Convention on the Rights of the Child: The case of Eastern Cape province, South Africa* (PhD thesis, University of Pretoria).

CRC. (1989). *UN General Assembly, Convention on the Rights of the Child, United Nations, Treaty Series, 1577(3)*. Retrieved October 26, 2024, from http://www.refworld.org/docid/3ae6b38f0.html

Creswell, J. W., & Creswell, J. D. (2022). *Research design: Qualitative, quantitative and mixed methods approaches*. Sage.

Cristobal-Fransi, E., Montegut-Salla, Y., Ferrer-Rosell, B., & Daries, N. (2020). Rural cooperatives in the digital age: An analysis of the internet presence and degree of maturity of agri-food cooperatives' commerce. *Journal of Rural Studies, 74*(1), 55–66.

CRPD. (2006). *International Convention on the Rights of Persons with Disabilities*. United Nations.

CRPD. (2016). *General Comment No. 4 (2016), Article 24: Right to Inclusive Education*. Retrieved September 15, 2024, from https://www.refworld.org/legal/general/crpd/2016/en/112080

Daniels, B. (2010). Developing inclusive policy and practice in diverse contexts: A South African experience. *School Psychology International, 31*(6), 631–643.

Deghaye, N. (2021). *Disability support and accessibility in ordinary schools in South Africa*. Stellenbosch Economic Working Papers: WP05/2021. Research on Socio-Economic Policy.

Dehnhardt, A., Grothmann, T., & Wagner, J. (2022). Cost-benefit analysis: What limits its use in policy making and how to make it more usable? A case study on climate change adaptation in Germany. *Environmental Science & Policy, 137*(22), 53–60.

Department of Education. (2008). *National strategy on screening, identification, assessment and support*. Government Printers.

Department of Education. (2011). *Guidelines for responding to learner diversity in the classroom through curriculum and assessment policy statements*. Government Printers.

Donohue, D. K., & Bornman, J. (2014). The challenges of realising inclusive education in South Africa. *South African Journal of Education, 34*(2), 1–14.

Dror, Y. (2006). Training for policy markers. In M. Moran, M. Rein, & R. Goodin (Eds.), *The Oxford handbook of public policy*. Oxford University Press.

Du Plessis, P., & Mestry, R. (2019). Teachers for rural schools a challenge for South Africa. *South African Journal of Education, 39*(4), 1–9.

Dube, B. (2020). Rural online learning in the context of COVID-19 in South Africa: Evoking an inclusive education approach. *Multidisciplinary Journal of Educational Research, 10*(2), 135–157.

Duri, B., & Luke, R. (2022). Transport barriers encountered by people with disability in Africa: An overview. *Journal of Transport and Supply Chain Management, 16*(2), 9–14.

Dyson, A. (2001). Special needs education as the way to equity: An alternative approach. *Support for Learning, 16*(3), 99–104.

Engelbrecht, P., Nel, M., Nel, N., & Tlale, D. (2015). Enacting understanding of inclusion in complex contexts: classroom practices of South African teachers. *South African Journal of Education, 35*(3), 1–10.

Equal Education. (2016). *Planning to fail: A report on equal educations visit to Eastern Cape province*. Equal Education Press.

Forsberg, J. T., & Schultz, J. (2023). Educational and psychosocial support for conflict-affected youths: The effectiveness of a school-based intervention targeting academic underachievement. *International Journal of School & Educational Psychology, 11*(2), 145–166.

Giandhari, J., Pillay, S., Wilkinson, E., Tegally, H., Sinayskiy, I., Schuld, M., Lourenco, J., Chimukangara, B., Lessells, R., Moosa, Y., Gazy, I., Fish, M., Singh, L., Khanyile, K. S., Fonseca, V., Giovanetti, M., Alcantara, L. C., Petruccione, F., & de Oliveira, T. (2021). Early transmission of SARS-CoV-2 in South Africa: An epidemiological and phylogenetic report. *International Journal of Infectious Diseases, 103*(2021), 234–241.

Grimes, P., Cruz, A., Soliman, D. M., Nabata, J. L., Cruz, K. L. S., & Vocales, S. I. (2023). *Mapping of the progress towards disability inclusive education in Eastern and Southern African countries, good practices and recommendations*. UNICEF.

Gumede, V. (2008). Public policy making in a post-apartheid South Africa—A preliminary perspective. *Africanus Journal of Development Studies*, (38)2, 7–23.

Hall, K. (2019). *Child-only households. Statistics on children in South Africa*. Children's Institute.

Hall, A., & Midgley, J. (2004). *Social policy for development*. SAGE Publications.

Herr, S. (1980). Rights of disabled persons. *Columbia Human Rights Law Review, 12*(1), 1–55.

Hill, M. (2007). *The public policy process*. Liber AB.

Hudson, B., Hunter, D., & Peckham, S. (2019). Policy failure and the policy-implementation gap: Can policy support programs help? *Policy Design and Practice, 2*(1), 1–14.

Human Rights Watch. (2021). *South Africa, events of 2020*. Retrieved August 16, 2024, from https://www.hrw.org/world-report/2021/country-chapters/south-africa

Jackson, D., McDowall, D., Mackenzie-Davey, A., & Whiting, R. (2016). *Principles of applied research methods*. SAGE Publications Inc.

Kanga, O. (1992). The politics of universalism: The case of Finnish Sickness Insurance. *Journal of Social Policy, 21*(1), 25–52.

Kett, M., Cole, E., & Turner, J. (2020). Disability, mobility and transport in low—And middle-income countries: A thematic review. *Sustainability, 12*(2), 1–18.

Khumalo, S., & Hodgson, T. F. (2017). The right to basic education for children with disabilities. *In Basic education rights handbook*. Retrieved November 16, 2024, from https://section27.org.za/wp-content/uploads/2017/02/Chapter-5.pdf

Kirchherr, J., & Charles, K. (2018). Enhancing the sample diversity of snowball samples: Recommendations from a research project on anti-dam movements in Southeast Asia. *PLoS One, 13*(8), 264–282.

Kirst-Ashman, K. K. (2010). *Empowerment series: Introduction to social work and social welfare*. Brooks/Cole Publishers.

Kota, Z., Hendricks, M., Matambo, E., & Naidoo, V. (2021). Provincial governance of education: The Eastern Cape experience. In B. Levy, R. Cameron, U. Hoadley, & V. Naidoo (Eds.), *The politics and governance of basic education: A tale of two South African provinces*. Oxford University Press.

Kumar, R. (2019). *Research methodology: A step-by-step guide for beginners*. Sage Publications Limited.

Kuper, H., Banks, L. M., Bright, T., Davey, C., & Shakespeare, T. (2020). Disability-inclusive COVID-19 response: What it is, why it is important and what we can learn from the United Kingdom's response. *Wellcome Open Research, 5*(2), 79–81.

Lang, R., Kett, M., Groce, N., & Trani, J. (2011). Implementing the United Nations Convention on the rights of persons with disabilities: Principles, implications, practice and limitations. *ALTER, European Journal of Disability Research, 5*(3), 206–220.

Magedi, M. J., Rakgogo, T. J., Mnguni, O. S., Segabutla, M. H., & Kgwete, L. (2023). COVID-19 and its impact on students with disabilities: A social justice expression at a South African university. *Transformation in Higher Education, 8*(2), 1–14.

Magnússon, G. (2019). An amalgam of ideals and images of inclusion in the Salamanca Statement. *International Journal of Inclusive Education, 23*(7), 677–690.

Mahlaule, A., McCrindle, C., & Napoles, L. (2024). Inclusive education and related policies in special needs schools in South Africa. *African Journal of Disability, 13*(2), 1–10.

Makoelle, T. M., & Burmistrova, V. (2020). The challenges of realising inclusive education in South Africa. *South African Journal of Education, 40*(4), 1–14.

Manyire, H., & Asingwire, N. (1998). The state of dynamics of social policy practice and research in Uganda. *Journal of Social Development in Africa, 13*(2), 75–83.

Mizunoya, S., Mitra, S., & Yamasaki, I. (2018). Disability and school attendance in 15 low-and middle-income countries. *World Development, 14*(3), 388–403.

Mkandawire, T. (2001). *Social policy and development programme*. United Nations Research Institute for Social Development.

Moosa-Tayob, S., & Risenga, P. R. (2022). Challenges of caregivers providing care to children with disabilities at non-governmental organisations in Tshwane townships, South Africa. *African Journal of Disability, 11*(2), 930–936.

Mpu, Y., & Adu, E. O. (2021). The challenges of inclusive education and its implementation in schools: The Southern African perspective. *Perspectives in Education, 39*(2), 225–238.

Murungi, L. (2015). Inclusive basic education in South Africa: Issues in its conceptualisation and implementation. *Potchefstroom Electronic Law Journal, 18*(1), 3160–3195.

Musenyente, E., Han, M. L., & Knigge, M. (2022). Implementation of UN Convention on the Rights of Persons with Disabilities in public and private schools in three districts of Uganda. *African Journal of Disability, 11*(2), 908–915.

Muthukrishna, N., & Schoeman, M. (2010). From 'Special Needs' to 'Quality Education for All': A participatory, problem-centred approach to policy development in South Africa. *International Journal of Inclusive Education, 4*(4), 315–335.

Ndlovu, S. (2023). Indigenous knowledge systems in Africa exclusion of students with disabilities in South African higher education. *Curriculum Perspect, 43*(1), 127–140.

Ngumbela, K. G. (2021). Unique challenges of the poverty dilemma in the Southern Cape province of South Africa. *African Journal of Public Affairs, 12*(1), 1–18.

Nilholm, C., & Göransson, K. (2017). What is meant by inclusion? An analysis of high impact research in North America and Europe. *European Journal of Special Needs Education, 32*(3), 437–451.

Odeku, O. K. (2022). Critical analysis of school pit toilet system as an impediment to the right to access quality education in South Africa. *African Journal of Public Affairs, 13*(1), 97–102.

Pillay, J., Patel, L., & Setlhare-Kajee, R. (2023). Teacher awareness of psychosocial support available as per the Integrated School Health Policy in South Africa. *South African Journal of Childhood Education, 13*(1), 1–9.

Pillay-van Wyk, V., Bradshaw, D., Groenewald, P., Seocharan, I., Manda, S., Roomaney, R. A., Awotiwon, O., Nkwenika, T., Gray, G., Buthelezi, S. S. & Mkhize, Z. L. (2020). *COVID deaths in South Africa: 99 days since South Africa's first death*. Retrieved November 16, 2024, from https://doi.org/10.7196/SAMJ.2020.v110i11.15249

Polat, F. (2011). Inclusion in education: A step towards social justice. *International Journal of Educational Development, 31*(1), 50–58.

Ramango, S. P., & Naicker, S. R. (2022). Conceptualising a framework for school leaders as they foster an inclusive education culture in schools. *Journal of Education, 8*(6), 85–106.

Ransom, B. (2008). *Missing voices: Children with disabilities in Africa*. The African Child Policy Forum.

Republic of South Africa. (1996). *The South African constitution: Chapter 2 The bill of rights*. Government Printers.

Republic of South Africa. (2013). *Regulations relating to minimum uniform norms and standards for public school infrastructure*. GN R920/2013. Government Gazette.

Republic of South Africa. (2020). *Address by President Cyril Ramaphosa on South Africa's response to the Coronavirus pandemic*. Retrieved November 16, 2024, from https://dirco.gov.za/message-by-president-cyril-ramaphosa-on-covid-19-pandemic-thursday-9-april-2020/

Savio, N. D. (2010). *Forecasting the effectiveness of policy implementation strategies*. University of Manchester.

Sewpaul, V., Lombard, A., Louw, L., & Noyoo, N. (1999). Social policy training and social work: The South African context. *Journal of Social Development in Africa, 14*(1), 5–18.

Shikwambana, M. M., & Fourie, J. V. (2023). Supporting a Tsonga learner living with Bardet-Biedl syndrome, a rare complex disability. *African Journal of Disability, 12*(1), 181–123.

Shukia, R. (2020). Fee-free basic education policy implementation in Tanzania: A 'phenomenon' worth rethinking'. *Huria Journal, 27*(1), 115–138.

Simo Fotso, A., Solaz, A., Diene, M., & Tsafack Nanfosso, R. (2018). Human capital accumulation of children in Cameroon: Does disability really matter? *Education Economics, 26*(3), 305–320.

Stats SA. (2020). *Child poverty in South Africa: A multiple overlapping deprivation analysis*. Retrieved November 19, 2024, from https://www.unicef.org/southafrica/media/4241/file/ZAF-multidimensional-child-poverty-analysis-policy-brief-07July-2020.pdf

Stats SA. (2022). *Census 2022: Statistical release*. Retrieved January 11, 2025, from https://census.statssa.gov.za/assets/documents/2022/P03014_Census_2022_Statistical_Release.pdf

Strohwald, A. (2021). *The child's rights to, in and through basic education: An analysis of South Africa's international obligations* (PhD thesis, Stellenbosch University).

Sumbane, G., Mothiba, T., Modula, M., Mutshatshi, T., & Manamela, L. (2023). The McKinsey's 7-S model framework for assessment of challenges faced by teachers of children with autism spectrum disorders in the Limpopo province, South Africa. *South African Journal of Childhood Education, 13*(1), 1–11.

Swart, E., & Pettipher, R. (2005). *A framework for understanding inclusion*. Van Schaik.

Tremmel, P., Myers, R., Brunow, D. A., & Hott, B. L. (2020). Educating students with disabilities during the COVID-19 Pandemic: Lessons learned from Commerce Independent School District. *Rural Special Education Quarterly, 39*(4), 201–210.

UNESCO. (1994). *The Salamanca statement and framework for action on special needs education*. Paris: UNESCO. Retrieved November 16, 2024, from https://unesdoc.unesco.org/ark:/48223/pf0000098427

UNESCO. (2020). *Inclusion and education: All means all*. Global Education Monitoring Report. Retrieved November 16, 2024, from https://www.right-toeducation.org/sites/right-to-education.org/files/resource-attachments/GEM_Report_Inclusion_2020_En.pdf

United Nations. (1948). *United Declaration for Human Rights*. Retrieved September 15, 2024, from https://www.un.org/sites/un2.un.org/files/2021/03/udhr.pdf

United Nations. (2015a). *Sustainable Development Goal 4 (SDG 4)*. Retrieved November 18, 2024, from https://www.globalgoals.org/goals/4-quality-education/

United Nations. (2015b). *Transforming our world: 2030 agenda for sustainable development*. United Nations.

Walton, E., & Engelbrecht, P. (2022). Inclusive education in South Africa: Path dependencies and emergences. *International Journal of Inclusive Education, 4*(2), 1–19.

WHO. (2020). *Disease burden and mortality estimates: Cause-specific mortality, 2000–2016*. Retrieved October 28, 2024, from https://www.who.int/data/gho/data/themes/mortality-and-global-health-estimates/ghe-leading-causes-of-death

Wihlborg, E. (2010). *The implementation loop—On Projects as an organizational form in the context of Governance*. Linköping University.

Yaacob, W. N. W., Yaacob, L. H., Muhamad, R., & Zulkifli, M. M. (2021). Behind the scenes of parents nurturing a child with autism: A qualitative study in Malaysia. *International Journal of Environmental Research and Public Health, 18*(16), 853–860.

Zwane, S. L., & Malale, M. M. (2018). Investigating barriers teachers face in the implementation of inclusive education in high schools in Gege branch, Swaziland. *African Journal of Disability, 6*(7), 391–404.

Part IV
Vulnerable Populations and Social Justice

Chapter 17
The Interface of Social Development and Drug Abuse Among Youths in Southern Africa: A Rehabilitative Approach

Edward Muzondo, Wilson Zvomuya, Ntokozo Nyathi, and Tariro Katsomba

17.1 Introduction

Contemporary Africa's transformative blueprint, Agenda 2063, which was enacted by the African Union (AU) in January 2013, envisions continent-wide inclusive growth and social development (AU, 2020). This blueprint, also dubbed "The Africa We Want," was conceived as a grand plan to achieve sustainable development and growth in Africa and transform it into a global powerhouse by 2063. Within the context of contemporary African social policy discourse, this transformative vision hinges on the active participation of young people, who comprise a sizeable portion of Africa's population. However, the alarming rise of drug and substance use among youths poses a significant threat to the attainment of this milestone. This challenge exemplifies the dynamic interconnected nature of contemporary African development issues that require integrated social policy responses, as highlighted in recent analyses of the continent's development trajectory (Adesina, 2025; Chipenda, 2023).

Drug use and addiction negatively impact development at personal, community, and national levels (Muzondo, 2022; Box, 2022; Jones, 2023). Young people struggling with addiction are more likely to engage in criminal activities, experience poor academic performance, engage in harmful sexual behaviors, and practice self-neglect (Box, 2022; Muzondo, 2023). The scourge is made even worse by

E. Muzondo (✉)
Department of Social Work, Eswatini Medical Christian University, Mbabane, Eswatini

W. Zvomuya
University of Zimbabwe, Harare, Zimbabwe

N. Nyathi
University of Ankara, Ankara, Turkey

T. Katsomba
Department of Social Work, University of Zimbabwe, Harare, Zimbabwe

E. Ndhlovu, C. Chipenda (eds.), *Redefining the Social Policy Discourse*,
https://doi.org/10.1007/978-3-032-24067-5_17

punitive, rather than rehabilitative, approaches when working with youths abusing drugs and/or substances (Muzondo, 2022; Box, 2022). These punitive approaches reflect broader challenges in contemporary African social policy, where external influences and donor-driven agendas often prioritize narrow, residual interventions over comprehensive, transformative approaches (Adesina, 2021; Chipenda, 2023). This not only undermines youths' potential but also negatively impacts their meaningful contribution to national and regional socio-economic development. The persistence of such approaches highlights the urgent need for redefining social policy discourse in contemporary Africa to prioritize locally informed, holistic solutions that address root causes rather than symptoms.

Despite the growing recognition of drug and substance use as a significant development challenge in Africa, there remains a critical gap in scholarship that examines this issue through a social development lens rather than conventional criminal justice or medical frameworks. Existing literature predominantly focuses on enforcement and clinical treatment approaches, with limited attention to community-based, rehabilitative interventions that prioritize youth empowerment and social reintegration. In addition, there is insufficient theoretical engagement with how drug rehabilitation can be conceptualized as a vehicle for achieving broader transformative development goals outlined in Agenda 2063.

This chapter contributes to the broader project of redefining social policy discourse in contemporary Africa by examining drug and substance use through a social development lens, thereby addressing this knowledge gap. The aim of this chapter is to provide a comprehensive analysis of the multifaceted issues associated with drug and substance use in the Kingdom of Eswatini, South Africa, and Zimbabwe. The objectives of the study included examining the prevalence of drug and substance use across Africa, thereby providing a nuanced understanding of the problem. The study also sought to identify key determinants of drug and substance use in Southern Africa, exploring the complex interplay of factors that contribute to the initiation and continuation of substance use among youths in the region. The study also further investigated the flow of illicit drugs in the region, mapping key drug trafficking routes, networks, and actors involved. The effects of drug and substance use have also been examined with the aim of developing a rehabilitative approach for drug and substance use. Importantly, this chapter fills the knowledge gap by proposing a decolonized, culturally grounded rehabilitation framework that moves beyond Western-centric approaches to embrace African philosophical traditions and community-based solutions. These objectives provide guidance in the development of a social development-inspired approach to rehabilitating youths showing symptoms of drug addiction in Southern Africa and beyond. By situating this analysis within contemporary African social policy discourse, this chapter demonstrates how targeted interventions can contribute to broader transformative development goals while addressing the structural constraints that have historically limited policy effectiveness across the continent.

The authors argue that a restorative approach focusing on rehabilitation, rather than punishment, has the potential to lead to improved outcomes towards the attainment of Agenda 2063. This perspective aligns with contemporary calls for more integrated and adaptive social policy frameworks that can address multiple development challenges simultaneously while building resilience and sustainability (UNRISD, 2023; Redaelli & Arevalo, 2023). Rehabilitation of youths who are victims of drug and substance use has the potential to increase their self-esteem and mental health, increase youth participation in community activities, and promote accelerated progress towards Africa's economic and social development. Such an approach exemplifies the kind of transformative social policy thinking needed in contemporary Africa, one that moves beyond narrow, punitive responses to embrace comprehensive, empowering interventions that address root causes while building human capital for sustainable development. By addressing drug and substance use through holistic rehabilitation programs, Africa can foster resilient and healthy youth who can provide future leadership, promote inclusive social development, drive economic growth, and achieve Agenda 2063 objectives.

17.2 Theoretical Framework: Social Development and Youth Empowerment

17.2.1 Conceptualizing Social Development in Contemporary Africa

The concept of social development, while defying precise definition due to its multifaceted nature, provides a crucial lens through which to examine drug and substance use among African youths (Midgley, 1995; Lombard, 2017; Raenarri & Stazzari, 2023). Rather than viewing social development as a static concept, this chapter adopts a dynamic understanding that emphasizes improving the well-being of individuals and their communities by reducing poverty, promoting economic growth and participation, and increasing access to social services. This approach is relevant in contemporary Africa, where traditional Western-centric development models have often failed to address the dynamic, interconnected challenges facing the continent's youth.

Social development represents a dynamic way of organizing resources and human interactions to create opportunities through which the potential of all people, individually and collectively, can be fully developed (Midgely, 2014; Lombard, 2017; Patel, 2005). This definition is significant because it shifts the focus from deficit-based approaches that pathologize youth drug use to asset-based frameworks that recognize the inherent potential within affected communities. It is apparent from the above definition that the concept of social development can be analyzed from different social, political, and economic perspectives. However, for the purposes of this chapter, the emphasis is placed on how social development principles can inform culturally grounded, community-based responses to drug and substance use that align with African philosophical traditions and values.

From a developmental perspective, relevant to the African context, the social development approach involves the use of locally available resources to empower people, including the youths, to solve their own social, economic, and political challenges with a human face (Todaro & Smith, 2012). This understanding is crucial because it challenges the dependency paradigm that has historically characterized many interventions in Africa. Instead, it promotes self-reliance and indigenous solutions that respect local knowledge systems and cultural practices.

17.2.2 Africa Agenda 2063: A Social Development Framework for African Transformation

Understanding Agenda 2063 as a social development endeavor requires examining its historical context and its potential to address contemporary challenges such as youth drug and substance use. Africa Agenda 2063 is a strategic framework to guide Africa's development and transformation over a 50-year period from 2013 to 2063. While the Organization of African Unity (OAU) focused on decolonization of Africa and performed very well in that regard (Twala, 2014), its successor, the African Union (AU), is focusing on development on a broader front: economic, social, political, scientific, and cultural. This shift represents a maturation in African development thinking, moving beyond political independence to encompass comprehensive human development that includes addressing social challenges such as drug and substance use among youth.

The development of the Africa Agenda 2063 was built on existing African frameworks, programs, and declarations. According to DeGhetto et al. (2016), consultations were made with a broad spectrum of African stakeholders at the grassroots level, involving the synthesis of 35 national and Regional Economic Communities (RECs) strategic and action plans and situational analysis of global mega trends. This participatory approach is significant because it demonstrates a commitment to locally informed policy development that contrasts with externally imposed development frameworks. Agenda 2063 was adopted by the African Union Summit in January 2015, and its First Ten Year Implementation Plan (FTYIP) was adopted in June 2015, predating the global Agenda 2030 and Sustainable Development Goals (SDGs) of September 2015. This timeline is important because it establishes Africa's agency in defining its own development trajectory, rather than merely responding to global initiatives.

It is noteworthy that the global Agenda 2030 and its SDGs were heavily influenced by the African Union's Common Position on Post 2015 Development Agenda (CAP), with Africa being the only region to submit a well-articulated position in writing. The UN Open Working Group (OWG) and formal inter-governmental negotiations relied heavily on CAP (African Union, 2020). This influence demonstrates Africa's growing capacity to shape global development discourse, and it provides a foundation for arguing that African-led solutions to challenges such as drug and substance use should be taken seriously by the international community. CAP was adopted by the AU Summit in Addis Ababa in January 2015 and promulgated in Ndjamena in February 2015.

Most relevant to this chapter's focus, Agenda 2063 sets out youth empowerment as one of its key program areas to achieve a prosperous, peaceful, and integrated Africa (AU, 2020). The initiative acknowledges the potential of Africa's young population in driving transformation and development across the continent. African countries have made a pledge to promote education and skills development, employment, and entrepreneurial opportunities and ensure active youth participation for Africa's economic growth (Akumbu & Jukanovich, 2024; Raphael & Winter-Ebmer, 2001). However, the reality that this same sub-population is increasingly falling victim to the scourge of drug and substance use presents a significant contradiction that requires urgent attention through innovative social policy approaches. It is clear from the above that Agenda 2063 puts youths at the center of a prosperous and sustainable Africa's future, yet the same sub-population is falling victim to the scourge of drug and substance use. This paradox highlights the need for integrated approaches that address drug and substance use not as isolated problems but as barriers to achieving broader development goals.

17.2.3 The Global Context of Youth Drug and Substance Use: Implications for Africa

To understand the significance of drug and substance use among African youths, it is essential to situate this challenge within the broader global context while recognizing the unique factors that shape the African experience. Drug use refers to the misuse or overindulgence of psychoactive substances that are illegal and addictive for various purposes, including recreational (Baumgartner, 2020; Jones, 2023). According to the United Nations Office on Drugs and Crime (UNODC) (2020), drug use is a global pandemic with more than 35 million people suffering from substance use disorders. The increase in drug and substance use across the globe has been attributed to the growing production, distribution, promotion, and easy availability of illicit substances (Sahu & Sahu, 2016). However, the manifestation of this global phenomenon in Africa is shaped by unique factors, including historical trauma, structural inequalities, limited access to mental health services, and the erosion of traditional community support systems.

The clinical definition of substance abuse, as provided by the American Psychiatric Association (2013, p. 483), describes it as a maladaptive pattern of substance use that leads to significant impairment or distress, as manifested by one or more of the following within a 12-month period: recurrent substance use resulting in failure to fulfil major role obligations, recurrent substance use in situations in which it is physically hazardous, recurrent substance-related legal problems, or continued substance use despite persistent or recurrent social or interpersonal problems. While this definition provides important clinical parameters, it fails to capture the dynamic social, cultural, and structural factors that contribute to substance use in African contexts, thereby reinforcing the need for more culturally grounded understanding and interventions.

The global prevalence of youth drug and substance use is indicative of the magnitude of this challenge. The maladaptive use of drugs and substances is a concern not only in Africa but across the entire world. This is evident in various global statistics, including the United Nations Office on Drugs and Crime (UNODC, 2020) report that 13.9% of youths aged 15–24 used illicit drugs in 2019 in the United Kingdom. Similarly, the National Survey on Drug Use and Health (NSDUH, 2020) found that 31.4% of American youths aged 12–17 used illicit drugs in 2020, with cannabis and prescription painkillers being the most used substances. In Europe, the European Monitoring Centre for Drugs and Drug Addiction (EMCDDA, 2020) reported that 15% of youths aged 15–24 used cannabis in 2019, with amphetamines and cocaine also being widely used. Additionally, studies in India (Ray et al., 2017) and Australia (Australian Institute of Health and Welfare, 2020) revealed that 28.5% and 22.2% of youths aged 18–25 and 12–17, respectively, used substances. These statistics collectively reveal that drug and substance abuse is indeed a global pandemic, but they also highlight the need for context-specific responses that acknowledge the unique circumstances facing African youth.

The factors contributing to youth substance use, as identified by the World Health Organization (WHO, 2019), include peer pressure, family dynamics, mental health issues, and social media influence. However, in the African context, these factors intersect with additional challenges such as poverty, unemployment, political instability, and the breakdown of traditional social structures. WHO recommended that countries put in place effective prevention and intervention strategies to address substance use among youths. UNODC (2020) notes that school-based programs, family-based interventions, and community-based initiatives have shown promise in reducing youth substance use. While these recommendations provide valuable guidance, their implementation in African contexts requires adaptation to local cultural norms, resource constraints, and the specific structural challenges facing the continent's youth. This adaptation process forms the foundation for the culturally grounded, social development approach proposed in this chapter.

17.3 Methodology

This study used a scoping review analysis to gather and synthesize information. Scoping review analysis is an approach whereby a body of literature is identified and synthesized to reach a conclusion on a certain topic (Logan et al., 2024). Scoping review uses a systematic and iterative approach to identify and synthesize an existing or emerging body of knowledge on a given topic (Munn et al., 2022). Researchers conducted an internet search to investigate the prevalence and effects of drug and substance use among youths in Africa and further delved into Southern

Africa and subsequently, the Kingdom of eSwatini, South Africa, and Zimbabwe. A total of 43 sources of information (21 journals, 15 book chapters, and seven reports) were retrieved and analyzed to address the study objectives.

This study focuses on Southern Africa, a region characterized by diverse cultures, languages, and socio-economic contexts (AU, 2020). Specifically, it zooms in on Eswatini, South Africa, and Zimbabwe, which share similar historical, cultural, and socio-economic contexts that contribute to the prevalence of drug and substance use among youths. The Kingdom of Eswatini and Zimbabwe are both landlocked countries, with South Africa being the southernmost country on the African continent. According to World Bank (2023) statistics, Eswatini has approximately 1.2 million people, Zimbabwe has approximately 16.4 million, and South Africa has approximately 60.6 million people.

The corresponding author facilitated team reflexivity through formal and informal group discussions with all three authors. This collaborative approach aimed to establish a shared understanding of team practices and research quality. The authors employed a relational approach (Munn et al., 2022), which considers authors' perspectives, relationships, and realities. Through this approach, authors contributed relevant literature, including their own published works and known references. The authors' diverse expertise provided comprehensive insights, with authors 1 and 2 focusing on Zimbabwe and eSwatini, and authors 3 and 4 exploring Zimbabwe and South Africa.

Availability sampling was used to identify studies focusing on drug and substance use among youths (15–35 years) in Africa. The authors targeted articles published after 2014 so that contemporary information is used. Search engines were used to quickly locate relevant studies using keywords that include "drug abuse," "substance use," "youths," "Africa," and "drug use and social development."

Library searches were downloaded into EndNote software, and relational articles were hand-entered by the lead author. The EndNote files were uploaded into the COVIDENCE online software, where auto-generated duplicates were removed. Title and full article screening were done by authors 1, 2, and 3. Out of 43 articles initially identified, 36 articles were finally synthesized for data analysis on the prevalence of drug and substance use in Southern Africa, effects of drug and substance use among the youth, analysis of illicit drug supply networks, and the applicability of social development in the rehabilitation of drug and substance use.

Thematic analysis was used to sort the narrative summary of the findings into key themes related to drug and substance use among youths in southern Africa. Key sub-themes were also identified to help readers quickly understand and navigate important findings from the study. The presentation and discussion of study findings were done simultaneously to promote a coherent understanding of drug and substance use in the region, its impact on national economies, and Agenda 2063, as well as gaps in the current intervention strategies. The authors used the study findings to propose a social development-inspired rehabilitation approach, "The ARISE Cycle," to deliver effective drug and substance users in a developmental manner to achieve Agenda 2063.

The authors were aware of the potential bias brought by scoping review; therefore, data extraction was shared among authors, and analysis was discussed in our weekly meetings. As such, a number of ethical considerations were observed in the process. These included data authenticity by verifying the source of extracted articles or sources of information, their publisher, and the ethical values of the retrieved article. Data acknowledgement and confidentiality were maintained throughout the compilation of this chapter. This helped to check and reflect on any potential bias or error during the collection and presentation of findings.

17.4 Study Findings and Discussion

17.4.1 Drug and Substance Use in Africa

Africa's position as a major transit route in the global illicit drug trade has transformed the continent's drug landscape, creating unprecedented challenges for youth development and social policy implementation (AU, 2020). The continent's strategic geographic location, combined with porous borders and governance challenges, has established what the African Union describes as the "path of least resistance" in international drug trafficking networks. This positioning has not only increased the availability and accessibility of drugs but has also exposed young people, the demographic central to Agenda 2063's transformative vision, to significant risks that threaten their potential contributions to continental development. The contemporary drug situation in Africa is characterized by both external trafficking networks and domestic cultivation, creating a complex supply environment that challenges traditional enforcement approaches. According to the World Drug Report, cannabis cultivation occurs throughout the continent, while key trafficking and transit countries include Nigeria and Ghana in West and Central Africa, Mozambique, Eswatini, and Malawi in Southern Africa, and Tanzania, Uganda, and Kenya in East Africa. In North Africa, Morocco serves as a significant hub for illicit drug activities. This geographical spread demonstrates how drug and substance use has evolved from isolated incidents to a continent-wide development challenge that requires integrated social policy responses aligned with Agenda 2063's holistic approach to transformation.

The emergence of New Psychoactive Substances (NPS) represents a concerning trend that exemplifies the dynamic nature of contemporary drug challenges in Africa. Despite limited empirical research on this phenomenon, media reports from countries including Zimbabwe, Eswatini, and South Africa suggest widespread NPS usage (AU, 2020). This gap in research reflects broader challenges in contemporary African social policy, where external influences often prioritize Global North experiences over locally informed understanding of African realities (Dumbili et al., 2021). The rapid introduction of NPS into African markets creates regulatory challenges and complicates rehabilitation efforts, highlighting the need for adaptive, culturally grounded approaches that can respond to evolving substance use patterns while maintaining focus on youth empowerment and social reintegration.

Poly-drug use patterns among African youth further complicate the relationship between substance use and development outcomes, requiring comprehensive social policy responses that address multiple interconnected challenges simultaneously. The African Union (2020) reports high alcohol usage rates in Zimbabwe (58%), Lesotho (54%), and Rwanda (76%), with addiction rates reaching 30% in Ethiopia, 17.4% in Rwanda, and 0.3% in Madagascar. This pattern of multiple substance use among youth directly contradicts Agenda 2063's vision of empowered young people driving continental transformation. Instead, it creates dynamic health risks and treatment challenges that require integrated, community-based interventions rather than punitive approaches.

17.4.2 Structural Determinants of Youth Drug Use in Southern Africa

The examination of drug use determinants through a social development lens reveals how systemic challenges create vulnerabilities that undermine youth potential and threaten Agenda 2063's transformative goals. Rather than viewing drug use as individual moral failings, this analysis demonstrates how structural factors: unemployment, poverty, corruption, social fragmentation, and peer influence create environments where substance use becomes a rational response to limited opportunities and systemic exclusion.

17.4.2.1 Unemployment as a Development Challenge

Youth unemployment in Africa represents both a cause and consequence of drug-related problems, creating cycles that perpetuate exclusion from meaningful participation in continental transformation. With approximately 13 million unemployed youth across the continent and an additional 60 million not in education, employment, or training, the scale of youth marginalization directly challenges Agenda 2063's emphasis on youth empowerment and participation. Specific country contexts illustrate this challenge: Zimbabwe's youth unemployment rate of 90%, Eswatini's overall unemployment rate of 58%, and South Africa's youth unemployment rate of 39% (WHO, 2021) create conditions where young people face limited legitimate opportunities for economic participation and social contribution.

This unemployment crisis creates dual pathways to drug involvement that reflect broader failures in contemporary African social policy approaches. On one hand, unemployed youth may engage in the drug trade as an alternative economic strategy, increasing the availability and accessibility of substances. On the other hand, unemployment exacerbates feelings of helplessness and frustration, leading to drug use as escapism from systemic exclusion (Lee et al., 2019). This analysis demonstrates how addressing youth drug use requires comprehensive social development approaches that create legitimate opportunities for economic participation rather than merely focusing on enforcement or individual treatment.

17.4.2.2 Poverty and Economic Marginalization

Poverty in Southern Africa represents a significant barrier to youth development that intersects with drug use in ways that require social development responses rather than punitive approaches. In Eswatini, 54% of youth live in poverty, with 42% experiencing extreme poverty and significant urban-rural income disparities (Taylor et al., 2020; WHO, 2021). These conditions create vulnerability to both drug use as a coping mechanism and involvement in the drug trade as a survival strategy, reflecting broader failures in contemporary African social policy to address the root causes of youth marginalization.

The relationship between poverty and drug use exemplifies how individual problems reflect systemic challenges that require comprehensive social policy responses. When young people without formal education become primarily responsible for drug-related offences (SAPS, 2021), this pattern reveals how educational and economic exclusion create pathways to criminalization rather than empowerment. This perspective supports the argument for rehabilitation over punishment, as addressing poverty and educational exclusion through social development approaches can prevent drug involvement while building human capital for continental transformation.

17.4.2.3 Corruption and Governance Challenges

Corruption and porous border facilities represent governance failures that facilitate drug trafficking while undermining the rule of law and social cohesion necessary for Agenda 2063's transformative vision. The easy accessibility of illicit substances, including heroin, cannabis, and codeine-containing medications, through corrupted entry points exposes young people to substances that compromise their developmental potential (AU, 2020). This situation demonstrates how governance challenges create environments where youth face increased risks rather than expanded opportunities, directly contradicting Agenda 2063's emphasis on good governance and youth empowerment. The proliferation of unlicensed alcohol outlets and local substance production further illustrates how governance failures create systematic vulnerabilities for youth. Examples include the widespread availability of *musombodiya* in Zimbabwe, an illicitly produced, highly intoxicating beverage, and unlicensed alcohol outlets (shebeens) in South Africa (Ayuka et al., 2014; Mukwenha et al., 2021). These conditions reflect broader challenges in contemporary African social policy, where weak implementation of regulations creates environments that undermine youth development rather than supporting their potential contributions to continental transformation.

17.4.2.4 Social Fragmentation and Cultural Challenges

The decay of traditional social structures and cultural practices represents a challenge that requires culturally grounded, community-based responses rather than externally imposed solutions. Contemporary sociocultural practices have evolved in ways that inadvertently promote substance use, with young people viewing entertainment at social events as opportunities for drug experimentation (Mabasa, 2018; Matutu & Mususa, 2019). This transformation reflects broader challenges in contemporary African social policy, where rapid social change has outpaced the development of supportive cultural frameworks that could guide youth development positively. Specific examples from the region illustrate how cultural practices have evolved in ways that create risks rather than opportunities for youth development. In Eswatini, young people increasingly overindulge in alcohol at traditional ceremonies (Langa, 2019), while in Zimbabwe, "*vuuzu*" parties expose school-age children to substance use and risky sexual behaviors (Kurevakwesu et al., 2023). These patterns demonstrate how addressing youth drug use requires culturally sensitive approaches that strengthen positive cultural practices while addressing harmful adaptations. This aligns with social development principles that emphasize community-based solutions and cultural appropriateness.

17.4.2.5 Peer Influence and Developmental Vulnerabilities

Peer pressure represents a significant factor in youth substance use that requires understanding within broader developmental and social contexts rather than individual pathology frameworks. Research consistently identifies peer influence as a predictor of youth substance use (Mishra, 2015; Kurevakwesu et al., 2023), with school students particularly vulnerable to experimentation and peer pressure. This pattern reflects normal developmental processes identified in Erik Erikson's psychosocial development theory, where adolescents naturally seek independence while being influenced by expanding social networks. Understanding peer influence through a social development lens reveals how creating positive peer networks and community support systems can redirect youth energy toward constructive activities that contribute to Agenda 2063's transformative goals. Rather than criminalizing peer interaction, effective social policy approaches should focus on creating environments where peer influence supports positive development through education, skills development, and meaningful participation in community activities.

17.4.3 Consequences of Drug Use: Barriers to Youth Development and Continental Transformation

The effects of drug and substance use among youth represent barriers to achieving Agenda 2063's vision of empowered young people driving continental transformation. These consequences extend beyond individual health impacts to affect educational outcomes, family stability, and community development, creating cyclical patterns that perpetuate exclusion and undermine social cohesion necessary for sustainable development.

17.4.3.1 Health and Cognitive Implications

Drug addiction's impact on brain circuits involved in learning, memory, and behavioral control directly undermines youth capacity for educational achievement and meaningful participation in development processes. This neurological dimension of addiction supports arguments for rehabilitation over punishment, as addiction represents a complex health condition requiring comprehensive treatment rather than criminal justice responses. The recognition that some individuals face greater vulnerability due to genetic makeup, age of exposure, and environmental influences emphasizes the need for preventive approaches that address environmental risk factors through social development interventions.

17.4.3.2 Educational Disruption and Human Capital Development

The relationship between substance use and poor educational performance represents a threat to human capital development necessary for achieving Agenda 2063's transformative goals. Cognitive impairment, memory loss, and decreased concentration among substance users seriously affect their capacity to acquire and retain knowledge (Muoti, 2014). This educational disruption creates long-term consequences that extend beyond individual students, affecting community development and national progress. Reduced educational attainment limits young people's capacity to contribute to economic growth and social development. The UN (2015) report's findings that adolescent substance users experience cognitive and behavioral issues that disrupt classroom environments demonstrate how individual drug use creates broader educational challenges. This pattern illustrates how addressing youth drug use requires comprehensive approaches that support both affected individuals and educational environments, aligning with social development principles that emphasize community-wide interventions and systemic solutions.

17.4.3.3 Family and Social Disruption

Drug use among youth creates family instability that undermines social cohesion and community development essential for Agenda 2063's transformative vision. Substance use leads to family distancing, disrupted relationships, and increased risk of divorce and child maltreatment (Swartbooi, 2017; Matutu & Mususa, 2019). These family disruptions create intergenerational cycles of vulnerability that perpetuate social problems and limit community capacity for positive development. The impact of youth addiction on parental marriages and family stability (Barnard, 2015; Velleman et al., 2015) demonstrates how individual substance use creates broader social consequences that require community-based responses rather than individualized treatment approaches. This analysis supports arguments for family-inclusive rehabilitation programs that address system-wide impacts and build collective resilience for sustainable recovery and development.

17.4.3.4 Economic and Criminal Behavior Consequences

The relationship between drug use and criminal behavior represents a barrier to youth participation in legitimate economic activities and social development processes. Substance use frequently leads to maladaptive behaviors, including asset liquidation, workplace problems, and property crimes (Muzondo, 2022; Van Wormer & Davies, 2008). These patterns create cycles where substance use leads to economic marginalization, which in turn increases vulnerability to continued drug involvement and criminal activity. The South African Police Service (2021) found that young people without formal education are primarily responsible for drug-related crimes, illustrating how educational exclusion intersects with substance use to create pathways to criminalization rather than empowerment. This pattern demonstrates how addressing youth drug use requires comprehensive social development approaches that address educational access, economic opportunities, and social inclusion simultaneously.

17.4.4 The Illicit Drug Supply Network: Understanding Structural Challenges

The illicit drug supply network in Southern Africa represents a dynamic system that requires comprehensive social development responses rather than enforcement-only approaches. This network comprises interconnected actors, drug traffickers, distributors, sellers, and users who operate within broader social, economic, and political contexts that facilitate drug availability and accessibility among youth. These are shown in Fig. 17.1 below:

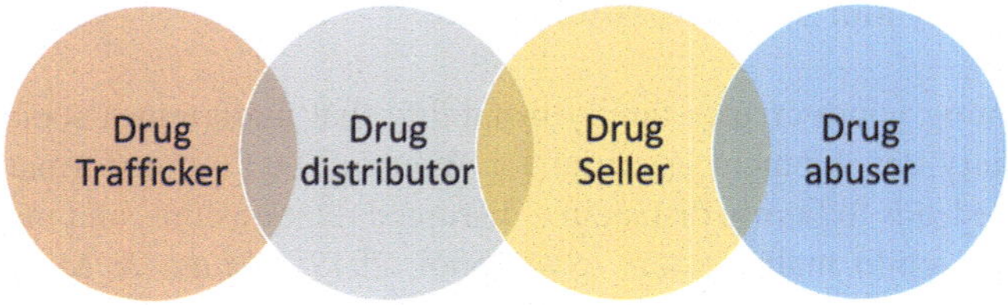

Fig. 17.1 The illicit drug network in Africa. (Source: Developed by Authors)

The hierarchical nature of this network, from drug traffickers and distributors to sellers and users, demonstrates how addressing youth drug use requires interventions at multiple levels of the supply chain. Drug traffickers operate through corrupted ports of entry, while distributors include influential community members who collect drugs from traffickers and supply them to local sellers. Drug sellers serve as foot soldiers who maintain direct contact with youth users, creating personal relationships that facilitate continued drug access and addiction. Understanding this network through a social development lens reveals how structural vulnerabilities create opportunities for drug involvement at each level. Rather than criminalizing participants, effective social policy approaches should focus on addressing the underlying conditions, poverty, unemployment, corruption, and social exclusion that make drug network participation attractive or necessary for survival.

The UNODC (2023) suggests that dismantling this network requires empowering key actors using social development principles, which aligns with this chapter's argument for rehabilitation over punishment. By addressing root causes through comprehensive social development approaches, African countries can create legitimate opportunities for youth participation in development processes rather than illicit activities that undermine continental transformation goals. This analysis demonstrates how contemporary African social policy must move beyond narrow enforcement approaches to embrace comprehensive, community-based interventions that address the structural determinants of drug network participation while building positive alternatives for youth engagement and empowerment.

17.5 The ARISE Cycle: A Transformative Framework for Youth Rehabilitation

The structural and systematic nature of drug use in southern Africa necessitates comprehensive, multidimensional responses that transcend traditional punitive approaches. Building on the social development principles outlined in this chapter and the empirical evidence from Eswatini, South Africa, and Zimbabwe, the authors propose the ARISE Cycle as a transformative framework that addresses the root causes of substance abuse while promoting sustainable rehabilitation outcomes aligned with Agenda 2063's vision of youth empowerment. The ARISE Cycle represents a paradigmatic shift from Western-centric, deficit-based interventions to African-centered, asset-based approaches that recognize the inherent potential within affected communities. This framework acknowledges that effective drug rehabilitation requires addressing multiple individual needs beyond drug use itself, bridging the micro-macro divide between individual users and their social environments while respecting cultural values and traditional knowledge systems. The cycle's five interconnected phases collectively address the complex nature of youth substance abuse through locally informed, community-driven solutions.

Associating with drug users represents a reorientation in perspective, viewing young people who use drugs as victims of systemic failures and illicit networks rather than criminals deserving punishment. This phase emphasizes decriminalization and treatment of users as "individuals in need of care," challenging the traditional punitive paradigm that has proven counterproductive. The evidence from the three countries demonstrates that criminalization approaches deter rehabilitation while often targeting users, leaving traffickers free. This highlights the need for resource mobilization focused on empowering drug users through job creation, social fabric reorganization, and dedicated support networks that reduce dependency on substances through meaningful engagement.

Research forms the critical knowledge foundation for effective, culturally appropriate intervention. This phase involves comprehensive information gathering about drug and substance use patterns among young people, utilizing strength-based approaches that respect African epistemological traditions to understand user ambitions and needs. Researchers must discover and disseminate insights on production and distribution networks while employing bio-psycho-social, cultural, and spiritual frameworks to develop contextually relevant harm reduction strategies. This research must be conducted by African scholars and practitioners who understand local contexts, building community awareness about new psychoactive substances and their health implications while fostering indigenous knowledge systems.

Introducing community support recognizes that positive change is central to social development and must be rooted in African communal values and practices. This phase facilitates improvements in tangible living conditions, including employment opportunities, healthcare access, literacy, and socioeconomic status, through community-led initiatives. When people are positively occupied and working toward their dreams within supportive cultural contexts, their need for drugs diminishes significantly. Social development's preventive nature requires early-stage empowerment that builds on traditional African concepts of ubuntu and collective responsibility, enabling individuals to contribute toward sustainable societal improvements.

Securing their future addresses the comprehensive needs of individuals within their social contexts, emphasizing the interconnectedness central to African worldviews. This phase establishes supportive environments connecting individual drug users with family, social groupings, community, and society at large through culturally appropriate mechanisms. By linking individual micro-needs with available macro capabilities while respecting traditional healing practices and community structures, rehabilitation services address medical, psychological, social, vocational, and legal aspirations within frameworks that honor age, gender, ethnicity, and cultural considerations.

Empowering institutions promotes independence and community-based self-sufficiency while strengthening governance structures necessary for sustainable development. This phase addresses systemic corruption through competitive remuneration of police and immigration personnel, creating viable institutional and employee welfare arrangements that support rather than undermine community well-being. Genuine community participation enables collective action against drug networks, with clear protocols and referral pathways for identifying and reporting drug suppliers that build on traditional conflict resolution mechanisms. Grassroots programs are essential for reducing drug supply at community levels where drug hubs operate, utilizing traditional leadership structures and community mobilization approaches that align with cultural practices. Below is a clear illustration of the ARISE Cycle proposed by the authors to rehabilitate young people from drug and substance use:

The ARISE Cycle framework offers a comprehensive, culturally grounded approach to youth substance abuse rehabilitation that addresses both individual needs and systemic challenges while contributing to broader Agenda 2063 objectives (see Fig. 17.2). Its implementation requires coordinated efforts between government agencies, educational institutions, healthcare providers, traditional leaders, and community organizations within frameworks that respect African values and knowledge systems. Success depends on sustained commitment to evidence-based

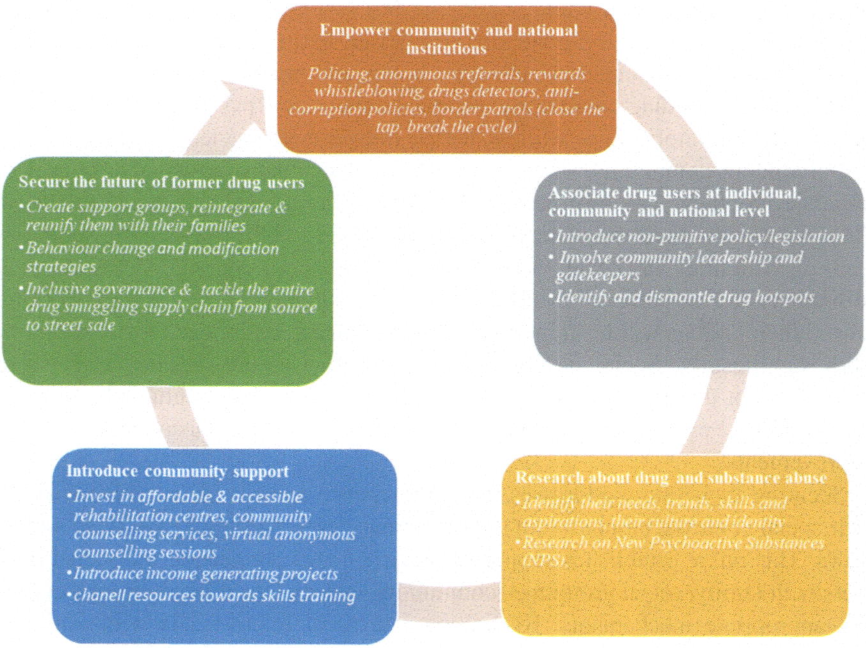

Fig. 17.2 The ARISE cycle. (Source: Developed by Authors)

practices that incorporate indigenous knowledge, adequate resource allocation that prioritizes community ownership, and continuous monitoring and evaluation of intervention outcomes that measure both individual rehabilitation and community transformation.

17.6 Conclusion

This chapter has established that drug and substance use has evolved into a global pandemic affecting diverse age groups across social, cultural, economic, and psychological dimensions. In southern Africa, this crisis has been exacerbated by structural factors including unemployment, poverty, corruption, porous border controls, social fabric deterioration, and peer pressure. The proliferation of drug use among African youth poses a direct threat to the achievement of Agenda 2063, undermining the continent's developmental aspirations and contradicting the framework's emphasis on youth empowerment as a cornerstone of transformation.

The analysis reveals that the complexity of drug supply networks significantly complicates regional efforts to combat substance abuse, while current interventions remain predominantly reactive and punitive, rendering them ineffective and inadequate. These challenges are further compounded by corruption, selective law enforcement, and political interference that inadvertently facilitate drug distribution networks among youth populations. The evidence presented demonstrates that traditional approaches that criminalize drug users while failing to address underlying structural causes have proven counterproductive, necessitating a paradigm shift toward rehabilitative, social development-oriented interventions.

Through examining the experiences of Eswatini, South Africa, and Zimbabwe, this chapter has demonstrated that effective intervention requires embracing social development principles that recognize drug users as victims of systemic failures rather than criminals deserving punishment. The research has shown that comprehensive education programs within schools and communities, when implemented through culturally grounded approaches, can increase awareness about drug-related risks while equipping young people with the necessary information and skills to resist peer pressure. These programs must promote healthy coping mechanisms, decision-making skills, and positive self-esteem through collaborative efforts that respect African philosophical traditions and community-based solutions.

The chapter has highlighted the critical importance of developing and strengthening support networks for youth struggling with substance use disorders through enhanced access to counselling services and specialized treatment centers that emphasize multidisciplinary approaches. Such interventions must integrate therapy, medication-assisted treatment, and family and community involvement within frameworks that bridge individual needs with broader social transformation goals. Continuous monitoring and evaluation mechanisms are essential to ensure intervention effectiveness, utilizing data and feedback to progressively refine and improve programs while maintaining cultural relevance and community ownership.

The research findings highlight that ongoing investigation into drug and substance use issues is crucial for informing evidence-based policy development and implementation that aligns with Agenda 2063 objectives. Data collection on trends, prevalence rates, and risk factors enables policymakers and communities to tailor prevention strategies to specific contexts while building local capacity for sustainable responses. Adequate funding allocation for research initiatives is crucial for fostering evidence-based practices and sustainable interventions that contribute to broader human development goals.

The adoption of the ARISE Cycle as a vehicle for change represents a paradigmatic shift toward African-centered, social development-based solutions that can contribute significantly to achieving Agenda 2063's developmental goals while addressing the youth substance abuse crisis in southern Africa and beyond. This framework demonstrates how locally informed, holistic approaches can transform social policy discourse in contemporary Africa, moving beyond narrow, externally imposed interventions to embrace comprehensive, empowering solutions that build human capital while addressing structural constraints. By implementing the ARISE Cycle, African nations can foster resilient and healthy youth who can provide future leadership, promote inclusive social development, drive economic growth, and achieve the transformative vision embodied in "The Africa We Want."

References

Adesina, J. (2021). Introduction: Social policy in the African context. In J. Adesina (Ed.), *Social policy in the African context* (pp. 1–12). CODESRIA.

Adesina, J. (2025). *Meeting the Cesare challenge: Locally grounded, globally relevant scholarship (in postgraduate training).* Paper presented at the UNISA College of Graduate Studies Research and Innovation Week, 19 May.

African Union. (2020) *Addressing drug use in sub-Saharan Africa: Policy recommendations for effective prevention and intervention*. Retrieved December 12, 2023 from https://au.int/

Akumbu, R. V., & Jukanovich, J. (2024). African Union's youth education challenge: A critical review of agenda 2063. *The Scholarship Without Boarders Journal, 2*(2), 1.

American Psychiatric Association. (2013). *Diagnostic and statistical manual of mental disorders* (5th ed.). American Psychiatric Publishing.

Australian Institute of Health and Welfare. (2020). *National Drug Strategy Household Survey 2019*. https://www.aihw.gov.au/reports/illicit-use-of-drugs/national-drug-strategy-household-survey-2019/contents/summary

Ayuka, F., Barnett, R., & Pearce, J. (2014). Neighbourhood availability of alcohol outlets and hazardous alcohol consumption in New Zealand. *Health & Place, 29*, 186–199.

Barnard, M. (2015). *Drugs in the family: The impact on parents and siblings*. Joseph Rowntree Foundation.

Baumgartner, C. (2020). Exploring the effects of legalizing Marijuana: A comparative analysis of Colorado and Washington state. *Journal of Drug Policy Analysis, 12*(1), 56–74.

Box, W. (2022). The increase in women who use drugs in Zimbabwe. In J. Buxton, G. Margo, & L. Burger (Eds.), *The impact of global drug policy on women: Shifting the needle* (pp. 141–148). Emerald Publishing Limited.

Chipenda, C. (2023). Rethinking the role of social policy in responding to the Covid-19 pandemic in South Africa: Critical reflections using the intellectual contribution of Thandika

Mkandawire. In V. Gumede & E. Ndhlovu (Eds.), *Development in Africa: Remembering Thandika Mkandawire* (pp. 197–231). Ace Press.

DeGhetto, K., Gray, J., & Kiggundu, M. N. (2016). The African Union's Agenda 2063: Aspirations, challenges and opportunities for management research. *Africa Journal of Management, 2*(1), 93–116. https://doi.org/10.1080/23322373.2015.1127090

Dumbili, E., Ebuenyi, I. D., & Ugoeze, K. C. (2021). New psychoactive substances in Nigeria: A call for more research in Africa. *Emerging Trends in Drugs, Addictions, and Health, 1*, 100008. ISSN 2667-1182.

European Monitoring Centre for Drugs and Drug Addiction. (2020). *European Drug Report 2020.* https://www.euda.europa.eu/edr2020_en

Jones, J. (2023). The role of adverse childhood experiences in drug use and addiction: A systematic review. *Journal of Substance Use Treatment, 15*(4), 285–302.

Kurevakwesu, W., Masona, G., Mabeza, T., Zengeni, F., & Chiweshe, C. C. (2023). Factors responsible for substance use among youths in urban areas and social work-related interventions. *Journal of Social Work Practice in the Addictions.* https://doi.org/10.1080/1533256X.2023.2212475

Langa, I. (2019). Wastewater analysis of psychoactive drugs: Non-enantioselective vs enantioselective methods for estimation of consumption. *Forensic Science International, 325*, 110873.

Lee, J. S., Mukhopadhyay, P., Matyas, C., Trojnar, E., Paloczi, J., Yang, Y. R., & Lohoff, F. W. (2019). Inhibition as a novel therapeutic target for alcoholic liver disease. *Scientific Reports, 9*(1), 17167.

Logan, J., Webb, J., Singh, N. K, & Ayala, A. P. (2024). Scoping review researches in social sciences: A scoping review. *Research Synthesis Methods*. Retrieved January 14, 2025, from https://doi.org/10.1002/jrsm.1742

Lombard, B. (2017). Understanding the link between mental health disorders and substance use: A comprehensive review. *Journal of Psychiatric Research, 25*(3), 123–140.

Mabasa, M. A. (2018). *Impact of socio-cultural practices on substance use amongst the rural youth: Towards the development of a school-based intervention programme*. University of Limpopo.

Matutu, V., & Mususa, D. (2019). Drug and alcohol use among young people in Zimbabwe: A crisis of morality or public health problem. *SSRN Electronic Journal.*

Midgely, J. (2014). *Social development: Theory and practice*. SAGE.

Midgley, C. (1995). *Social development: The developmental perspective in social welfare*. SAGE.

Mishra, S. (2015). *An investigation into extent of drug use among youth in Africa.* Shreemati Nathibai Damodar Thackersey (SNDT) Women's University Mumbai.

Mukwenha, S., Murewanhema, G., Madziva, R., Dzinamarira, T., Herrera, H., & Musuka, G. (2021). Increased illicit substance use among Zimbabwean adolescents and youths during the COVID-19 era: An impending public health disaster. *Addiction (Abingdon, England), 117*(4), 1177–1178. https://doi.org/10.1111/add.15729

Munn, Z., Pallock, D., Hannan, K., & Alexander, L (2022). What are scoping reviews? Providing a formal definition of scoping review as a type of evidence synthesis. *JBI Evidence Synthesis*. Retrieved January 12, 2025, from https://www.researchgate.net/publication/359060419

Muoti, S. K. (2014). *Effects of drug and substance use on academic performance among secondary school students.* Kathonzweni District, Makueni County, Kenya.

Muzondo, E. (2022). Accessibility of human rights and sexual & reproductive health services by female drug users in Zimbabwe: Challenges and opportunities. *Journal of Social Issues in Non-Communicable Conditions & Disability, 1*(1), 38–50.

Muzondo, E. (2023). Women and illegal drug use in Zimbabwe: Causes and effects. *Sprin Journal of Arts, Humanities and Social Sciences, 2*(07), 11–25.

National Survey on Drug Use and Health. (2020). *2020 National Survey on Drug Use and Health*. https://www.samhsa.gov/data/release/2020-national-survey-drug-use-and-health-nsduh-releases

Patel, L. (2005). *Social welfare and social development* (2nd ed.). Oxford University Press.

Raenarri, L., & Strazzari, F. (2023). The data that we do not have: Studying drug trafficking and organized crime in Africa. *Trends in Organized Crime, 26*, 358–378. https://doi.org/10.1007/s12117-023-09482-5
Raphael, S., & Winter-Ebmer, R. (2001). Identifying the effect of unemployment on crime. *The Journal of Law and Economics, 44*(1), 259–283. Retrieved November 19, 2023, from https://doi.org/10.1086/320275
Ray, R., Mudholkar, S., & Saxena, S. (2017). Substance use in India: A review. *Journal of Substance Use, 22*(4), 361–366.
Redaelli, C., & Arevalo, C. (2023). Targeting drug lords: Challenges to IHL between Lege Lata and Lege Ferenda. *International Review of the Red Cross, 105*(923), 652–673.
Sahu, R., & Sahu, R. (2016). Stress and drug use: A review of current knowledge and future direction. *Journal of Stress and Health, 18*(2), 87–104.
South Africa Police Services. (2021). *Crime statistics*. Government Printer.
Swartbooi, C. (2017). *A phenomenological study on parents' experiences of their adolescent's substance use* (M.A Psychology Dissertation, University of Western Cape).
Taylor, S., Paluszek, M. M., Rachor, G. S., McKay, D., & Asmundson, G. J. G. (2020). Substance use and use, COVID-19-related distress, and disregard for social distancing: A network analysis. *Addictive Behaviors, 114*, 106754.
Todaro, M. P., & Smith, S. C. (2012). Economic Development (11th ed., 801 p.). Harlow: Addison-Wesley, Pearson. https://shahroodut.ac.ir/fa/download.php?id=1111128678.
Twala, C. (2014). The 50 years struggle of the Organization of African Unity (OAU): A search for a developmental paradigm in Africa*? Journal of Social Sciences, 39*(1), 23–38.
United Nations. (2015). *World Drug Report*. Oxford University Press.
United Nations Office on Drugs and Crime. (2020). *World Drug Report 2020*. https://wdr.unodc.org/wdr2020/en/index2020.html
United Nations Office on Drugs and Crime (UNODC). (2023). *Global Drug Report 2023: Understanding the impact of drug use on health and society*. https://www.unodc.org/drugs/
UNRISD. (2023). *Crises of inequality: Shifting power for a new eco-social contract. Research and Policy Brief 39*. UNRISD.
Van Wormer, K., & Davis, D. R. (2008). *Addiction treatment: A strengths perspective* (2nd ed.). Cengage.
Velleman, R. B., Templeton, L. J., & Copello, A. G. (2015). The role of the family in preventing and intervening with substance use and misuse: A comprehensive review of family interventions, with a focus on young people. *Drug and Alcohol Review, 24*, 93–109.
World Bank. (2023). *World development indicators*. Retrieved January 21, 2025, from World Development Indicators | DataBank
World Health Organization. (2019). *Substance use*. Retrieved October 14, 2024, from https://www.who.int/topics/substance_use/en-
World Health Organization. (2021). *World Drug Report*. Oxford University Press.

Chapter 18
Transformative Pathways to Children's Rights: A Social Policy Analysis in Rural Zimbabwe

Primrose Hove and Clement Chipenda

18.1 Introduction

Children's rights in rural Zimbabwe remain significantly compromised despite decades of policy interventions and international commitments. This situation reveals the inadequacy of conventional approaches to address systemic inequalities affecting vulnerable populations. The focus of this study is premised on three key research questions. First, how do existing social policy frameworks and institutional arrangements in Mazvihwa (a district in rural Zimbabwe) impact the realisation of children's rights through the lens of the multiple instruments of Transformative Social Policy (TSP), including production, reproduction, redistribution, protection, and social cohesion? Second, what are the structural barriers and opportunities for transforming social policies to better advance children's rights, especially considering local sociocultural dynamics and economic constraints? Third, how can TSP approaches be effectively integrated with rights-based frameworks to enhance children's well-being and promote sustainable social change in rural Zimbabwe?

As Africa confronts an unprecedented convergence of global disruptions and local imperatives (see Chipenda's Introduction in this volume), children's rights in rural Zimbabwe must be understood within this broader continental context of social policy transformation needs. In rural Zimbabwe, children's rights must be engaged with within the socio-economic and cultural context to develop effective interventions that address their unique challenges. Poverty, limited access to

P. Hove (✉)
Department of Community and Social Development, University of Zimbabwe, Harare, Zimbabwe

C. Chipenda
South African Research Chair in Social Policy, College of Graduate Studies, University of South Africa, Pretoria, South Africa
e-mail: chipec@unisa.ac.za

E. Ndhlovu, C. Chipenda (eds.), *Redefining the Social Policy Discourse*,
https://doi.org/10.1007/978-3-032-24067-5_18

education and healthcare, and cultural practices that often undermine children's rights are prevalent in these areas. These challenges reflect the continent-wide pattern where narrowly conceived social policies have proven inadequate in addressing systemic inequalities affecting children's well-being (see Chipenda & Tom in this volume). By placing children's rights in this context, the study can identify systemic barriers and design targeted strategies that reflect local values and practices. This holistic understanding not only informs policy frameworks but also fosters community engagement, ensuring that interventions are culturally sensitive and sustainable.

The current global landscape, characterised by interconnected crises including climate change, rising inequality and democratic instability, has exposed the limitations of traditional social policy approaches in advancing children's rights, especially in rural African contexts. To address the aforementioned research questions, the study aims to achieve three primary objectives: first, to examine and analyse the intersection between existing social policy frameworks and the realization of children's rights in rural Zimbabwe; second, to identify and critically assess the structural barriers and opportunities for a TSP approach; and third, to contribute to the development of a framework that integrates TSP approaches with rights-based mechanisms to enhance children's well-being and promote sustainable social change in rural Zimbabwe.

While continental achievements in reducing child mortality and expanding primary education access demonstrate progress, entrenched poverty and escalating inequality highlight the imperative for holistic children's rights protection strategies. Despite the existence of various social policy frameworks, there remains a significant gap in the literature regarding their effective application to children's rights, especially in rural contexts. This scholarly void is compounded by research that emphasises abstract frameworks while neglecting implementation obstacles and contextual factors that obstruct rights fulfilment in rural African communities, necessitating localised analytical approaches.

This investigation confronts a significant academic lacuna by analysing children's rights through social policy perspectives, questioning prevailing methodologies rooted in neo-liberal paradigms. This study goes beyond being an academic exercise; it advocates for progressive policy frameworks that address the challenges posed by the current children's rights discourse, which often overlooks issues of inequality and inequity. African children's rights discourse demonstrates what Fischer (2018) characterises as 'stratified, segmented and segregated' methodologies, functioning within inequitable fiscal arrangements that compromise holistic rights achievement. It seeks to propose quality livelihood outcomes that reduce market-driven child protection mechanisms and enhance state and citizen resilience against systemic shocks, emphasising the need for a holistic understanding of local contexts and community engagement.

This research's theoretical innovation centres on synthesising Transformative Social Policy (TSP) principles with Human Rights-Based Approach (HRBA) methodologies, acknowledging that enduring children's rights advancement demands both normative frameworks and transformational strategies. The TSP framework emphasises the necessity of addressing systemic inequalities and structural barriers that impede children's rights, ensuring that interventions are grounded in normative standards and accountability. This synthesised methodology addresses continental demands for social policy architectures that concurrently advance SDG targets while establishing foundations for post-2030 development agendas emphasising resilience, sustainability, and inclusive transformation.

The COVID-19 pandemic's impact has revealed the inadequacies of restrictive social policy designs, demonstrating the hazards of separating economic strategies from social welfare considerations in safeguarding children's rights. These frameworks are essential for effectively addressing the socio-economic and cultural challenges that children in rural Zimbabwe face, especially considering Africa's contemporary circumstances, where conventional social policy methodologies have failed to address children's rights obstacles.

To strengthen the analytical framework and policy relevance, this chapter incorporates comparative perspectives from South Africa, utilising secondary evidence to examine similar challenges and policy responses in rural Southern African contexts. The realisation of children's rights in rural areas of Southern African countries, especially in South Africa, is influenced by several socio-economic factors, including poverty, legislative gaps, and historical inequalities, which collectively hinder the effective implementation of children's rights (Mirugi-Mukundi, 2010; Khangala, 2024). In South Africa, despite having a progressive constitution, the socio-economic rights of children remain largely unmet due to widespread poverty and unemployment, which limit families' ability to provide adequate care for their children (Mirugi-Mukundi, 2010; Khangala, 2024). A significant proportion of children in South Africa live in poverty, with estimates indicating that 60–75% of the youth population is affected (Khangala, 2024). Poverty exacerbates the vulnerability of children, limiting access to essential services such as healthcare and education (Mirugi-Mukundi, 2010).

The South African experience reveals critical legislative and policy gaps that parallel Zimbabwe's challenges. South Africa has ratified the UN Convention on the Rights of the Child and enacted several laws to protect children's rights, yet gaps remain in comprehensively addressing children's healthcare rights (Buchner-Eveleigh & Nienaber, 2012). There is a lack of legislation specifically protecting the health needs of children with disabilities, highlighting a critical area for improvement (Buchner-Eveleigh & Nienaber, 2012). The legacy of apartheid in South Africa has resulted in enduring socio-economic challenges, including disparities in education and healthcare access, which continue to affect children's rights (Khangala, 2024). These historical inequalities contribute to poor service delivery in rural areas, further marginalising children from disadvantaged backgrounds (Mirugi-Mukundi, 2010). These comparative perspectives from South Africa inform the study's analysis of how similar structural barriers manifest in rural Zimbabwe and provide lessons for transformative social policy approaches.

This research makes four key scholarly contributions to African social policy discourse. First, it bridges the theoretical gap between social policy and rights-based approaches by demonstrating how TSP can enhance rights realisation in rural contexts. Second, it challenges the dominant neo-liberal paradigm in children's rights discourse by proposing alternative frameworks that prioritise structural transformation over residual interventions. Third, it provides empirical evidence from rural Zimbabwe that can inform policy recommendations for rights-based transformative interventions across similar African contexts. Fourth, it offers comparative analysis drawing from South African experiences to identify regional patterns and policy lessons for Southern African rural contexts.

The study's relevance is acute given Zimbabwe's ongoing social and economic challenges and the broader continental need for sustainable and contextually appropriate solutions to children's rights violations. This approach is relevant given Zimbabwe's ongoing social and economic challenges and the need for sustainable and contextually appropriate solutions. By examining children's rights through the TSP lens while incorporating comparative perspectives from South Africa, this study contributes to the broader African social policy discourse that seeks to move beyond residual approaches toward comprehensive frameworks that address systemic inequalities and promote inclusive development.

18.2 Literature Review

Children's rights in rural Zimbabwe face significant challenges from poverty, climate change and child maltreatment due to cultural differences (Ringson, 2019). These contribute to severe rights violations, necessitating a comprehensive understanding to inform policy and community responses. The inadequacy of conventional approaches to address systemic inequalities affecting vulnerable populations reflects broader continental patterns. As argued earlier, these narrowly conceived social policies have proven inadequate in addressing children's well-being challenges across rural African contexts. Child poverty and deprivation studies in Zimbabwe by Musiwa (2020, 2021) reveal 78% of rural children experience severe early childhood deprivation, with significant deficits in water (46%), healthcare (44%) and sanitation (40%) (Musiwa, 2021). Rural children are disproportionately affected, highlighting the urgent need for targeted interventions. This pattern resonates with South African contexts where estimates indicate 60–75% of the youth population is affected by poverty, exacerbating vulnerability and limiting access to essential services (Khangala, 2024).

Climate change exacerbates vulnerabilities, leading to food insecurity, decreased school attendance and increased risks of child labour and exploitation (Chikoko & Chihiya, 2023). Government investment in climate adaptation strategies is essential to protect children's rights. Moreover, the interconnected crises, including climate change, rising inequality and democratic instability, have exposed the limitations of traditional social policy approaches in advancing children's rights across rural African contexts. In this complex landscape, traditional disciplinary methods often infringe on children's rights, with corporal punishment and deprivation of basic needs being prevalent (Mafa & Simango, 2022). Cultural beliefs prioritise traditional values over children's rights frameworks, leading to violations of autonomy and protection (Mangena & Ndlovu, 2014). This highlights cultural conflict between child discipline and rights, necessitating a shift towards rights-oriented approaches. Similar tensions exist in South Africa, where historical inequalities contribute to poor service delivery in rural areas, further marginalising children from disadvantaged backgrounds (Mirugi-Mukundi, 2010).

Children's rights are challenged by poverty, the HIV/AIDS epidemic and inadequate community support, necessitating collaborative efforts among local and international organisations (Madziva, 2017). According to Midzi (2015), rights are significantly impacted by harmful cultural practices, parental separation and ignorance of children's law. Rights are severely compromised by poverty, inadequate education and high malnutrition rates, exacerbated by long-term crisis and lack of focused governmental policies (Munro, 2015). The COVID-19 pandemic's impact has revealed inadequacies of restrictive social policy designs, demonstrating the hazards of separating economic strategies from social welfare considerations in safeguarding children's rights.

Despite frameworks like the UNCRC, practical application remains limited. Legal frameworks like the 2013 Constitution and Children's Act [Chapter 5.06] support children's rights, yet challenges persist due to inadequate resources, poverty and lack of services (Mutale, 2015). Rights are influenced by international treaties and local laws, yet challenges persist due to socio-economic factors and limited access to education and healthcare (Ngwende, 2016). The Children's Act [Chapter 5: 06] lacks alignment with international standards, particularly in child protection and participation (Bhaiseni, 2016). A discernible gap exists between human rights frameworks and the realisation of children's rights in rural areas. South Africa's experience parallels these challenges—despite ratifying the UNCRC in 1995 and having a progressive Constitution (1996) with Section 28 specifically addressing children's rights, gaps remain in comprehensively addressing children's healthcare rights, especially for children with disabilities (Buchner-Eveleigh & Nienaber, 2012).

Children's rights issues are rarely examined through a social policy lens, with dominant approaches primarily anchored in social protection logic that perpetuates neo-liberal-oriented social policy architecture (see Tekwa, Mokone & Ndhlovu in this volume). This architecture, as Fischer (2018) argues, remains 'stratified, segmented and segregated,' operating within an unfair fiscal contract that undermines comprehensive children's rights realisation. This reflects methodologies functioning within inequitable fiscal arrangements that compromise holistic rights achievement across African contexts. The Transformative Social Policy Approach (TSP) adopted in this study as the theoretical and evaluative framework deliberately opposes this limited perspective to examine systemic barriers and disparities that prevent children from realising their rights. This study uniquely integrates the TSP framework and the Human Rights Based Approach (HRBA) and posits that sustainable advancement of children's rights requires both normative standards and transformative mechanisms. This chapter bridges the gap between social policy and rights-based approaches, advancing theoretical knowledge and practical implementation by examining how TSP can enhance rights realisation in rural regions. This approach addresses continental demands for social policy architectures that concurrently advance SDG targets while establishing foundations for post-2030 development agendas emphasising resilience, sustainability and inclusive transformation. This approach is relevant given Zimbabwe's ongoing social and economic challenges and the need for sustainable and contextually appropriate solutions.

18.2.1 Children's Rights in the Global Rural Context

International dialogue regarding children's rights has markedly progressed, especially in rural contexts (Fottrell, 2021). International frameworks, including the UNCRC, emphasise equitable access to education, healthcare and social services for all children regardless of geographical location (Tanveer, 2024). Realisation frequently encounters obstacles in rural areas due to poverty, cultural norms and inadequate infrastructure (Bartlett et al., 2016). Studies demonstrate that rural children globally experience disproportionate deprivation, exhibiting substantial deficiencies in essential services (Shamrova & Lampe, 2020). The convergence of traditional practices and contemporary rights-based frameworks creates a multifaceted environment where children's rights may be simultaneously promoted and obstructed (Dexit et al., 2024; Joudaki, 2024). Current literature indicates that although international commitments exist, local realities frequently differ, resulting in a disconnect that necessitates comprehensive strategies involving local communities and stakeholders (Broberg & Sano, 2018; Cloward, 2014). The International Labour Organisation (2010) indicates globally, one in six children is engaged in labour, with 126 million children working in hazardous conditions. The highest prevalence is in sub-Saharan Africa, where 26% of children (49 million) are involved in work (Dinbabo, 2013).

Global child protection across Europe, Asia, Australia and America illustrates varied landscapes influenced by differing legal systems, cultural practices and socio-economic factors. In Europe, strong policies and international agreements support child protection initiatives; however, child poverty and the situation of migrant children remain prevalent (Iusmen, 2020). Asia exhibits swift economic expansion alongside substantial challenges, including child labour and trafficking, especially in South and Southeast Asia, requiring customised interventions (ILO, 2025; Han 2024). Australia possesses robust legal frameworks; however, Indigenous children disproportionately encounter systemic challenges, and this highlights the necessity for culturally sensitive methodologies (Glaberson, 2019). These regions collectively exhibit dedication to promoting children's rights; however, the efficacy of protection mechanisms frequently depends on tackling socio-economic inequalities and incorporating cultural viewpoints into policy frameworks.

18.2.2 Children's Rights in the Rural African Context

The African continent is characterised by rich cultural diversity, economic challenges and varying political stability. Children's rights are codified in international treaties, such as the UNCRC and the African Charter on the Rights and Welfare of the Child (Umaru, 2019). Implementation is inconsistent, exhibiting considerable disparities between urban and rural regions. In Sub-Saharan Africa's rural areas, children face poverty, lack of access to education and healthcare disparities, leading to the highest rates of child malnutrition and mortality globally (Marume et al., 2025). Despite initiatives such as the African Union's Agenda 2063 intended to enhance children's rights, challenges remain (Union, 2016). Conventional practices frequently clash with contemporary legal systems, hindering children's rights actualisation.

In West Africa, socio-economic disparities and political instability have significantly affected children's rights (Tuwor & Sossou, 2008). Despite legislative advancements in countries like Nigeria and Ghana, enforcement remains inadequate, with persistent child labour and trafficking issues (Okunade & Shulika, 2021). ECOWAS and Southern Africa have enacted regional policies to tackle these challenges; however, numerous rural communities persist in facing difficulties due to insufficient service delivery (Ojo & Adedayo, 2022). Regional entities such as SADC have launched initiatives to enhance children's welfare; however, difficulties persist in reconciling traditional practices with contemporary rights-based frameworks (Riggirozzi, 2017). Cultural, political and socio-economic factors in North and East Africa impact children's rights. Countries like Egypt, Morocco, Kenya and Uganda have made legislative advances, yet enforcement and cultural norms often hinder rights realisation (Mbaku, 2019). Child marriage (Walker, 2012) and healthcare disparities are prevalent issues, exacerbated by political instability impacting service delivery. Child protection framework execution is inconsistent, with traditional practices impacting children's rights, especially for girls (Ndondo, 2024). Substantial execution deficiencies remain.

18.2.3 Comparative Analysis: Children's Rights in Southern African Rural Contexts

The Southern African region presents a dynamic landscape of children's rights realisation, with South Africa and Zimbabwe sharing similar challenges despite different legislative trajectories. Both countries demonstrate the gap between progressive legal frameworks and practical implementation, especially in rural areas.

18.2.3.1 The Legislative Framework and Implementation Challenges

South Africa's children's rights framework is anchored in the Constitution (1996), with Section 28 specifically addressing children's rights and establishing that a child's best interests are paramount in all matters. The country ratified the UNCRC in 1995 and the African Charter on the Rights and Welfare of the Child in 2000, demonstrating commitment to international standards. The comprehensive Children's Act (Act 38 of 2005, implemented 2010) replaced earlier legislation, covering parental responsibilities, children's courts, adoption and child protection mechanisms. The Child Justice Act (Act 75 of 2008, implemented 2010) created a separate criminal justice system for children, emphasising restorative justice and rehabilitation. Despite this progressive legislative framework, implementation challenges persist. The socio-economic rights of children remain largely unmet due to widespread poverty and unemployment, limiting families' ability to provide adequate care (Mirugi-Mukundi, 2010; Khangala, 2024). The legacy of apartheid has resulted in enduring socio-economic challenges, including disparities in education and healthcare access, continuing to affect children's rights (Khangala, 2024). Critical legislative gaps exist in comprehensively addressing children's healthcare rights, particularly for children with disabilities (Buchner-Eveleigh & Nienaber, 2012).

Zimbabwe's legal framework evolution demonstrates parallel challenges with different trajectories. The foundational Children's Protection and Adoption Act of 1972 and Children's Act [Chapter 5.06] underwent critical revisions in 1996 and 2001, establishing basic child protection mechanisms, significant after independence in 1980. Subsequent developments included the Legal Age of Majority Act (1982), Education Act (1987, amended 2006), and Social Welfare Assistance Act (1988). The 2013 Constitution marked a watershed moment by incorporating children's rights into the supreme law, providing a constitutional anchor for existing and future legislation. Recent developments include the Child Justice Bill (2020), aimed at establishing juvenile justice systems focused on restoration and rehabilitation, demonstrating ongoing efforts to address contemporary challenges (Oyewale, 2022). Zimbabwe ratified the UNCRC in 1990 and the African Charter on the Rights and Welfare of the Child in 1995, effectively aligning domestic legislation with international standards (Bhaiseni, 2016).

Both countries face similar implementation challenges despite different resource capacities. In South Africa, estimates indicate 60–75% of the youth population is affected by poverty, exacerbating vulnerability and limiting access to essential services (Khangala, 2024). Zimbabwe's evidence reveals 78% of rural children experience severe early childhood deprivation, with significant deficits in water (46%), healthcare (44%) and sanitation (40%) (Musiwa, 2021). Both contexts demonstrate how poverty perpetuates service access inequalities that disproportionately affect rural children. The strain on traditional care systems, combined with inadequate formal social services, creates compound effects impacting vulnerable children in both countries. In South Africa, historical inequalities contribute to poor service delivery in rural areas, further marginalising children from disadvantaged backgrounds (Mirugi-Mukundi, 2010). Zimbabwe's traditional care systems, historically resilient in rural communities, experience unprecedented strain due to socio-economic challenges and demographic shifts (Madziva, 2017).

18.2.4 Children's Rights in Zimbabwe's Rural Context

The discourse on children's rights in rural Zimbabwe reveals a complex nexus of international frameworks, local realities and structural challenges hindering effective implementation. This complexity reflects broader Southern African patterns where progressive legal frameworks struggle with practical implementation challenges, especially in rural areas. Zimbabwe's legislative trajectory parallels South Africa's progression from basic protection mechanisms to comprehensive rights-based frameworks. The mid-1990s saw further important developments with the Guardianship of Minors Act [Chapter 5.08] of 1995 (amended in 2022) and the Maintenance Act [Chapter 5.09] of 1996, which provided comprehensive guidelines for child custody, support and parental responsibility (GoZ, 1996, 2022). A significant policy milestone was the Zimbabwe National Orphan Care Policy of 1999, developed in response to the HIV/AIDS epidemic's impact on orphaned and vulnerable children (Kurevakwesu & Chizasa, 2020). This policy framework laid the groundwork for subsequent initiatives, including the National Action Plan for Orphans and Vulnerable Children (2004–2010). The children's rights landscape was further strengthened by the National Case Management System for Child Protection (2012) and the Protocol on Multi-Sectoral Management of Child Sexual Abuse (2012) (Nhapi, 2020).

Despite comprehensive legal and institutional mechanisms and supportive legal structures, implementation remains constrained by limited resources and capacity, and gaps remain between international commitments and the lived experiences of children in rural areas (Mutale, 2015). This mirrors South Africa's experience where, despite progressive legislation like the Children's Act (2005) and Child Justice Act (2008), socio-economic factors continue to undermine effective implementation. Evidence reveals stark deprivation in rural contexts. The cascading effects of structural constraints in Zimbabwe parallel those documented in South Africa, where poverty limits families' ability to provide adequate care despite legislative protections. Chinyoka (2014) notes that 44% lack healthcare services, 46% lack clean water and 40% lack adequate sanitation facilities. Koyratty et al. (2022) observes a correlation between food insecurity and compromised infant growth. Over 50% of children lack access to safe drinking water and primary healthcare services (Musiwa, 2020, 2021). These challenges are particularly acute in rural areas with underdeveloped economic infrastructure, characterised by high poverty levels and inadequate social service provision. Social service provision emerges as a critical dimension in children's rights realisation. While the UNCRC (Article 28) establishes education and social services as fundamental rights, scholarly attention to service delivery in rural Zimbabwe remains limited (UNICEF, 1989). The interaction between formal social services and traditional care systems creates a dynamic landscape where children's rights often hang in the balance, similar to South Africa, where historical inequalities continue to affect service delivery patterns.

Protection and social security mechanisms reveal significant implementation gaps. Fischer (2018) identifies a fragmented approach to protection strategies that fails to address structural inequalities, and we can argue that this is applicable to both Zimbabwe and South African contexts. This can be exemplified by disability rights, where both countries struggle with implementation despite having relevant legislation—Zimbabwe's Disability Act lacks formal implementation strategies (Nhapi, 2020; Mtetwa, 2018), while South Africa lacks comprehensive legislation specifically protecting health needs of children with disabilities (Buchner-Eveleigh & Nienaber, 2012).

The relationship between social cohesion and children's rights reveals underlying tensions. Mafa and Simango (2022) and Mangena and Ndlovu (2014) document tensions between traditional practices and rights-based approaches. This tension exists across Southern African contexts where cultural practices often clash with contemporary legal frameworks, requiring culturally sensitive approaches that respect traditional values while protecting children's rights. In this context, resource redistribution emerges as a critical dimension in both countries. The UNCRC's Articles 2 and 4 mandate non-discriminatory rights protection and appropriate resource allocation (UNICEF, 1989). However, the effectiveness of redistributive mechanisms in rural contexts remains understudied, perpetuating service access inequalities that disproportionately affect rural children (Chipenda & Cochrane, 2024). South Africa's experience with social grants demonstrates both the potential and limitations of redistributive mechanisms in advancing children's rights, providing lessons for Zimbabwe's policy development.

This comparative literature review reveals significant research gaps at the intersection of TSP and children's rights in Southern African rural contexts. While existing scholarship documents implementation barriers, limited attention has been paid to how TSP's multiple functions, production, reproduction, redistribution, protection, and social cohesion, interact in rural settings. The study of Mazvihwa aims to contribute to this comparative understanding by examining how local sociocultural dynamics inform the intersection of social policy frameworks and children's rights in Zimbabwe, with implications for broader Southern African contexts. The ultimate goal is to contribute theoretical and practical evidence-based recommendations that can advance children's rights through transformative interventions, building community resilience against systemic challenges and addressing the dynamics between structural constraints and local capacities for rights realisation.

18.3 Theoretical Framework

This study's theoretical foundation integrates the Transformative Social Policy (TSP) approach with the Human Rights-Based Approach (HRBA), creating an analytical lens that combines TSP's emphasis on structural transformation and multidimensional social policy functions (Mkandawire, 2004; UNRISD, 2023) with HRBA's focus on rights, empowerment and accountability mechanisms (Vandenhole, 2024). The TSP framework provides critical perspectives for addressing systemic inequalities through its five key functions: production, reproduction, protection, redistribution, and social cohesion, while the HRBA complements this by emphasising rights-holders' agency and duty-bearers' obligations. This theoretical synthesis enables a comprehensive examination of both structural dimensions of social policy and rights-based imperatives in advancing children's wellbeing in rural Zimbabwe, transcending conventional social protection paradigms while ensuring rights-based principles inform TSP interventions.

18.3.1 The Transformative Social Policy Framework

The TSP framework emerged from the United Nations Research Institute for Social Development's seminal project *Social Policy in a Development Context (2000–2006)* (Mkandawire, 2004; UNRISD, 2006), challenging conventional development paradigms by offering a comprehensive understanding of social policy's role in societal transformation. This framework critically responds to neoliberal approaches that have historically dominated African social policy discourse, characterised by a narrow focus on targeted interventions and minimal safety nets for the 'deserving poor,' which have proven inadequate in addressing societal challenges (Meagher, 2022; Adesina, 2020).

Distinguished African scholars like Thandika Mkandawire and Jimi Adesina have significantly contributed to this theoretical development (see Chipenda & Tom in this volume). Mkandawire (2004:1) articulates social policy as encompassing 'collective interventions in the economy to influence access to and the incidence of adequate and secure livelihoods and income,' while Adesina (2009:38) expands this to include 'collective public efforts aimed at protecting the well-being of people in a given territory.' These definitions establish TSP's intellectual foundation, highlighting a departure from conventional approaches that merely address market failures.

The framework's transformative potential lies in integrating five essential functions: production, reproduction, protection, redistribution, and social cohesion/nation building (UNRISD, 2023). This multidimensional approach represents what Adesina (2015) terms the 'wider vision' of social policy, moving beyond palliative interventions to address structural inequalities and systemic challenges. The framework emphasises policies that actively shape economic structures, enhance human capabilities, transform social relations and strengthen institutional frameworks (Adesina, 2020). Crucially, TSP advocates for a shift from externally determined, neoliberal-oriented policies to locally grounded, transformative approaches. This reorientation transcends limitations of anti-poverty schemas that fail to address root causes of poverty, inequality and marginalisation (Chipenda, 2023, 2024; Tom, 2024), instead promoting comprehensive societal transformation through integrated policy mechanisms (Meagher, 2022). By emphasizing interconnectedness of economic and social policies, TSP provides a robust framework for achieving sustainable development outcomes extending beyond poverty alleviation to encompass broader societal transformation and solidarity.

18.3.2 The Human Rights-Based Approach

The HRBA represents a transformative framework that reconceptualises development processes through human rights principles and standards (Gready, 2019). This approach shifts from needs-based or charity-oriented development paradigms to one recognising rights holders as active agents of change rather than passive beneficiaries (Cornwall & Nyamu-Musembi, 2004). Anchored in international human rights standards, HRBA analyses and addresses structural inequalities and power dynamics perpetuating development challenges (UN, 2024).

The theoretical underpinning rests on key pillars including universality, indivisibility, participation, accountability and non-discrimination (Vandenhole, 2024), providing both normative guidance and operational frameworks for development interventions. A distinctive feature is its dual focus on empowering rights-holders while strengthening duty-bearers' capacity and accountability (Oestreich, 2023). This is significant in addressing power asymmetries and institutional barriers impeding rights realisation. However, Gatenio Gabel (2024a, 2024b) notes that this binary categorisation may oversimplify complex webs of responsibilities and relationships in rights implementation, especially in contexts where state capacity is limited or traditional authorities play significant roles.

Critics highlight several challenges. Pantzerhielm (2023) argues that Western-centric conceptualisation of rights may not fully align with diverse cultural contexts, values and practices, as well as alternative approaches to justice and well-being. The disparity between legal promises and real-world practices raises concerns about HRBA effectiveness in addressing local contexts (De Man, 2018). This focus on legal frameworks may neglect grassroots movements and necessitate collaboration with local civil society (Schmitz, 2012). Effective HRBAs require a thorough understanding of local power dynamics and sociopolitical contexts, often overlooked (Yamin & Cantor, 2014), creating dilemmas, especially in health-related applications, where navigating power relations is essential.

18.3.3 Framework Integration

Despite these challenges, HRBA remains influential in shaping development discourse and practice. The integration of HRBA with TSP addresses structural inequalities more effectively, offering possibilities for bridging implementation gaps between rights rhetoric and practical outcomes (Chambers & Vastardis, 2021). This theoretical synthesis provides a unique analytical perspective that maintains core human rights principles while addressing contemporary challenges, including climate change, digital transformation, and growing global inequality (UNRISD, 2023). The evolution of both frameworks continues to be shaped by emerging challenges, requiring adaptive approaches that can address contemporary contexts while maintaining theoretical rigour and practical relevance for advancing children's well-being in rural settings.

18.4 Research Context and Methodology

18.4.1 Study Setting

The study was undertaken in the Mazvihwa community, located in the Southeast Zvishavane District, approximately 20 km from Zvishavane mining town in Zimbabwe's Midlands Province (Zela, 2023). The area faces significant socio-economic challenges, characterised by high temperatures and low rainfall patterns severely impacting its agro-based economy, leading to food insecurity and poverty (Chidakwa et al., 2020; Manyakaidze, 2023). Most households survive on less than US$2 per day, limiting their ability to provide basic needs for children's health and education (Mwapaura et al., 2022). The community's infrastructure is notably inadequate (Nciizah et al., 2021). Educational facilities are limited (Manjengwa, 2012), with few primary and secondary schools serving a large population of school-age children. Healthcare services are restricted to three understaffed rural health centres, contributing to elevated child mortality and morbidity rates (Mwapaura et al., 2022; Mhaka-Mutepfa et al., 2014). Economic activities primarily involve smallholder farming, livestock rearing and informal trading, with increasing migration of working-age adults to urban areas and neighbouring countries (Chidakwa et al., 2020). Recurrent droughts have led to migration patterns disproportionately affecting vulnerable family members. Men often migrate seeking employment, leaving women, children and the elderly to manage family farms and depend on remittances (Muzari et al., 2014), resulting in increased child-headed, women-headed, and elderly-headed households caring for grandchildren (Chidakwa et al., 2020).

18.4.2 Research Methodology

According to Gaus (2017), clear research methodology frameworks are important for delineating, elucidating and forecasting phenomena. This study adopted an interpretivist philosophy, prioritising understanding the subjective significance of social phenomena from individual perspectives (Ryan, 2018). This mixed-methods approach incorporates both primary qualitative data collection and secondary literature analysis to comprehensively explore how existing social policy frameworks and institutional arrangements impact children's rights realisation through TSP instruments. It also aims to identify structural barriers and opportunities for transformative approaches, as well as examine integration possibilities between TSP and rights-based frameworks.

The secondary literature component involved systematic literature searches across multiple databases, including JSTOR, Google Scholar, EBSCO and academic repositories, using keyword combinations such as 'children's rights', 'rural contexts', 'social policy', 'transformative approaches' and 'Southern Africa' and others. Special attention was given to research undertaken in South Africa on children's rights realisation in rural settings to strengthen the analytical framework and policy relevance through comparative perspectives from similar Southern African contexts. Literature was grouped thematically, focusing on structural barriers, policy gaps and transformative interventions, and subjected to word search analysis to identify recurring patterns and themes. Thematic coding of secondary literature was conducted to complement primary data analysis, enabling triangulation of findings and a broader regional understanding of children's rights challenges.

Primary data collection followed a qualitative methodology, utilising multiple methods including key informant interviews, in-depth interviews and focus group discussions. As noted by Denzin (2017), qualitative data collection aims to obtain subjective viewpoints and contextual narratives. Key informant interviews were conducted with a Chief, a school headmaster and headmistress and a Social Development Officer. In-depth interviews involved community stakeholders, including schoolteachers, childcare workers, village heads, a Councillor and local NGO representatives, totalling nine comprehensive interviews. Sampling employed a non-probability purposive approach, as described by Campbell et al. (2020), selecting participants with specific characteristics relevant to research objectives. Focus group discussions were structured across four distinct guardian categories: elderly-headed families, biological parents, child-headed families and general community members. Ethical considerations were paramount, drawing on Hallberg's (2013) definition of ethics as criteria governing conduct. The study adhered to key ethical principles, including informed consent, anonymity and confidentiality. As O'Neill (2017) notes, informed consent ensures participants are fully apprised of the study's nature, purpose, risks and potential outcomes. Confidentiality, as defined by Vainio (2013), protects participants' personal information, while anonymity involves eliminating identifiable traits from collected data.

Data analysis utilised thematic analysis (TA) for both primary and secondary data. Brown and Clarke (2022) recommend TA for identifying and interpreting patterns of meaning within qualitative data. This approach allowed systematic exploration of themes emerging from collected information, providing a structured framework for analysing complex interactions between social policies, institutional arrangements and children's rights. The integration of primary findings with secondary literature analysis from South African contexts enhanced the analytical depth. It provided comparative insights into regional patterns of children's rights realisation, informing policy recommendations for transformative interventions across similar Southern African rural contexts. This comprehensive methodological approach, combining rigorous qualitative methods with comparative secondary analysis from South African contexts, aimed to capture diverse perspectives while generating comprehensive insights into complex social dynamics affecting children's rights. It was considered effective in contributing to broader African social policy discourse that seeks to move beyond residual approaches toward comprehensive frameworks addressing systemic inequalities.

18.5 Research Findings and Discussion

18.5.1 Policy Implementation Challenges: A Cross-Regional Analysis

Existing social policy frameworks and institutional arrangements in Mazvihwa demonstrate significant gaps between policy intent and practical implementation in children's rights protection. This reveals how TSP instruments of production, reproduction, redistribution, protection and social cohesion face systematic undermining. The study reveals significant gaps between policy frameworks and their practical implementation in children's rights protection across the Southern African region. While Zimbabwe's child rights legislation and policies like the National Orphan Care Policy offer comprehensive protections, local execution in Mazvihwa remains inadequate. This challenge mirrors the South African experience, where, despite extensive constitutional protections described as the most extensive constitutional protection for children anywhere (Matadi, 2019), implementation challenges persist across provinces. The South African Constitution's entrenchment of justiciable socio-economic rights for children has enriched jurisprudence and earned international recognition, yet inequalities in access to life's essentials continue to affect how children access opportunities for rights fulfilment (Matadi, 2019). These implementation challenges reflect the inadequacy of conventional approaches to address systemic inequalities affecting vulnerable populations. All respondents in Mazvihwa acknowledged constitutional provisions for children's rights to education, healthcare and protection from abuse. However, community awareness remains critically limited, with one community leader explaining.

> People know children have rights, but they don't know exactly what these rights mean in practice and how children should realize them exactly. Even with the existing policy frameworks with regard to child protection, most people in this community are not aware of them.

Documentation of child protection cases lacks consistency, with limited follow-up mechanisms, primarily due to manual systems that are cumbersome given the caseload volume. This was corroborated by the Social Development Officer from the Social Welfare Department, who noted that '…in the era of digitalisation, a lot of work is done manually which is cumbersome given the number of cases'.

The Chief highlighted the dysfunction of local structures, stating.

> Local structures for child protection and monitoring children's rights are non-functional. The Social Welfare office, responsible for overseeing policy implementation, is based in town and rarely visits the area. They mostly appear when implementing the drought relief programme.

This inadequate coverage results in vulnerable children, especially those in child-headed and elderly-headed families, being excluded from welfare programs. Similarly, in South Africa's Eastern Cape, despite constitutional guarantees, implementation gaps persist with inadequate infrastructure and service delivery challenges. South African evidence demonstrates how poverty affects 60% of children below 6 years, with half experiencing food poverty (Chirowamhangu, 2024). Historical inequalities, on the other hand, contribute to poor service delivery in rural areas, further marginalising children from disadvantaged backgrounds (Mirugi-Mukundi, 2010). Research indicates that several factors affect the welfare of the South African child, ranging from poverty and inequality to preventable diseases, nutrition challenges and healthcare access issues (Matadi, 2019). The implementation of children's rights provisions faces ongoing challenges despite the solid foundation provided by the 1996 Constitution, the Children's Act and the Child Justice Act.

18.5.2 Resource Constraints and Systemic Underfunding

Inadequate economic production limits families' capacities to provide for children, directly impacting access to education and healthcare while undermining TSP's production and redistribution instruments. Severe resource limitations significantly impact children's rights implementation across both countries. Health centres operate with minimal staff and supplies, while schools face critical infrastructure challenges. A primary school headmaster in Mazvihwa provided detailed insight into educational resource constraints.

> At my school children share textbooks with four or more students as to one textbook. In this era where primary education curriculum has been changed and resources such as computers are needed but are not available at my rural school presents a challenge to students. Only a few if any households can provide these for their children. This means that rural pupils are lagging behind and will continue to lag behind due to lack of adequate resources in terms of the education being delivered to them.

Limited transportation resources hinder Child Protection Officers from reaching remote areas in a timely manner. Social Development Officers, based in Zvishavane town, cover extensive rural areas with insufficient personnel and resources. One Child Care Worker (CCW) explained the operational challenges.

> As CCWs, the frontline workers for child protection, we are incapacitated. We lack transportation to cover our area and are not compensated for our voluntary work. The Offices for reporting child cases are based in Zvishavane town, requiring transport money that we do not have. Consequently, we often have to ignore child protection issues because we cannot address them.

The Basic Education Model (BEAM), previously covering fees for vulnerable children, has become ineffective. A schoolteacher detailed the impact.

> The Basic Education Model (BEAM), which previously covered fees for vulnerable children, is no longer effective as schools now require fees from all students. The government has delayed releasing BEAM funds to school accounts, impacting administration. As a result, those relying on BEAM are left vulnerable, with some dropping out and being denied access to education at a young age.

This echoes challenges in South Africa where the National School Nutrition Programme faces significant funding backlogs, affecting meal provision for vulnerable children. Research indicates that 66% of the participants indicated that there is a backlog of financial assistance on the feeding programmes initiated as part of the NSNP between 2019 and 2022 in the Eastern Cape province (Chirowamhangu, 2024). Despite government efforts to provide nutritious meals to combat malnutrition, participants revealed ongoing challenges, with one stating, 'I cannot say we do not get any support... we have been given so many promises, but those promises are not putting anything on the table' (Chirowamhangu, 2024).

Institutional weaknesses permeate all levels of child protection. In rural Zimbabwe, the Social Development Officer noted the overwhelming nature of their responsibilities.

> Since the department office is in town and lacks suboffices in rural areas like Mazvihwa, it is overwhelmed and unable to address child rights issues effectively. Officers struggle to monitor children's rights to health and education due to limited community contact and resources for regular visits.

These institutional weaknesses demonstrate how narrowly conceived social policies have proven inadequate in addressing systemic inequalities affecting children's well-being. This validates arguments for transformative approaches that address structural barriers. In South Africa, similar institutional capacity challenges persist despite progressive legislation. Research shows that a wealth of legislation and policies exists, with abundant case law demonstrating that all human rights are justiciable, yet poverty and social exclusion are factors that still hamper a full realisation of children's survival rights (Matadi, 2019). The challenge lies not in legal frameworks but in implementation capacity and resource allocation.

18.5.3 Structural Barriers to Rights Realisation

Analysis of structural barriers reveals how local sociocultural dynamics and economic constraints create systematic impediments to TSP approaches, affecting TSP's instruments of production, reproduction, and social cohesion.

18.5.3.1 Economic Limitations and Intergenerational Poverty

Economic hardship emerges as the most critical barrier to children's rights realisation, directly challenging TSP's production and redistribution instruments. This situation also reflects broader continental crises, including climate change and rising inequality, which expose the limitations of traditional social policy approaches. Chronic poverty, unemployment and inadequate income-generating opportunities severely limit families' capacity to provide for their children's basic needs. One parent in Mazvihwa illustrated this challenge comprehensively.

> I am not employed; I'm depending on piece jobs which are not enough to raise money to send my children to school. I used to depend on a long-based income but due to low rainfall and high temperatures every year I'm not getting anything from the fields. So, I work so that I feed the family and the pay I get is insufficient. All my children are now school dropouts because I cannot afford to pay school fees of $30 USD per term for each child out of the 5, I have.

The intersection of poverty and climate change exacerbates these challenges, as families struggle with declining agricultural productivity. This economic vulnerability mirrors experiences in South Africa's Eastern Cape, where 60% of children under six live in poverty, with half experiencing food insecurity (Chirowamhangu, 2024). The COVID-19 pandemic further intensified these vulnerabilities, revealing the inadequacies of restrictive social policy designs and demonstrating the dangers of separating economic strategies from social welfare considerations in safeguarding children's rights. In South Africa, research indicated that nearly 3.2 million children under five lacked access to any support program, while many others had access only to poorly resourced, low-quality, improvised services (Fredman et al., 2022). Consequently, the pandemic primarily affected the most vulnerable, who have historically faced significant challenges in accessing quality services, perpetuating their vulnerability (Fredman et al., 2022).

18.5.3.2 Cultural Tensions and Deep-Rooted Practices

Cultural practices significantly impact TSP's social cohesion and reproduction instruments, revealing tensions between traditional norms and formal rights frameworks. This situation highlights the necessity for culturally sensitive interventions that respect local values while advancing children's rights. Traditional norms present significant challenges to policy implementation, especially regarding child marriage and gendered labour roles. Despite constitutional prohibitions, child marriage remains prevalent, driven by economic hardships and cultural acceptance. In Mazvihwa, one community member articulated this perspective:

> Let girls go and get married and start their home. What else can they do if they are not going to school and sitting at home? We did the same during our time and we are still in our homes today. There is nothing called a child as long as one is able to do chores and what she is expected of.

A headmaster provided additional insight into the educational impact of cultural acceptance of child marriage.

> Acceptance of child marriages by the community, mostly parents, is affecting children's education negatively as few of our students are pursuing education up to the highest level. Mostly boys go far and for girls is otherwise.

These cultural dynamics highlight the importance of a holistic understanding of local contexts and community engagement in developing effective interventions that address systemic challenges. Gender inequalities significantly impact rights realisation, with girls facing additional barriers to education as families often prioritise boys' schooling when resources are limited. Female-headed households highlighted their vulnerability, affecting children's access to rights. These gender dynamics resonate with South African experiences where discrimination affects access to basic education. Despite having a progressive constitution, the socio-economic rights of children remain largely unmet due to widespread poverty and unemployment, which limit families' ability to provide adequate care (Mirugi-Mukundi, 2010; Khangala, 2024). Research indicates that a greater proportion of girls than boys are likely to be enrolled in home-based playgroups. However, boys are more likely than girls to attend Grade R in a primary school setting, suggesting that girls may be kept home while boys are sent to school at this early age (Fredman et al., 2022). The study found that children living in poorer households are less likely to be exposed to high-quality early learning programmes, with almost half of the children in lower household-income quintiles not attending any educational centre (Fredman et al., 2022).

18.5.3.3 Infrastructure and Service Delivery Challenges

Infrastructure deficits directly undermine TSP's protection instrument, revealing how inadequate service delivery systematically compromises children's rights realisation across both countries. Water and sanitation challenges significantly impact children's rights realisation. In South Africa's Eastern Cape, participants highlighted that 'you would never say that we have schools that have pit latrine toilets,' expressing concern about the state of sanitation infrastructure. The province still has the highest number of pit latrine toilets at 1585, despite government initiatives to eliminate them by 2023 (Chirowamhangu, 2024). One participant noted: 'But we are still sitting with over 1000 schools with pit latrines.' The second timeline was on school infrastructure: 'we still have hundreds of schools with improper infrastructure, with poor quality materials from zinc, asbestos, and all those things' (Chirowamhangu, 2024). Safety infrastructure concerns compound these challenges. Research indicates that the position was that by 29 November 2016, all schools were supposed to have access to some form of power supply, water supply and sanitation. Schools entirely made of inappropriate materials, such as mud, metal, asbestos or wood, were to be replaced with new schools (Chirowamhangu, 2024). However, implementation has remained inadequate with targets not being met (Chirowamhangu, 2024).

18.5.3.4 Power Relations and Political Dynamics

Local power structures significantly impact TSP's social cohesion instrument. Political influences affect resource distribution and service access while traditional authority systems sometimes conflict with formal child protection mechanisms. One focus group participant in Mazvihwa detailed the political dimensions.

> This area is a stronghold of ZANU PF, everything that comes or that is being done is being politicized and party members or their families mostly benefit than the general public. Most of those in power are members of the ruling party and things are done as if everything belongs to the party.

Another participant highlighted collaboration challenges.

> We face collaboration challenges with local government and NGOs, as initiatives are often politicised. High political patronage makes everything appear tied to the ZANU PF party, causing community members to withdraw or be excluded, with only ruling party supporters benefiting. This politicisation hinders effective collaboration.

These political dynamics reflect broader challenges of moving beyond residual approaches toward comprehensive frameworks that address systemic inequalities and promote inclusive development. Similar challenges exist in South Africa where racial discrimination continues to affect access to basic education. The legacy of apartheid has resulted in enduring socio-economic challenges, including disparities in education and healthcare access, which continue to affect children's rights (Khangala, 2024). This is indicative of how historical inequalities shape contemporary policy implementation.

18.5.4 Transformative Opportunities and Community Resilience

Despite significant structural barriers, analysis reveals promising opportunities for transforming social policies through local innovations and community mobilisation. This demonstrates how TSP approaches can be effectively integrated with rights-based frameworks to enhance children's well-being and promote sustainable social change. In Mazvihwa, community-driven projects like Muonde Trust provide agricultural employment opportunities, while food-for-work programmes offer alternative income sources. These community-driven initiatives demonstrate movement beyond market-driven child protection mechanisms toward enhanced state and citizen resilience against systemic shocks. Village savings groups enable women to support their children's education, and community gardens provide supplementary nutrition.

Traditional courts demonstrate potential for integrating customary practices into formal frameworks. One village head provided an example of effective integration.

> Last month, two child abuse cases were reported, resolved amicably in the traditional court, and referred to the police for action against the perpetrators. The police then involved social welfare officers, who accompanied the child survivors to formal court. Addressing child protection issues at the village level ensures cases are identified and resolved locally before escalating, fostering transformation through village heads.

This integration of traditional and formal systems illustrates culturally sensitive and sustainable interventions that respect local values while advancing children's rights. Community mobilisation emerges as crucial for enhancing rights implementation. Faith-based organisations coordinate support for vulnerable children, while traditional leaders engage in child protection advocacy. Youth serve as effective peer educators, raising awareness about children's rights and health issues. This community-based approach aligns with South African experiences where NGOs and communities play significant roles in addressing education challenges (Chirowamhangu, 2024).

18.6 Integrating TSP and HRBA for Sustainable Children's Rights Realisation

18.6.1 Theoretical Innovation and Policy Integration

The integration of TSP and HRBA emerges as a crucial strategy for enhancing children's wellbeing, addressing how TSP approaches can be effectively integrated with rights-based frameworks to promote sustainable social change while responding to critiques of prevailing neo-liberal methodologies. This integration addresses the limitations of rights-based approaches that may oversimplify power relations and adopt Western-centric perspectives, while TSP's focus on structural transformation addresses systemic barriers to rights realisation. The integration represents theoretical innovation in synthesising TSP principles with HRBA methodologies, acknowledging that enduring children's rights advancement demands both normative frameworks and transformational strategies. Policy integration potential exists through harmonising traditional and formal child protection mechanisms. The Social Development Officer in Mazvihwa emphasised.

There is need to integrate children's rights into all community development initiatives. This integration will prioritise children's needs and create a conducive environment for realising their rights.

This integrated approach addresses continental demands for social policy architectures that concurrently advance SDG targets while establishing foundations for post-2030 development agendas emphasising resilience, sustainability and inclusive transformation. In South Africa, similar integration challenges exist despite constitutional provisions ensuring that the best interests of the child are paramount and extensive legal frameworks. Research indicates that the fulfilment of children's rights requires a broad strategy involving the promulgation of relevant laws, the design and implementation of suitable programmes and policies, and the provision of services. The challenge rests with translating constitutional rights into practical outcomes where poverty and social exclusion remain significant barriers to full rights realisation.

18.6.2 Structural Barriers and Transformative Opportunities

Findings reveal that existing social policy frameworks in both countries partially influence children's rights realisation. Effectiveness remains hampered by structural inefficiencies that validate arguments for TSP frameworks to address socio-economic and cultural challenges facing children in rural contexts where conventional methodologies have failed. Inadequate economic production limits families' capacities to provide for children, directly impacting access to education and healthcare. This aligns with Mkandawire's (2004, 2007) assertion that social policy should encompass collective interventions for securing adequate livelihoods. The nexus between social reproduction and cultural norms complicates the children's rights landscape. Traditional gender roles confining girls to domestic responsibilities result in forced early marriages and high out-of-school rates. This cultural acceptance undermines social cohesion and perpetuates poverty cycles, highlighting the need for policies that enhance human capabilities while transforming social relations (Adesina, 2015).

Comparative analysis demonstrates how similar structural barriers manifest across Southern African contexts, strengthening policy relevance for regional approaches. The South African experience demonstrates that even progressive constitutional frameworks face implementation challenges. Despite having a system of laws and programmes to ensure basic support for children and millions of children benefiting from child support grants, inequalities in access to the essentials of life still exist. The country's experience with justiciable socio-economic rights provides important lessons for other African contexts, showing that legal frameworks alone are insufficient without adequate implementation mechanisms and resource allocation.

Economic hardships, characterised by unemployment and limited opportunities, severely restrict families' ability to provide for their children's needs. This aligns with Chipenda's (2023, 2024) critique of neoliberal policies in African contexts that fail to address poverty and inequality root causes, necessitating locally grounded, transformative approaches. The COVID-19 pandemic highlighted existing vulnerabilities and inequalities in both countries, with the chasm between the 'haves and the have-nots' getting ever deeper, especially when considering the socio-economic inequalities. Rural learners and teachers were particularly affected, with structural issues like limited online network access in rural areas exacerbating educational disparities.

18.6.3 Stakeholder Collaboration and Implementation Challenges

Stakeholder collaboration emerges as essential for effective children's rights implementation, demonstrating contributions to broader African social policy discourse that seeks comprehensive frameworks addressing systemic inequalities and promoting inclusive development. In Mazvihwa, participants emphasised the need for stronger partnerships between local government, traditional leaders and NGOs. However, politicisation of NGO initiatives creates collaboration challenges, with participants noting that high political patronage makes initiatives appear tied to ruling parties, causing community withdrawal. NGO representatives highlighted successful rights-based programming models from other areas that could be adapted locally. One village head supported enhanced monitoring systems.

> Strong monitoring will address loopholes and ensure that systems work effectively for children to realize their rights. Those responsible must take this seriously and ensure all structures function properly. The government should also improve its service delivery, as it appears reluctant regarding children's rights in rural areas.

Similarly, in South Africa, collaboration challenges persist despite extensive constitutional and legal frameworks. The country's experience with early childhood care and education demonstrates that the central role of private providers complicates service delivery pictures, with glaring inequalities between wealthier and poorer communities. Services in poorer communities are frequently characterised by overcrowding, poor infrastructure and an informal, poorly paid and under-qualified workforce. These collaboration challenges reflect recognition that effective interventions must be grounded in normative standards and accountability while addressing systemic inequalities and structural barriers. Capacity-building initiatives emerge as crucial for effective implementation. Participants in both countries emphasised the need for training workshops for social workers, teachers, community child workers and community leaders on child protection laws and practices. These initiatives should clarify each stakeholder's role in child protection and ensure effective policy implementation.

18.6.4 Towards Comprehensive Rights Realisation

Findings highlight that children's rights realisation requires comprehensive approaches. These must address structural barriers while leveraging transformative opportunities. They provide empirical evidence that can inform policy recommendations for rights-based transformative interventions across similar African contexts. Findings also indicate that inadequate economic production and persistent cultural norms significantly restrict families' capacities to provide for children, negatively impacting education and healthcare access. South African experiences validate the comparative approach by demonstrating regional patterns and policy lessons for Southern African rural contexts. It demonstrates that constitutional protections and progressive legislation, while necessary, are insufficient without adequate implementation mechanisms. The country's experience with justiciable socio-economic rights provides important lessons, showing that the Constitutional Court develops ways for the implementation of socioeconomic rights, specifically in the context of children's rights, while ensuring access to legal remedies. However, success requires addressing structural inequalities and ensuring culturally relevant, locally driven approaches. Integration of TSP with HRBA emerges as a vital strategy representing core theoretical innovation, demonstrating how synthesised methodologies can empower rights holders as active change agents while navigating complex local power dynamics. This synergy potentially alleviates poverty while fostering accountability and cultural relevance in policy design.

Evidence demonstrates opportunities for creating quality livelihood outcomes that reduce market-driven child protection mechanisms and enhance state and citizen resilience against systemic shocks. Both countries demonstrate that opportunities exist for creating equitable and sustainable environments conducive to children's rights realisation. However, success requires comprehensive approaches that address structural barriers, leverage community resilience, and ensure culturally sensitive policy implementation that integrates formal and traditional systems effectively. The experiences of both Zimbabwe and South Africa highlight the importance of moving beyond legal frameworks to address dynamic socio-economic and cultural factors that shape children's lived experiences and rights realisation, validating holistic approaches to children's rights in rural contexts.

The study validates arguments that synthesised methodologies address continental demands for social policy architectures that concurrently advance SDG targets while establishing foundations for post-2030 development agendas. Evidence validates emphasis on sustainable and contextually appropriate solutions given Zimbabwe's ongoing challenges and broader continental needs for addressing children's rights violations.

18.7 Conclusion

This study examined transformative pathways to children's rights through integrated social policy analysis in Mazvihwa, rural Zimbabwe, revealing the complex nexus between TSP instruments and children's rights realisation in resource-constrained environments. The research demonstrates that effective children's rights protection requires comprehensive approaches transcending conventional legal frameworks through systematic integration of TSP's five instruments with HRBA methodologies. The investigation reveals substantial implementation gaps between progressive policy intentions and lived realities in rural contexts, highlighting how structural barriers, including economic hardship, cultural tensions, and institutional weaknesses, systematically undermine children's rights realisation across Southern African rural contexts.

The theoretical innovation of synthesising TSP and HRBA emerges as crucial for addressing these challenges, demonstrating that sustainable children's rights advancement demands both normative standards and transformational strategies. Comparative analysis with the South African context validates regional patterns while strengthening policy relevance for Southern Africa. The study illustrates that transformative opportunities exist through community-driven initiatives, traditional-formal system integration, and stakeholder collaboration, yet success requires addressing structural inequalities while leveraging local resilience and cultural sensitivity.

The findings contribute significantly to African social policy discourse by challenging neo-liberal paradigms and proposing alternative frameworks prioritising structural transformation over residual interventions. This research provides empirical evidence informing policy recommendations for rights-based transformative interventions applicable across similar African contexts, addressing continental demands for social policy architectures that concurrently advance SDG targets while establishing foundations for post-2030 development agendas that emphasise resilience, sustainability, and inclusive transformation.

The study advocates comprehensive policy frameworks recognising children as active rights-holders while addressing systemic barriers through locally grounded, transformative approaches. It calls upon policymakers, researchers and community leaders to develop contextually appropriate strategies that integrate formal and traditional systems, strengthen institutional capacities and foster meaningful community engagement. This research serves as a foundation for rethinking child protection approaches in rural African contexts, emphasising that sustainable change emerges from an understanding of local dynamics, authentic community participation and systematic addressing of structural challenges affecting children's rights realisation in contemporary Southern Africa.

References

Adesina, J. (2009). Social policy in sub-Saharan Africa: A glance in the rear-view mirror. *International Journal of Social Welfare, 18*(1), 37–51.

Adesina, J. (2015). Return to a wider vision of social policy: Rereading theory and history. *South African Review of Sociology, 46*(3), 99–119.

Adesina, J. (2020). Policy merchandising and social assistance in Africa: Don't call dog monkey for me. *Development and Change, 51*(2), 561–582.

Bartlett, S., Hart, R., Satterthwaite, D., de la Barra, X., & Missair, A. (2016). *Cities for children: Children's rights, poverty and urban management.* Routledge.

Bhaiseni, B. (2016). Zimbabwe Children's Act alignment with international and domestic legal instruments: Unravelling the gaps. *African Journal of Social Work, 6*(1), 3–6.

Braun, V. & Clarke, V., (2022). Conceptual and design thinking for thematic analysis. *Qualitative psychology, 9*(1), p.3.

Broberg, M., & Sano, H. O. (2018). Strengths and weaknesses in a human rights-based approach to international development: An analysis of a rights-based approach to development assistance based on practical experiences. *The International Journal of Human Rights, 22*(5), 664–680.

Buchner-Eveleigh, M., & Nienaber, A. (2012). Gesondheidsorg vir Kinders: Voldoen Suid-Afrikaanse Wetgewing Aan die Land se Verpligtinge Ingevolge die Konvensie Oor die Regte van die Kind en die Grondwet? *Potchefstroom Electronic Law Journal, 15*(1), 102–146. https://doi.org/10.4314/PELJ.V15I1.4

Campbell, S., Greenwood, M., Prior, S., Shearer, T., Walkem, K., Young, S., Bywaters, D., & Walker, K. (2020). Purposive sampling: Complex or simple? Research case examples. *Journal of Research in Nursing, 25*(8), 652–661.

Chambers, R., & Vastardis, A. Y. (2021). Human rights disclosure and due diligence laws: The role of regulatory oversight in ensuring corporate accountability. *Chicago Journal of International Law, 21*(2), 4. https://chicagounbound.uchicago.edu/cjil/vol21/iss2/4

Chidakwa, P., Mabhena, C., Mucherera, B., Chikuni, J., & Mudavanhu, C. (2020). Women's vulnerability to climate change: Gender-skewed implications on agro-based livelihoods in rural Zvishavane, Zimbabwe. *Indian Journal of Gender Studies, 27*(2), 259–281.

Chikoko, W., & Chihiya, P. (2023). Climate change and vulnerabilities of children in rural Zimbabwe: The case of ward 14 of Bikita District, Zimbabwe. *African Journal of Social Work, 13*(2), 78–86.

Chinyoka, K., (2014). Causes of school drop-out among ordinary level learners in a resettlement area in Masvingo, Zimbabwe. *Journal of Emerging Trends in Educational Research and Policy Studies, 5*(3), pp. 294–300.

Chipenda, C. (2023). Rethinking the role of social policy in responding to the Covid-19 pandemic in South Africa: Critical reflections using the intellectual contribution of Thandika Mkandawire. In V. Gumede & E. Ndhlovu (Eds.), *Development in Africa: Remembering Thandika Mkandawire* (pp. 197–231). Ace Press.

Chipenda, C. (2024). A transformative social policy perspective on land and agrarian reform in Zimbabwe. *Africa Spectrum, 59*(1), 89–118.

Chipenda, C., & Cochrane, L. (2024). Out of school children in the context of new policy trajectories and interlocking crises in Zimbabwe: A transformative social policy perspective. *SN Social Sciences, 4*(131), 1–30.

Chirowamhangu, R. (2024). Rights-based analysis of basic education in South Africa. *Journal for Juridical Science, 49*(3), 110–146.

Cloward, K., (2014). False commitments: Local misrepresentation and the international norms against female genital mutilation and early marriage. *International Organization, 68*(3), pp.495–526.

Cornwall, A., & Nyamu-Musembi, C. (2004). Putting the 'rights-based approach' to development into perspective. *Third World Quarterly, 25*(8), 1415–1437. https://doi.org/10.1080/0143659042000308447

De Man, A. (2018). Critiques of the human rights framework as the foundation of a human rights-based approach to development. *Journal for Juridical Science, 43*(1), 84–116. https://doi.org/10.18820/24150517/JJS43.v1.5

Denzin, N. K. (2017). Critical qualitative inquiry. *Qualitative inquiry, 23*(1), pp.8–16.

Dexit, A. K., Tiwari, N., & Yadav, G. (2024). Evolution of human rights: Trend challenges and new horizons. AADARSH Technosoft Publishers.

Dinbabo, M., (2013). Child rights in sub-Saharan Africa: A call for a rights-based global research agenda. *Social Work/Maatskaplike Werk, 49*(3). https://doi.org/10.15270/49-3-47

Fischer, A. M. (2018). Poverty as Ideology: Rescuing Social Justice from Global Development Agendas. London: Zed Books.

Fottrell, D. (Ed.). (2021). *Revisiting children's rights: 10 years of the UN Convention on the Rights of the Child*. Brill.

Fredman, S., Donati, G., & Naicker, S. (2022). New beginnings: The right to equality and early childhood care and education. *South African Journal on Human Rights, 38*(3–4), 167–191.

Gatenio Gabel, S. (2024a). *A human rights-based approach to justice in social work practice*. Oxford Academic. https://doi.org/10.1093/oso/9780197570647.001.0001

Gatenio Gabel, S. (2024b). Applying a rights-based approach to justice in social work. In *A human rights-based approach to justice in social work practice* (online ed.). Oxford Academic. https://doi.org/10.1093/oso/9780197570647.003.0012

Gaus, N. (2017). Selecting research approaches and research designs: A reflective essay. *Qualitative Research Journal, 17*(2), 99–112.

Glaberson, S. (2019). Coding over the cracks: Predictive analytics and child protection. *Fordham Urban Law Journal, 46*(2) Available at SSRN: https://ssrn.com/abstract=4448913

Government of Zimbabwe. (1996). Maintenance Act [Chapter 5.09]. Harare: Government Printers.

Government of Zimbabwe. (2022). Guardianship of Minors Act [Chapter 5:08]. Harare: Government Printers.

Gready, P., (2019). Reflections on a human rights decade, near and Far. J*ournal of Human Rights Practice, 11*(2), pp. 422–437.

Hallberg, L. (2013). Quality criteria and generalization of results from qualitative studies. *International Journal of Qualitative Studies on Health and Well-being, 8*.

Han, L. (2024). Addressing Child Trafficking Adopting a Human Rights-Based Approach. In: Child Trafficking in China. Springer, Singapore. https://doi.org/10.1007/978-981-97-3717-8_2

International Labour Organization. (2010). ILO facts on child labour 2010 [Report]. ILO. https://www.ilo.org/sites/default/files/2025-05/Facts_on_CL_2010_En.pdf

International Labour Organization (ILO) (2025). Child labour in Asia and the Pacific: Regional technical brief (ILO Brief, December 2025). International Labour Organization. https://www.ilo.org/sites/default/files/2026-01/Child%20Labour%20Brief%20-%20Asia-Pacific.pdf

Iusmen, I. (2020). Whose Children? Protecting Unaccompanied Migrant Children in Europe: A Case of Diffused Responsibility?. *The International Journal of Children's Rights, 28*(4), 925–949. https://doi.org/10.1163/15718182-28040006

Joudaki, S. (2024). Exploring children's experiences in the digital environment: A narrative inquiry based on an interdisciplinary approach to child rights (Doctoral dissertation, University of New Brunswick).

Khangala, T. K. (2024). An analysis of South Africa's constitutional and UNCRC-imposed obligations to achieve children's socio-economic rights: A critique. *Athens Journal of Humanities & Arts, 11*(4), 387–402. https://doi.org/10.30958/ajha.11-4-3

Koyratty, N., Ntozini, R., Mbuya, M. N., Jones, A. D., Schuster, R. C., Kordas, K., Li, C. S., Tavengwa, N. V., Majo, F. D., Humphrey, J., & Smith, L. E. (2022). Growth and growth trajectory among infants in early life: Contributions of food insecurity and water insecurity in rural Zimbabwe. *BMJ Nutrition, Prevention & Health, 5*(2), 332–343. https://doi.org/10.1136/bmjnph-2022-000470

Kurevakwesu, W., & Chizasa, S. (2020). Ubuntu and child welfare policy in Zimbabwe: A critical analysis of the national orphan care policy's six-tier system. *African Journal of Social Work, 10*(1), 89–100. https://doi.org/10.31920/2409-5605/2020/v10n1a7

Madziva, C. (2017). Community responses to vulnerable children in rural Zimbabwe: Lessons from a partnership case study. *Development in Practice, 27*(1), 37–52.

Mangena, T., & Ndlovu, S. (2014). Reflections on how Selected Shona and Ndebele Proverbs Highlight a Worldview that Promotes a Respect and/or a Violation of Children's Rights. *The International Journal of Children's Rights, 22*(3), 660–671. https://doi.org/10.1163/15718182-02203003

Manjengwa, J. (2012). Environment, natural resources and poverty reduction: Capturing and sharing the gains of natural resources exploitation. In *Understanding poverty, promoting wellbeing and sustainable development* (p. 79).

Manyakaidze, P. (2023). *Assessing climate change compatible smallholder water management strategies in southern dry lands of Zimbabwe* (Doctoral dissertation).

Marume, A., Kasanzu, S. & Chirenda, J. Socio-economic and rural-urban. (2025). disparities in the double burden of childhood malnutrition in sub-Saharan Africa. *Journal of Health, Population and Nutrition, 44*, 335. https://doi.org/10.1186/s41043-025-01075-y

Matadi, M. T. (2019). The realisation of children's survival rights in South Africa, Kenya and the Democratic Republic of Congo: A comparative analysis. *Comparative and International Law Journal of Southern Africa, 52*(3), 352–388.

Mbaku, J. M. (2019). The Rule of Law and the Exploitation of Children in Africa, 42 Hastings Int'l & Compar. L. Rev. 287. Available at: https://repository.uclawsf.edu/hastings_international_comparative_law_review/vol42/iss2/2.

Meagher, K. (2022). Crisis narratives and the African paradox: African informal economies, COVID-19 and the decolonization of social policy. *Development and Change, 53*(6), 1200–1229.

Midzi, N. A. (2015). *Civil society and children rights in Zimbabwe: Case of Justice for Children organization in Budiriro district in Harare 2013 to present* (Doctoral dissertation).

Mirugi-Mukundi, G. (2010). Reaffirming the social security rights of children in South Africa with particular reference to the child support grant. *ESR Review: Economic and Social Rights in South Africa, 11*(1), 7–9.

Mhaka-Mutepfa, M., Maree, J. G. & Chiganga, G. (2014). Towards respecting children's rights, obligations and responsibilities: The Zimbabwean case. *School Psychology International, 35*(3), pp. 241–252.

Mkandawire, T. (2004). Social Policy in a Development Context: Introduction. In: Mkandawire, T. (eds) Social Policy in a Development Context. Social Policy in a Development Context. London: Palgrave Macmillan. https://doi.org/10.1057/9780230523975_1

Mkandawire, T. (2007). Transformative Social Policy and Innovation in Developing Countries. *The European Journal of Development Research, 19*(1), 13–29. https://doi.org/10.1080/09578810601144236

Mtetwa, E. (2018). Rural social security for Zimbabwe: Challenges and opportunities for state and non-state actors. *Journal of Social Development in Africa, 33*(1). https://doi.org/10.4314/jsda.v33i1

Munro, L. T. (2015). Children in Zimbabwe after the long crisis: Situation analysis and policy issues. *Development Southern Africa, 32*(4), 477–493.

Musiwa, A. S. (2020). Extent and depth of child poverty and deprivation in Zimbabwe: A multidimensional deprivation approach. *Child Indicators Research, 13*, 885–915. https://doi.org/10.1007/s12187-019-09656-0

Musiwa, A. S. (2021). A rights-based approach to child poverty measurement and child rights realization in Zimbabwe. *The International Journal of Children's Rights, 29*(1), 148–198. https://doi.org/10.1163/15718182-29010007

Mutale, Q. (2015). Challenges facing school children in rural Zimbabwe: A case of Tyunga and Luunga wards of Binga District. *Research on Humanities and Social Sciences, 5*(9), 32–39.

Muzari, W., Muvhunzi, S., Soropa, G., & Kupika, O. L. (2014). Impacts of climate variability and change and farmers' responsiveness in the agricultural sector in Zimbabwe. *International Journal of Science and Research, 3*(9), 1726–1731.

Mwapaura, K., Chikoko, W., Nyabeze, K., Kabonga, I. & Zvokuomba, K. (2022). Provision of child protection services in Zimbabwe: review of the human rights perspective. *Cogent Social Sciences, 8*(1), p.2136606.

Nciizah, T., Nciizah, E., Mubekaphi, C. & Nciizah, A. D. (2021). Role of small grains in adapting to climate change: Zvishavane District, Zimbabwe. In African handbook of climate change adaptation (pp. 581–599). Cham: Springer International Publishing.

Ndondo, S. (2024). *Public perceptions of child protection programs in Zimbabwe: Towards strategies to improve the full realization of children's rights* (Master's Dissertation, University of Cape Town, Cape Town). http://hdl.handle.net/11427/40383

Ngwende, O. F. (2016). Child Rights in Zimbabwe: A Review of Literature. *Imperial Journal of Interdisciplinary Research, 3*(1), 828–837.

Nhapi, T. (2020). Zimbabwe's national case management system for child protection and enhanced rights realization for children with disabilities. *African Journal of Social Work, 10*(2), 45–58. https://doi.org/10.31920/2409-5605/2020/v10n2a5

O'Neill, O. (2017). Some limits of informed consent. In *The elderly* (pp. 103–106). Routledge.

Oestreich, J. E. (2023). Closing the circle of implementation: The sustainable development goals, universal periodic review, and the rights-based approach to development. *The International Journal of Human Rights, 28*(1), 69–88. https://doi.org/10.1080/13642987.2023.2232998

Ojo, O. O. P. & Adedayo, A. M. (2022). Economic Community of West African States (ECOWAS) in the Maintenance of Peace and Security of West-African Countries: Prospects and Challenges. *American Journal of Multidisciplinary Research & Development (AJMRD), 4*(09), pp. 30–39.

Okunade, S. K., & Shulika, L. S. (2021). The dynamics of child trafficking in West Africa. *African Human Mobility Review, 7*(3), 116–140.

Oyewale, A. (2022). Yoruba Opera: A practice-based research enquiry into the idea of vocal composition in a non-European cultural framework. Doctoral Dissertation, University of Salford.

Pantzerhielm, L. (2023). On multiple objects and ontic fixes: Human rights and the 'forgotten' politics of the United Nations' human rights-based approach. *Millennium, 51*(2), 519–551. https://doi.org/10.1177/03058298221146303

Republic of South Africa. (1996). Constitution of the Republic of South Africa, No. 108 of 1996. Cape Town: Government Gazette.

Riggirozzi, P. (2017). Regional integration and welfare: Framing and advocating pro-poor norms through southern regionalisms. *New Political Economy, 22*(6), 661–675.

Ringson, J. (2019). The impact of child rights cultural contestation in orphans and vulnerable children (OVC) support in Zimbabwe. *Child Abuse Research in South Africa, 20*(1), 51–61.

Ryan, G. (2018). Introduction to positivism, interpretivism and critical theory. *Nurse Researcher, 25*(4), 41–49.

Schmitz, H. P. (2012). A human rights-based approach (HRBA) in practice: Evaluating NGO development efforts. *Polity, 44*(4), 523–541. https://doi.org/10.1057/pol.2012.18

Shamrova, D., & Lampe, J. (2020). Understanding patterns of child material deprivation in five regions of the world: A children's rights perspective. *Children and Youth Services Review, 108*, 104595.

Simango, T., & Mafa, I. (2022). Child disciplinary practices versus child rights in Zimbabwe: Viewed through social work lenses. *Qualitative Sociology Review, 18*(1), 114–131.

Tanveer, F. (2024). The United Nations Convention on the Rights of the Child (UNCRC): Examining the tenets, practices, and related studies. *Library of Progress-Library Science, Information Technology & Computer, 44*(3).

Tom, T. (2024). Decades of land and agrarian reform in Zimbabwe: A review of transformative social policy lessons for other African countries. *The African Review.*

Tuwor, T. & Sossou, M. A. (2008). Gender discrimination and education in West Africa: strategies for maintaining girls in school. *International Journal of Inclusive Education, 12*(4), pp.363–379.

Umaru, S. (2019). *The role of UN in the protection of children's rights.* Kampala International: School of Law.

UN. (2024). *Human rights-based approach.* https://unsdg.un.org/2030-agenda/universal-values/human-rights-based-approach

UNICEF. (1989). *Convention on the Rights of the Child.* UNICEF.

Union, A. (2016). *Africa health strategy 2016–2030.* Addis Ababa: African Union.

UNRISD. (2006). *Transformative social policy: Lessons from UNRISD research.* UNRISD.

UNRISD. (2023). *Crises of inequality: Shifting power for a new eco-social contract.* Research and Policy Brief 39. UNRISD.

Vainio, A. (2013). Beyond research ethics: Anonymity as 'ontology', 'analysis' and 'independence'. *Qualitative Research, 13*(6), 685–698.

Vandenhole, W. (2024). Human rights-based approaches: Methodological challenges. In B. A. Andreassen, C. Methven O'Brien, & H.-O. Sano (Eds.), *Research methods in human rights* (pp. 321–338). Edward Elgar Publishing. https://doi.org/10.4337/9781803922614.00029

Walker, J. A. (2012). Early marriage in Africa: Trends, harmful effects and interventions. *African Journal of Reproductive Health, 16*(2), 231–240.

Yamin, A. E. & Cantor, R. (2014). Between insurrectional discourse and operational guidance: Challenges and dilemmas in implementing human rights-based approaches to health. *Journal of Human Rights Practice, 6*(3), pp.451–485.

Zela, K. (2023). Exchange rate forecasting with Artificial Intelligence. *Smart Cities and Regional Development (SCRD) Journal, 7*(1), pp.65–70.

Part V
Synthesis and Future Directions

Chapter 19
Conclusion: Towards Reframing Social Policy for Inclusive Development

Emmanuel Ndhlovu and Clement Chipenda

19.1 Introduction

Africa faces unprecedented new political, social, economic and environmental challenges. These challenges often mutually reinforce each other and may affect not only the income of households or of individuals (e.g., through divorce, the loss of a job and disability) but also the management of time as people become under pressure to successfully combine household roles and employment. These new challenges require that contemporary social policy move beyond the traditional conventions of focusing on a limited number of specific risks, mainly short-duration unemployment, disability during active age and resource inadequacy in childhood and during old age. The approach adopted within traditional social policy interventions was based on several assumptions: a clear identification of the various phases of human life (childhood, studies, activity retirement); firm allocation of the gender roles within the household (housewife, male breadwinner); the presence of strong ties within the household; and uninterrupted full-time work careers. In many contemporary societies, including Africa, these assumptions no longer match reality due to the emergence of new socio-economic, political and environmental risks. These new social risks are problematic and require urgent and innovative social policy responses. The new social risks call for the need for new social policies capable of responding to developing trends more effectively.

E. Ndhlovu (✉)
School of Public Management, Governance and Public Policy, College of Business and Economics, University of Johannesburg, Auckland Park, South Africa

C. Chipenda
South African Research Chair in Social Policy, College of Graduate Studies, University of South Africa, Pretoria, South Africa
e-mail: chipec@unisa.ac.za

E. Ndhlovu, C. Chipenda (eds.), *Redefining the Social Policy Discourse*,
https://doi.org/10.1007/978-3-032-24067-5_19

Africa is among the poorest global regions, with the economies of most countries based on agriculture (Ndhlovu, 2025). Africa is also home to most of the world's poor people. In 2024, Sub-Saharan Africa accounted for 16% of the world's population. Nonetheless, out of the 3.5 billion people (44% of the global population) living in poverty (measured by the $6.85 per day standard of upper-middle-income countries), 67% of them were in SSA (World Bank, 2024). This implies that two-thirds of the world's population living in poverty lives in Africa. Approximately 72% of the world's population living in extreme poverty are in countries that are eligible to receive assistance from the International Development Association (IDA) (World Bank, 2024). Therefore, since poverty and inequality are pervasive and watertight interventions elusive in Africa, it is vital to sustain the opportunities for discussion, experimentation, innovation, and learning in social development. Finding ways to develop watertight interventions for social policy-related challenges and to encourage a deeper appreciation for the contextual factors that shape development outcomes can represent a significant contribution.

Societies are also undergoing rapid transformation, promoting freedom and opening up new opportunities for the pursuit of optimal and cost-effective wellbeing choices and preferred values. Contemporary social policy, therefore, needs reframing to be able to help individuals and or whole groups positively respond to shocks that threaten their lives and livelihoods. This approach is supported by Banerjee and Duflo (2019, p. 322), who state that the '...goal of social policy, in these times of change and anxiety, is to help people absorb the shocks that affect them without allowing those shocks to affect their sense of themselves.' However, it is vital to understand that social policies may not always result in human welfare and may not always generate positive outcomes, since a social policy that may generate positive outcomes for one group may actually generate detrimental outcomes for another group. Hence, a critical lens is needed to understand the pertinent issues within social policy. Therefore, contemporary social policy needs to be reframed not only to respond to new shocks but also to be reframed in ways that align with the contexts of cultures and situations of particular individuals and or groups. While there are divergent views regarding the impact and implementation of social policies, there seems to be a general consensus that when social policies are formulated, designed and implemented appropriately, they have the potential to generate positive societal development impacts. Social policies can help society by making enhancements in a variety of areas, including education, citizenship, culture, income and economic status (Ndhlovu & Gumede, 2023). The chapters that comprise this book use several case studies across the African continent to demonstrate the various contexts of culture and situations within which social policy should be reframed.

This chapter concludes the book *Reframing Social Policy for African Futures: Harmonising International Discourse and Continental Imperatives* by revisiting and summarising the key issues raised in the chapters comprising the book. Based on the chapters in the book, the current chapter (i) reconceptualises social policy, with the intention to open up debates on how social policy can be reframed, (ii) outlines various aspects which policymakers can target in reframing social policy and (iii) provides examples of how social policy can be reframed. The multidisciplinary outlook and summary nature of this chapter readily commend it as compulsory reading for policymakers, donor communities, students and activists interested in emerging new social policy directions in African development thinking and research.

The chapter is organised as follows: After the introduction, the chapter outlines the research methodology adopted. This is then followed by a conceptualisation of the social policy framework, with a focus on Africa. Under separate sub-sections, the chapter also outlines various aspects that policymakers can target in reframing social policy. After that, examples of how social policy can be reframed are then outlined.

19.2 Materials and Methods

This chapter is based on the chapters that comprise this book. Therefore, the chapter utilises a critical document analysis to summarise the key issues raised in the book. This book comprises a dirty mixture of user-friendly, simple, but contextually rich chapters that are easy to read and informative. Collectively, the chapters deployed a multi-disciplinary approach, utilising historical, sociological and political economy to discuss prospects for reframing social policy in Africa. Some of the chapters used various research methods to help generate an in-depth understanding of the nature of current social policy interventions and how social policy can be reframed to satisfy new needs. This chapter used the content analysis approach to explore the key arguments and conclusions reached in the chapters. While this loose approach does not ensure generalisability, it provides an opportunity to present, in summary format, the key issues in the book. It is hoped that readers will read the book to gain a detailed understanding of the issues summarised in this chapter.

19.3 Social Policy Framework

The majority of chapters in this book agree that social policy refers to a set of guidelines and interventions deployed to change, maintain or create living conditions that are conducive to human welfare. Examples of social policy covered in the book include education, health, housing, employment and food for all people. Social policy is a segment of public policy that deals with social issues (Vargas-Hernandez et al., 2011). Citing the Malcolm Wiener Centre for Social Policy at Harvard University, Rittel and Webber (1973, p. 155) describe social policy as 'public policy and practice in the areas of health care, human services, criminal justice, inequality, education, and labour'. According to Rittel and Webber (1973), social policy often deals with 'wicked problems'. The chapters in this book also identify social policy as a distinct academic field which deals with the systematic assessment of societal responses to social needs. For instance, Chap. 11 recommends the need to adjust and redesign the university curriculum to include social policy courses as part of the effort towards social policy reframing. In addition, some chapters link social policy with the notion of social welfare and its connection with politics and society (see Chaps. 1–3).

For social policy to attain its objectives, it must have specific principles concerning the policies it determines. Human welfare, needs, treatment and social justice, equality and preference, altruism, reciprocity and duty and exclusion are some of the central values associated with social policy (Ndhlovu, 2024a; Oyelere et al., 2023; Papadakis & Tzagkarakis, 2025). The following are some of the principles of social policy:

- Achieving democracy through equality in resource distribution.
- Equal opportunity: Equal treatment of all classes of people. All people needed to be granted equal opportunities in resource access, utilisation, livelihood and income generation opportunities regardless of their gender or group. In addition, everybody should have the same benefits in the educational system and in the job market.
- Need: Caring and accommodation are also basic human needs. Needs have no limits, even though it is not clear which needs governments should focus on.
- Freedom and Rights: Social policy should also be able to ensure freedom and the rights of people in terms of civil rights (e.g., the lack of unjust imprisonment and incarceration) and the freedom to express any viewpoint without persecution. It should also promote democratic rights, including social welfare, social security, and the right to an education, voting and entering political parties.

By ensuring and promoting the achievement of various categories of rights, social policy then opens up opportunities for individual and collective development (of the type that delivers a progressive, sustained and sustainable transformation of lives and societies). With adequate and equal access to and utilisation of resources, people are able to pursue opportunities of their own choice to develop and sustain their socio-economic activities and thus improve their lives and communities. The concept of social welfare, which pervades the social policy discourse, refers to the basics required to provide individuals or communities with the services and the opportunities they need to live healthy and happy lives (Shirazi & Keivani, 2019). All the chapters in this book show that there is a strong bond between social welfare and social policy. It is posited that when implemented, social policy should result in the availability and access of welfare facilities in society. For this reason, social policy is an integral part of public policies. In Africa, although much of the social welfare activities are championed by donor communities because of government resource constraints (Ndhlovu, 2024a), it is always vital that governments play an active role in terms of providing frameworks on how social welfare activities can be rolled out.

This review shows that social policies wield much potential to change the living conditions of people who live in society. In Africa, where poverty statistics are high (World Bank, 2024) and where the majority of people depend on subsistence farming, social policies are vital in supporting people in dealing with challenges such as climate change, food insecurity and unequal access to education. For instance, in South-eastern Zimbabwe, Ndhlovu (2024b) found that most peasant farmers relied on social assistance for seed, agrochemicals and packaging resources to engage in their farming activities. In addition, social welfare actors, mainly the donor community, together with the government, promoted education access by paying school fees and buying school uniforms for underprivileged families (Ndhlovu, 2024b). In the long run, this will empower people to meet their food needs and exercise their human rights. In reframing social policy to respond to African needs, the chapters in this book have emphasised a focus on social security, social structures and social interactions (see Chaps. 20–23). The active and increasing presence of social services provision by aid donors, non-governmental organisations and private businesses has reinforced the role of social policy in human welfare. In these circumstances, the State's duty is to provide a conducive environment for private providers of social services while lowering its social spending and activities.

19.4 Towards Inclusive Development in Africa

Social policies need to be understood in the context of development (see Chap. 4). In conventional social policy discourses, social policy is viewed merely as a tool of governance with no clear and close link to development. This thought has limited social policy and social programmes to disaster response strategies. The thought is generated by a misconception that confines development to the acquisition of resources and capital without considering the value of non-monetary needs such as mental, spiritual and cultural liberation. This has generated a knowledge and conceptual gap that has clouded the understanding of African conditions.

Understanding the nature of social policy and social programmes and the development crisis in African countries requires, first, understanding the nature of politics, power relations, the type of social-political culture, and the interaction that exists between the state and civil society in post-independence Africa. In assessing the nature of politics and the political culture of a society, scholars such as Owolabi (1999) draw attention to the need to understand the role of the state and civil society. As a result, how development is conceived in Africa differs from how it is understood in other global regions, particularly the Euro-North American countries of the Global North.

In the Global North, development is conceptualised as an evolutionary process through which the countries in the North evolved and through which the countries in the Global South were still in the process of evolving (Ndhlovu & Mhlanga, 2023). This thinking, which was propagated by W.W. Rostov, an adviser in both the John F Kennedy and Lyndon Johnson administrations in the United States, was based on the belief that societies could attain development if they could only copy Western capital, technology and social organisation and values.

W.W. Rostov's conceptualisation of development is superficial and devoid of detail as it disregards factors such as culture and history, which are key factors in the development efforts of non-Western societies. For instance, Africa suffered colonialism and much resource exploitation by European countries, such that the history of exploitation cannot be ignored in any development efforts in the region. As a result, the majority of the chapters in this book touch on the impact of Africa's colonial past and how the past has influenced social policies and programmes (see Chaps. 12–14). Amin (1973) argued that Euro-North American societies achieved 'development' through primitive accumulation from other regions, especially Africa and that this development continues to be sustained through value flows from these regions, thus compromising any prospects for these regions to realise development. As a result, W.W. Rostov's conceptualisation of development is rejected by African scholars for its uselessness.

Due to Africa's colonial past, some African scholars define development as a 'liberatory human aspiration to attain freedom from political, economic, ideological, epistemological, and social domination…' (Ndlovu-Gatsheni, 2012, p. 2). Mensah (2019, p. 4) defined development as 'an evolutionary process in which the human capacity increases in terms of initiating new structures, coping with problems, adapting to continuous change, and striving purposefully and creatively to attain new goals'. Gumede (2019, p. 51) defines development as '…improvements in wellbeing, involving socio-economic progress', while Brobbey (2010, p. 1) understands development as 'the capacity of a state with the aim of achieving higher outcome of production for the satisfaction of citizens and empowering them to make demands'. Reyes (2001) avers that development entails a social condition whereby existing natural resources and methods satisfy people's needs. It is also understood as a practice that involves improving people's conditions (Todaro & Smith, 2006).

As shown in Chap. 4, what emerges from African scholars' definitions of 'development' is a clear rejection of the capitalistic features in W.W. Rostov's definition. African definitions revolve around the need for the well-being of all people—inclusive development. Inclusive development is a process that focuses on including marginalised people, sectors and countries in social, political and economic development (Gumede, 2019). It is a way to ensure that all people can participate in the development process. The goals of inclusive development include equity, which is the fair distribution of social and material benefits across society; *empowerment*, which is giving power and voice to marginalised groups; *sustainability*, which is attaining social and environmental sustainability; *and wellbeing*, which is increasing human welfare.

The inclusive development practices covered by the chapters in this book include issues of direct democracy with emphasis on civil, civic and political rights (see Chaps. 9–11). Some of the chapters also touch on the distribution of amenities, focusing on ensuring that everyone has adequate access to amenities such as education, infrastructure and healthcare. There is also an overwhelming call for the need to invest in human capital to counterbalance losses in natural capital. Achieving inclusive development goals through the pursuit of ideal practices necessitates reframing policies, particularly social policy, since most Africans cannot develop themselves without social support. Due to poverty, many people in Africa cannot afford access to amenities such as education, infrastructure and healthcare required in the development agenda. Therefore, reframing social policy is important as it opens up new categories of knowledge on how best to pursue development by drawing on local situations and needs.

19.5 Importance of Reframing Social Policy

The conventional conceptualisation of social policy mainly promotes human well-being through social services and support, but offers little prospects for development. With poverty on the rise in Africa (World Bank, 2024), it is vital that social policy be reframed so that it enables people to put their labour and passion to productive use to improve their livelihood conditions—such is the basis of the realisation of inclusive development (see Chap. 4). There are several factors that could be capitalised on to generate development once social policy is reframed and redesigned. For instance, Africa is basically an agrarian community, with over 50% of the population in many countries directly engaged in farming for livelihoods and income (Ndhlovu, 2025). Many indigenous Africans also possess significant traditional farming expertise. In reframing social policies, focus could be placed on how to support these people to improve and strengthen their farming activities, which provide a promising base for broad-based development.

Africa's demographic structure also offers much potential in shaping its development trajectory. Africa has a significant youthful population (World Bank, 2024). This youthful population holds enormous potential as it can serve as a catalyst for economic growth and societal transformation and an engine for sustainable development (Onyeiwu, 2024). Africa is the youngest continent in the world, with a youth population of over 420 million people aged 15–35 (World Bank, 2024). By 2050, this number is projected to more than double, exceeding 830 million, making Africa home to one of the fastest-growing youth populations globally (Food and Agriculture Organization (FAO), the United Nations Economic Commission for Africa (ECA), African Union Commission (AUC), and the World Food Programme (WFP), 2023). This demographic structure could be harnessed to drive development. Currently, youth unemployment is a significant challenge. In 2023, the youth not in employment, education, or training rate in Sub-Saharan Africa was at 21.9%, with wide variations (International Labour Organization (ILO), 2024). Harnessing this youthful population's contribution to regional development lies in the ability to reframe social policy and align social development to the growth of the youth population, with opportunities for meaningful economic engagement.

Reframing social policies allows for an opportunity to steer the African development agenda and addresses societal systematic inequalities, promotes social justice, increases public support, facilitates innovation and creativity, and promotes adaptation potential.

19.5.1 Addressing Systemic Inequalities

Inequality is documented as high in Africa where, on average in Africa, the wealthiest population group receives 50% of total income in most of Sub-Saharan Africa (Odusola et al., 2017; Chancel et al., 2019), with the percentage reaching 65% in South Africa (Zizzamia et al., 2021). In South Africa, for instance, Chatterjee et al. (2020, p. 2) found that

> The top 10 per cent own 86 per cent of aggregate wealth, and the top 0.1 per cent close to one-third of aggregate wealth. The top 0.01 per cent of the distribution (3,500 individuals) own 15 per cent of household net worth, more than the bottom 90 per cent as a whole. Such high levels of inequality can be accounted for in all forms of assets, including housing, pension funds, and other financial assets.

In a study on income and wealth inequality focused on Uganda, Malawi and Tanzania, De Magalhães and Santaeulàlia-Llopis (2018) found that there was a very low correlation between income and wealth inequality in Africa compared to the United States. The driving forces of inequality in Africa are a mix of historical and institutional factors that have ensured policy decisions (or lack thereof) (Zizzamia et al., 2021). Africa had a disruptive colonial past. Most of the countries in the region suffered colonial rule by Western countries such as Britain, Germany, Portugal and Belgium, for instance. At independence, most of these countries were faced with deep-rooted poverty amongst their vulnerable populace. Colonialism concentrated productive assets in the hands of a few colonisers and limited the mobility of the indigenous people. Today, these are Africa's key drivers of inequality (Alvaredo et al., 2018). The structure of African economies and their economic growth have also been implicated in generating inequality. Scholars such as Cornia and Martorano (2016) also point to the shifts in the percentage structure of value-added shares, from low-inequality to high-inequality sectors and vice versa, as drivers of inequality in Africa. For Zizzamia et al. (2021), the relative importance of the extractive sectors is another reason for high inequality, as it has a low value regarding employment and can have an adverse impact on investments in other sectors.

The advent of digital technology during the Fourth Industrial Revolution also resulted in increasing returns to high skills and, thus, an increase in wage inequalities (Mhlanga & Ndhlovu, 2023). While Rushambwa and Ndhlovu (2023) found varied effects across African countries, high informality levels across the continent suggest that technological change could significantly influence inequality in Africa. According to Zizzamia et al. (2021), this effect might be even stronger where strong institutions governing the labour market do not exist.

Another factor leading to inequality is discrimination. This results in unequal access to opportunities and resources for marginalised population categories. Some discriminatory tendencies are often the result of socio-cultural, economic and political practices in society, reproducing inequalities. Gender inequalities are examples of such social and cultural norms, and in Africa, they strongly correlate with high-income inequality (Zitha et al., 2024). Therefore, by re-examining and reframing social policies, policymakers and other interested stakeholders can identify and tackle systemic barriers that may be perpetuating inequalities, leading to more targeted solutions.

19.5.2 Promoting Social Justice

Social justice denotes the promotion of fairness, equality, equity and rights across various features of society, including economic, educational, and workforce opportunities (Killen et al., 2021; Nkhasho et al., 2024). Several studies have shown that in Africa, the condition of social justice is generally poor, marked by prevalent inequality, limited access to basic requirements such as healthcare and education and systemic discrimination based on race, gender and socio-economic status, particularly in areas with high rates of poverty and limited economic opportunities (ILO, 2019; Ndhlovu, 2022; Pillay, 2020). Countries such as Zimbabwe, South Africa, Mozambique, Ghana, Kenya and Uganda are examples of how settler colonialism, for instance, continues to impact social disparities. However, the chapters in this book demonstrate that social justice cannot be reached through adopting international conventions or declarations. All citizens need to take responsibility for protecting, advancing, and promoting the values, principles and ideals of social justice through their own initiatives. The ILO (2019) calls upon all stakeholders to take responsibility for building a just and equitable society through strategic partnerships at the national, regional and international levels with good-quality data for monitoring and evaluating advancement. Several scholars who contributed to this book have also argued that in order to increase social justice, it is essential to change attitudes and prejudices that preserve prejudicial and discriminatory treatment of others (see Chaps. 21 and 22).

Furthermore, it can be argued that realising social justice requires moving from individual blame and focus to wider societal factors, promoting policies that aim to create a more just and equitable society. This constitutes the reframing of social policy for inclusive development.

19.5.3 Increased Public Support

Reframing social policies to align with current needs and realities is also vital in mobilising public support for activities and commitments. Social policy and public support denote the rapport between government policies meant to deal with social issues such as poverty, healthcare, education and joblessness and the extent of public approval by citizens (van der Does, 2022). This essentially refers to the extent to which the general population supports a particular social policy implemented by the government (Lemanski, 2020). Public support and participation are important features of democratic agendas and commitments to inclusive development (Baloyi & Lubinga, 2017). Having first emerged as a response to increasing democratic discontent, public support is now widely promoted to address local development and public policy challenges (Hofer et al., 2024). For any society to achieve development, it requires the collective support of all citizens. On the other hand, citizens are more willing to support government policies if they feel that such policies are fair, transparent and meant for the collective good.

Public support is imperative for the effective execution of social policies because general approval can influence policy decisions, funding allocation, and political will to effect change (Ahrens, 2024). There are many factors that can impact public backing for a social policy, including the apparent usefulness of the policy, who benefits from it, the cost to taxpayers, and individual beliefs and values (Frank & Ahrens, 2020). Therefore, policymakers and governments can gain public support by reframing social policies to directly align with societal needs. Reframing social policies, therefore, can help promote public participation, foster peace, and encourage democratic practices in society. If a substantial percentage of the population perceives welfare programmes as vital for helping the poor, it may result in increased political pressure to maintain or expand those programmes. Conversely, if a proposed reframing of social policy is generally alleged as damaging or unfair, it may face substantial public resistance, possibly impeding its implementation.

19.5.4 Innovation and Creativity

Reframing is also important in that it can open up new avenues for policy development by promoting creative thinking and considering alternative approaches. Creativity enables policy stakeholders to understand and solve problems in more transparent ways and with innovation. Creativity opens the mind (Edler et al., 2024). A society with no creative attributes is considered an imprisoned society in that its people may be closed-minded for a long time (Ejoh & Omoile, 2023). Creativity broadens perspectives and can help individuals and whole communities to overcome prejudices.

19.5.5 Adapting to Changing Societal Needs

As societies evolve, existing policies may need to be revised as they become outdated. As time moves on, new needs emerge, necessitating a re-evaluation of needs and potential solutions to new problems. The adaptation of social policy to changing societal needs entails actively reviewing and modifying current social programmes to address new challenges, such as demographic shifts, technological advancements, economic fluctuations and changing social norms (Ndhlovu, 2024a). The aim is to ensure that social policies and programmes remain relevant and effective in supporting the well-being of people across different demographics.

Adapting social programmes and policies to changing needs requires regular monitoring of social trends, economic indicators and demographic shifts through data collection and assessment to identify emerging needs and areas requiring policy adjustments. The chapters in this book also mention the need for governments and other key stakeholders to carry out regular assessments of existing social programmes and policies to evaluate their effectiveness, identify gaps in service delivery, and identify areas for improvement (see Chaps. 21 and 22). It is also important that when designing policies and social programmes, flexibility and adaptability features are also built in. Designing policies and social programmes that have some built-in mechanisms to respond to changing situations, such as adjustable eligibility criteria or flexible programme delivery models (Ahrens, 2024).

Most of the chapters in this book also show that the challenges to the adaptation of social policies and politics include the lack of political will, data limitations, resource constraints and the complexity of social issues.

19.6 Towards Reframing Social Policy in Africa

There is no perfect-finished strategy for how social policies should be conceptualised. This is because countries and societies have specific challenges and issues they must deal with. In addition, societies evolve at different paces, thus experiencing different challenges at different times or stages of evolution. As a result, it is vital that in seeking to reframe social policy, policymakers and other relevant stakeholder first understand the contents of existing policies, assess their impact on society over time and their weaknesses, and also establish public views on such policies and programmes. This book demonstrates that this process should be data-driven to ensure that the envisaged changes will generate publicly accepted results.

In departing from conventional social policies that mainly focus on individual income levels to measure poverty, contemporary reframing might need to consider factors such as access to quality education, healthcare and affordable housing as key indicators. These factors represent the basic requirements for individuals and communities to improve their material conditions and initiate accumulation. Educated and healthy people stand a better chance of securing good jobs that may enable them to improve their welfare. Therefore, social policies that seek to address poverty need to transcend cash and food assistance and focus on addressing the root causes of poverty and its intergenerational prospects. In most poverty responses in Africa, the general trend has been to provide food parcels, roll out food-for-work activities, or provide cash assistance (Ndhlovu, 2024a; Nzabamwita & Ndhlovu, 2024). This has been successful in ensuring that the poor have meaningful access to food, but has had little impact on access to healthcare and education. As a result, such responses have had little impact on poverty; hence, there is a need to target the very factors that are actors and determinants of poverty.

Reframing social policy can also target unemployment, which is widespread in Africa, especially among youth (ILO, 2024). In responding to unemployment, the approach used by many African governments has been to provide benefits to the unemployed. For instance, in trying to mitigate the impact of COVID-19 on the unemployed, South Africa introduced a special COVID-19 Social Relief Distress (SRD) grant of R350 in May 2020. This was meant to ensure the well-being of its citizens through lessening the social and economic impacts of COVID-19 (Bhorat & Kohler, 2020). It was believed that the grant would be used mostly to purchase food (Ramaphosa, 2020). About 4,379,331 men and about 2,070,585 benefited from the grant (Ndhlovu & Gumede, 2023). In reframing social policy in relation to unemployment, the focus could shift from solely providing benefits to individuals to policies that address systemic issues like skills gaps and job market inequalities. When people have the skills required in the labour market, they are likely to secure jobs that could help them meet their needs without solely relying on social assistance.

19.7 Concluding Remarks

This chapter concluded the book *Reframing Social Policy for African Futures: Harmonising International Discourse and Continental Imperatives* by revisiting and summarising the key issues raised in the chapters comprising the book. The chapter showed that, unlike other global regions, such as Europe and North America, Africa has a history of colonialism, which continues to affect its development campaign. Africa is among the global regions with the highest number of poor people. The region also experiences pervasive food insecurity, inequality, and poor healthcare. As a result, many people on the continent rely on the State to provide them with these basic necessities. Unlike Europe and North America, which fast-tracked development through colonial exploitation, Africa has no region where it can go and colonise to achieve development through primitive accumulation. As a result, the continent cannot develop through the copy-and-paste development method suggested by W.W. Rostov. This implies that a different strategy is required. This book, therefore, showcased the diagnostic potential of reframing social policy as a development-for-all strategy—inclusive development.

References

Ahrens, L. (2024). The impact of public opinion on voting and policymaking: Is public opinion exogenous or endogenous? *Zeitschrift für Politikwissenschaft, 34*, 77–100. https://doi.org/10.1007/s41358-024-00366-w

Alvaredo, F., Chancel, L., Piketty, T., Saez, E., & Zucman, G. (Eds.). (2018). *World inequality report 2018*. Belknap Press of Harvard University Press.

Amin, S. (1973). *Neocolonialism in West Africa*. Harmondsworth: Penguin.

Baloyi, M. L., & Lubinga, E. N. (2017). After Izimbizo, what next? A participatory development communication approach to analysing feedback by the Limpopo Provincial Government to its citizens. *TD: The Journal for Transdisciplinary Research in Southern Africa, 13*(1), 1–7. https://doi.org/10.4102/td.v13i1.402

Banerjee, A., & Duflo, E. (2019). *Good economics for hard times*. Allen Lane.

Bhorat, H., & Kohler, T. (2020). *Lockdown economics in South Africa: Social assistance and the Ramaphosa stimulus package*. Retrieved December 13, 2020, from https://www.brookings.edu/blog/africa-in-focus/2020/11/20/lockdown-economics-in-south-africasocial-assistance-and-the-ramaphosa-stimulus-package/

Brobbey, P. (2010). *Theories of social development*. University of Ghana: Department of sociology.

Chancel, L., Cogneau, D., Gethin, A., & Myczkowski, A. (2019). How large are African inequalities? Towards distributional national accounts in Africa, 1990–2017. *WID World Working Paper* n° 2019/13.

Chatterjee, A., Czajka, L., & Gethin, A. (2020). Estimating the distribution of household wealth in South Africa. *WIDER Working Paper 45/2019*. UNI-Wider.

Cornia, G. A., & Martorano, B. (2016). *Building the IID-SSA inequality dataset and the seven sins of inequality measurement in sub-Saharan Africa*. UNDP, Regional Bureau for Africa, Working Paper Series on Inequality.

De Magalhães, L., & Santaeulàlia-Llopis, R. (2018). The consumption, income, and wealth of the poorest: An empirical analysis of economic inequality in rural and urban Sub-Saharan Africa for macroeconomists. *Journal of Development Economics, 134*, 350. https://doi.org/10.1016/j.jdeveco.2018.05.014

Edler, J., Ostertag, K., & Schuler, J. (2024). Social innovation, transformation, and public policy: Towards a conceptualization and critical appraisal. *Science and Public Policy, 51*(1), 80–88. https://doi.org/10.1093/scipol/scad054

Ejoh, E., & Omoile, P. A. (2023). Corporate social responsibility on innovation and creativity in the workplace. *Malikussaleh Social & Political Reviews, 4*(2), 66–71. https://doi.org/10.29103/mspr.v4i2.13610

FAO, AUC, ECA, & WFP. (2023). *Africa—Regional overview of food security and nutrition 2023: Statistics and trends*. FAO. https://doi.org/10.4060/cc8743en

Frank, B., & Ahrens, L. (2020). The impact of partisanship in the era of retrenchment: Insights from quantitative welfare state research. *Journal of European Social Policy, 30*, 34–47.

Gumede, V. (2019). Rethinking and reclaiming development in Africa. In S. J. Ndlovu-Gatsheni & B. Mpofu (Eds.), *Rethinking and unthinking development: Perspectives on inequality and poverty in South Africa and Zimbabwe*. Berghahn Books.

Hofer, K., Wicki, M., & Kaufmann, D. (2024). Public support for participation in local development. *World Development, 178*, 106569. https://doi.org/10.1016/j.worlddev.2024.106569

ILO. (2019). *Advancing social justice: Shaping the future of work in Africa: Report of the Director-General*. ILO.

ILO. (2024). *Share of young people not in education, employment or training* [dataset].

Killen, M., Yee, K. M., & Ruck, M. D. (2021). Social and racial justice as fundamental goals for the field of human development. *Human Development, 65*(5–6), 257–269. https://doi.org/10.1159/000519698

Lemanski, C. (2020). Infrastructural citizenship: The everyday citizenships of adapting and/or destroying public infrastructure in Cape Town, South Africa. *Transactions of the Institute of British Geographers, 45*(3), 589–605. https://doi.org/10.1111/tran.12370

Mensah, J. (2019). Sustainable development: Meaning, history, principles, pillars, and implications for human action: Literature review. *Cogent Social Sciences, 5*(1), 1–21. https://doi.org/10.1080/23311886.2019.1653531

Mhlanga, D., & Ndhlovu, E. (2023). Goals of social policy in the fourth industrial revolution from a developmental context in Africa. In D. Mhlanga & E. Ndhlovu (Eds.), *The fourth industrial revolution in Africa: Exploring the development implications of smart technologies in Africa*. Springer.

Ndhlovu, E. (2022). Neoliberalism, the COVID-19 pandemic, and the public healthcare crisis in Africa. *Journal of African Transformation, 7*(1), 80–94.

Ndhlovu, E. (2024a). Social protection instruments and strategies in Southern Africa. *International Journal of Innovation in Management Economics and Social Sciences, 4*(2), 44–59.

Ndhlovu, E. (2024b). Building farming resilience to climate change through social assistance in South-Eastern Zimbabwe. *International Journal of Research in Business and Social Science, 13*(10), 105–117.

Ndhlovu, E. (2025). *Agriculture, autonomous development, and prospects for industrialization in Africa*. Springer.

Ndhlovu, E., & Gumede, V. (2023). Reflections on South Africa's social policy response to the COVID-19 pandemic. In V. Gumede & E. Ndhlovu (Eds.), *Development in Africa: Remembering Thandika Mkandawire* (pp. 128–155). Africa Century Editions Press.

Ndhlovu, E., & Mhlanga, D. (2023). Decolonization of development in early post-independence Africa. In D. Mhlanga & E. Ndhlovu (Eds.), *Post-independence development in Africa decolonisation and transformation prospects* (pp. 63–81). Springer.

Ndlovu-Gatsheni, S. (2012, October 16). *Coloniality of power in development studies and the impact of global imperial designs on Africa*. Inaugural Professorial Lecture, University of South Africa.

Nkhasho, M., Rammala, R., & Benson, K. (2024). Social justice for peace: Towards the complementarity of criminal justice systems—The case of the San communities of the Northern Cape Province of South Africa. *Global Change, Peace & Security, 35*(2), 185–201. https://doi.org/10.1080/14781158.2024.2390458

Nzabamwita, J., & Ndhlovu, E. (2024). Social grants impact on household livelihoods: Empirical evidence from international migrants in Cape Town, South Africa. *E-Journal of Humanities, Arts and Social Sciences, 5*(14), 2521–2537.

Odusola, A. F., Cornia, G. A., Bhorat, H., & Conceição, P. (Eds.). (2017). *Income inequality trends in sub-Saharan Africa: Divergence, determinants and consequences*. United Nations Development Programme, Regional Bureau for Africa.

Onyeiwu, S. (2024). Population, youth bulge and economic development. In S. Onyeiwu (Ed.), *Emerging issues in contemporary African economies* (pp. 123–141). Palgrave Macmillan.

Owolabi, K. A. (1999). *The quest for development in Africa: A theoretical exploration*. Obaroh & Ogbinaka Publishers Ltd.

Oyelere, M., Olowookere, K., Oyelere, T., Opute, J., & Adisa, A. T. (2023). The conceptualisation of employee voice in Permacrisis: A UK perspective. In T. A. Adisa, C. Mordi, & E. Oruh (Eds.), *Employee voice in the global north: Insights from Europe, North America and Australia* (pp. 9–34). Springer.

Papadakis, N., & Tzagkarakis, S. I. (2025). Welfare state, social policy and social sustainability, within the context of the permacrisis. *Frontiers in Political Sciences, 6*, 1451406. https://doi.org/10.3389/fpos.2024.1451406

Pillay, J. (2020). Social justice implications for educational psychologists working with orphans and vulnerable children in South Africa. *School Psychology International, 41*(1), 37–52. https://doi.org/10.1177/0143034319893005

Ramaphosa, C. (2020) *President Cyril Ramaphosa: Additional Coronavirus COVID-19 economic and social relief measures*. https://www.gov.za/speeches/president-cyril-ramaphosa-additional-coronavirus-covid-19economic-and-social-relief

Reyes, G. E. (2001). Four main theories of development: Modernization, dependency, word-system, and globalization. *Nómadas. Revista Crítica de Ciencias Sociales y Jurídicas, 4*(2), 109–124.

Rittel, H., & Webber, M. (1973). Dilemmas in a general theory of planning. *Policy Sciences, 4*, 155–169.

Rushambwa, T., & Ndhlovu, E. (2023). Digital transformation and social change in Africa: Issues in technology governance and social participation. In D. Mhlanga & E. Ndhlovu (Eds.), *The fourth industrial revolution in Africa: Exploring the development implications of smart technologies in Africa*. Springer.

Shirazi, M. R., & Keivani, R. (2019). Social sustainability discourse. A critical revisit. In R. M. Shirazi & R. Keivani (Eds.), *Urban social sustainability. Theory, policy and practice* (pp. 1–26). Routledge. https://doi.org/10.4324/9781315115740-1

Todaro, M. P., & Smith, S. C. (2006). *Economic development* (8th ed.). Addison-Wesley.

van der Does, R. (2022). Citizen involvement in public policy: Does it matter how much is at stake? *Public Administration*, 1–16. https://doi.org/10.1111/padm.12846

Vargas-Hernandez, J., Noruzi, R. N., & Ali, I. F. N. (2011). What is policy, social policy and social policy changing? *International Journal of Business and Social Science, 2*(10), 287–291.

World Bank. (2024). *Poverty, prosperity, and planet report: Pathways out of polycrisis*. Retrieved February 20, 2025, from https://www.worldbank.org/en/publication/poverty-prosperity-and-planet?gad_source=1&gclid=Cj0KCQiAwtu9BhC8ARIsAI9JHal_SgkswKTMf3Sd9x2kfOtJgboEC4mpR7rIG-K2FovkYPNVjCqCu4kaAtE4EALw_wcB

Zitha, M. G., Lumadi, R. I., & Ndhlovu, E. (2024). Perceived impact of culture on the progress of women as school leaders in South Africa. *Social Sciences and Education Research Review, 11*(2), 135–145.

Zizzamia, R., David, A., & Leibbrandt, M. (2021). Inequality in sub-Saharan Africa: A review paper. *Working Paper Series*, no. 8. Retrieved February 24, 2025, from https://aceir.uct.ac.za/sites/default/files/content_migration/aceir_uct_ac_za/1639/files/WP8_Inequality-in-sub-Saharan-Africa.pdf

The manufacturer's authorised representative in the EU is Springer Nature Customer Service Centre GmbH, Europaplatz 3, 69115 Heidelberg, Germany. If you have any concerns regarding our products, please contact ProductSafety@springernature.com

Printed and bound by CPI Group (UK) Ltd, Croydon, CR0 4YY

17/09/2026

02218430-0015